Title 29 – Labor

Subtitle B - Regulations Relating to Labor

Chapter XVII - Occupational Safety and Health Administration, Department of Labor

Contents

Title 29 – Labor...1
Chapter XVII - Occupational Safety and 1
Health Administration, Department of Labor 1
   Subpart A - General 9
      § 1926.1 Purpose and scope..9
      § 1926.2 Variances from safety and health standards..9
      § 1926.3 Inspections - right of entry. 9
      § 1926.4 Rules of practice for administrative adjudications for enforcement of safety and health standards. 10
      § 1926.5 OMB control numbers under the Paperwork Reduction Act. 10
      § 1926.6 Incorporation by reference. 13
   Subpart B - General Interpretations 21
      § 1926.10 Scope of subpart..21
      § 1926.11 Coverage under section 103 of the act distinguished. 21
      § 1926.12 Reorganization Plan No. 14 of 1950. 22
      § 1926.13 Interpretation of statutory terms. 30
      § 1926.14 Federal contract for "mixed" types of performance. 31
      § 1926.15 Relationship to the Service Contract Act; Walsh-Healey Public Contracts Act. 31
      § 1926.16 Rules of construction. 32
   Subpart C - General Safety and Health Provisions 32
      § 1926.20 General safety and health provisions. 33
      § 1926.21 Safety training and education. 34
      § 1926.22 Recording and reporting of injuries. [Reserved]..34
      § 1926.24 Fire protection and prevention. 34
      § 1926.25 Housekeeping. 35
      § 1926.26 Illumination. 35
      § 1926.27 Sanitation. 35
      § 1926.28 Personal protective equipment. 35
      § 1926.29 Acceptable certifications. 35
      § 1926.30 Shipbuilding and ship repairing. 35
      § 1926.32 Definitions. 36
      § 1926.33 Access to employee exposure and medical records. 37
      § 1926.34 Means of egress. 37
      § 1926.35 Employee emergency action plans. 37
   Subpart D - Occupational Health and Environmental Controls 38
      § 1926.50 Medical services and first aid. 39
Appendix A to § 1926.50 - First aid Kits (Non-Mandatory) 40
      § 1926.51 Sanitation..40

Table D-2 - Permissible Noise Exposures	43
§ 1926.53 Ionizing radiation.	44
§ 1926.54 Nonionizing radiation.	44
§ 1926.55 Gases, vapors, fumes, dusts, and mists.	45
Table 1 to § 1926.55 - Permissible Exposure Limits for Airborne Contaminants	51
§ 1926.56 Illumination.	70
Table D-3 - Minimum Illumination Intensities in Foot-Candles	70
§ 1926.57 Ventilation.	70
Table D-57.1 - Grinding and Abrasive Cutting-Off Wheels	76
Table D-57.2 - Buffing and Polishing Wheels	76
Table D-57.3 - Horizontal Single-Spindle Disc Grinder	77
Table D-57.4 - Horizontal Double-Spindle Disc Grinder	77
Table D-57.5 - Vertical Spindle Disc Grinder	77
Table D-57.6 - Grinding and Polishing Belts	78
Standard Buffing and Polishing Hood	84
Table D-57.7 - Minimum Maintained Velocities Into Spray Booths	93
Table D-57.8 - Lower Explosive Limit of Some Commonly Used Solvents	94
Table D-57.9 - Determination of Hazard Potential	98
Table D-57.10 - Determination of Rate of Gas, Vapor, or Mist Evolution[1]	98
Table D-57.11 - Control Velocities in Feet Per Minute (f.p.m.) for Undisturbed Locations	99
Table D-57.12 - Minimum Ventilation Rate in Cubic Feet of Air Per Minute Per Square Foot of Tank Area for Lateral Exhaust	100
§ 1926.59 Hazard communication.	106
§ 1926.60 Methylenedianiline.	106
Appendix A to § 1926.60 - Substance Data Sheet, for 4-4'Methylenedianiline	126
Appendix C to § 1926.60 - Medical Surveillance Guidelines for MDA	126
§ 1926.61 Retention of DOT markings, placards and labels.	126
§ 1926.62 Lead.	126
Appendix A to § 1926.62 - Substance Data Sheet for Occupational Exposure to Lead	147
I. Substance Identification	147
II. Health Hazard Data	148
Appendix B to § 1926.62 - Employee Standard Summary	150
I. Permissible Exposure Limit (PEL) - Paragraph (C)	150
II. Exposure Assessment - Paragraph (D)	150
III. Methods of Compliance - Paragraph (E)	152
IV. Respiratory Protection - Paragraph (F)	152
V. Protective Work Clothing and Equipment - Paragraph (G)	153
VI. Housekeeping - Paragraph (H)	154
VII. Hygiene Facilities and Practices - Paragraph (I)	155

VIII.	Medical Surveillance - Paragraph (J)	155
IX.	Medical Removal Protection - Paragraph (K)	158
X.	Employee Information and Training - Paragraph (L)	159
XI.	Signs - Paragraph (M)	159
XII.	Recordkeeping - Paragraph (N)	160
XIII.	Observation of Monitoring - Paragraph (O)	161
XIV.	For Additional Information	161

Appendix C to § 1926.62 - Medical Surveillance Guidelines 161
    Introduction 161
    I.   Medical Surveillance and Monitoring Requirements for Workers Exposed toInorganic Lead....162
    II. Adverse Health Effects of Inorganic Lead 165
    III. Medical Evaluation 168
    IV. Laboratory Evaluation 171
        § 1926.64 Process safety management of highly hazardous chemicals. 173
        § 1926.65 Hazardous waste operations and emergency response. 174
    Table D-65.1 - Minimum Illumination Intensities in Foot-Candles 196
    Table D-65.2 - Toilet Facilities 196

Appendices to § 1926.65 - Hazardous Waste Operations and EmergencyResponse 208
    A. Totally-encapsulating chemical protective suit pressure test 208
CAUTION 210
    B. Totally-encapsulating chemical protective suit qualitative leak test 210
5.0 - Safety precautions 212
6.0 - Test procedure 212
CAUTION 215
Appendix B to § 1926.65 - General Description and Discussion of theLevels of Protection and Protective Gear 215
Appendix C to § 1926.65 - Compliance Guidelines 219
Appendix D to § 1926.65 - References 226
Appendix E to § 1926.65 - Training Curriculum Guidelines 227
    § 1926.66 Criteria for design and construction of spray booths. 245
    Subpart E - Personal Protective and Life Saving Equipment 253
    § 1926.95 Criteria for personal protective equipment. 253
    § 1926.96 Occupational foot protection. 254
    § 1926.97 Electrical protective equipment. 254
Table E-3 - Glove Tests - Water Level[12] 261
Table E-4 - Rubber Insulating Equipment, Voltage Requirements 261
Table E-5 - Rubber Insulating Equipment, Test Intervals 262
    § 1926.98 [Reserved] 262
    § 1926.101 Hearing protection. 263

§ 1926.102 Eye and face protection. 263
Table E-1 - Filter Lens Shade Numbers for Protection Against Radiant Energy 264
Table E-2 - Selecting Laser Safety Glass 265
  § 1926.103 Respiratory protection. 265
  § 1926.104 Safety belts, lifelines, and lanyards. 266
  § 1926.105 Safety nets. 266
  § 1926.106 Working over or near water. 267
  § 1926.107 Definitions applicable to this subpart. 267
  Subpart F - Fire Protection and Prevention 267
  § 1926.150 Fire protection. 267
  § 1926.151 Fire prevention. 270
  § 1926.152 Flammable liquids. 272
Table F-2 273
Table F-10 - Wetted Area Versus Cubic Feet (Meters) Free Air Per Hour 280
Table F-11 - Vent Line Diameters 288
Table F-3 301
Table F-31 301
  § 1926.154 Temporary heating devices. 302
Table F-4 303
  Subpart G - Signs, Signals, and Barricades 305
  § 1926.200 Accident prevention signs and tags. 305
  § 1926.201 Signaling. 307
  Subpart H - Materials Handling, Storage, Use, and Disposal 307
  § 1926.250 General requirements for storage. 307
  § 1926.251 Rigging equipment for material handling. 309
Table H-1 - Maximum Allowable Wear at any Point of Link 315
  § 1926.252 Disposal of waste materials. 315
  Subpart I - Tools - Hand and Power 316
  § 1926.300 General requirements. 316
  § 1926.301 Hand tools. 320
  § 1926.302 Power-operated hand tools. 320
  § 1926.303 Abrasive wheels and tools. 322
  § 1926.304 Woodworking tools. 324
  § 1926.305 Jacks - lever and ratchet, screw, and hydraulic. 325
  § 1926.306 Air receivers. 325
  § 1926.307 Mechanical power-transmission apparatus. 326
  Subpart J - Welding and Cutting 334
  § 1926.350 Gas welding and cutting. 334

§ 1926.351 Arc welding and cutting. 337
§ 1926.352 Fire prevention. 339
§ 1926.353 Ventilation and protection in welding, cutting, and heating. 340
§ 1926.354 Welding, cutting, and heating in way of preservative coatings. 342
Subpart K - Electrical 342
§ 1926.400 Introduction. 343
§ 1926.401 [Reserved] 343
§ 1926.402 Applicability. 343
§ 1926.403 General requirements. 344
Table K-1 - Working Clearances 345
Table K-2 - Minimum Depth of Clear Working Space in Front of ElectricEquipment 347
Table K-3 - Elevation of Unguarded Energized Parts Above Working Space 348
§ 1926.404 Wiring design and protection. 348
Table K-4 - Receptacle Ratings for Various Size Circuits 350
§ 1926.406 Specific purpose equipment and installations. 365
§ 1926.407 Hazardous (classified) locations. 367
§ 1926.408 Special systems. 369
§§ 1926.409-1926.415 [Reserved] 372
§ 1926.416 General requirements. 372
§ 1926.417 Lockout and tagging of circuits. 373
§§ 1926.418-1926.430 [Reserved] 373
§ 1926.431 Maintenance of equipment. 373
§ 1926.432 Environmental deterioration of equipment. 373
§§ 1926.433-1926.440 [Reserved] 374
§ 1926.441 Batteries and battery charging. 374
§§ 1926.442-1926.448 [Reserved] 374
§ 1926.449 Definitions applicable to this subpart. 374
Subpart L - Scaffolds 384
§ 1926.450 Scope, application and definitions applicable to this subpart. 385
§ 1926.451 General requirements. 390
§ 1926.452 Additional requirements applicable to specific types of scaffolds. 404
§ 1926.453 Aerial lifts. 413
§ 1926.454 Training requirements. 416

PART 1926 - SAFETY AND HEALTH REGULATIONS FORCONSTRUCTION

Authority: 40 U.S.C. 3704; 29 U.S.C. 653, 655, and 657; and Secretary of Labor's Order No. 12-71 (36 FR 8754), 8-76 (41 FR 25059), 9-83 (48 FR 35736), 1-90 (55 FR 9033), 6-96 (62 FR 111), 3-2000 (65 FR 50017), 5-2002 (67 FR 65008), 5-2007 (72 FR 31159), 4-2010 (75 FR 55355), 1-2012 (77 FR 3912), or 8-2020 (85 FR 58393), as applicable;

and 29 CFR part 1911, unless otherwise noted

Sections 1926.58, 1926.59, 1926.60, and 1926.65 also issued under 5 U.S.C. 553 and 29 CFR part 1911.

Section 1926.61 also issued under 49 U.S.C. 1801-1819 and 5 U.S.C. 553.

Section 1926.62 also issued under sec. 1031, Public Law 102-550, 106 Stat. 3672 (42 U.S.C. 4853).

Section 1926.65 also issued under sec. 126, Public Law 99-499, 100 Stat. 1614 (reprinted at 29 U.S.C.A. 655 Note)

and 5 U.S.C. 553.

Source: 44 FR 8577, Feb. 9, 1979; 44 FR 20940, Apr. 6, 1979, unless otherwise noted.

Editorial Notes: 1. At 44 FR 8577, Feb. 9, 1979, and corrected at 44 FR 20940, Apr. 6, 1979, OSHA reprinted without change the entire text of 29 CFR part 1926 together with certain General Industry Occupational Safety and Health Standards contained in 29 CFR part 1910, which have been identified as also applicable to construction work. This republication developed a single set of OSHA regulations for both labor and management forces within the construction industry.

2. Nomenclature changes to part 1926 appear at 84 FR 21597, May 14, 2019.

Subpart A - General

Authority: 40 U.S.C. 3701 *et seq.;* 29 U.S.C. 653, 655, 657; Secretary of Labor's Order No. 12-71 (36 FR 8754), 8-76 (41 FR 25059), 9-83 (48 FR 35736), 1-90 (55 FR 9033), 6-96 (62 FR 111), 3-2000 (65 FR 50017), 5-2002 (67 FR 65008), 5-2007 (72 FR 31160), 4-2010 (75 FR 55355), or 1-2012 (77 FR 3912), as applicable; and 29 CFR part 1911.

§ 1926.1 Purpose and scope.

(a) This part sets forth the safety and health standards promulgated by the Secretary of Labor under section 107 of the Contract Work Hours and Safety Standards Act. The standards are published in subpart C of this part and following subparts.

(b) Subpart B of this part contains statements of general policy and interpretations of section 107 of the Contract Work Hours and Safety Standards Act having general applicability.

§ 1926.2 Variances from safety and health standards.

(a) Variances from standards which are, or may be, published in this part may be granted under the same circumstances whereunder variances may be granted under section 6(b)(A) or 6(d) of the Williams-Steiger Occupational Safety and Health Act of 1970 (29 U.S.C. 65). The procedures for the granting of variances and for related relief under this part are those published in part 1905 of this title.

(b) Any requests for variances under this section shall also be considered requests for variances under the Williams-Steiger Occupational Safety and Health Act of 1970, and any requests for variances under Williams-Steiger Occupational Safety and Health Act with respect to construction safety or health standards shall be considered to be also variances under the Construction Safety Act. Any variance from a construction safety or health standard which is contained in this part and which is incorporated by reference in part 1910 of this title shall be deemed a variance from the standard under both the Construction Safety Act and the Williams-Steiger Occupational Safety and Health Act of 1970.

§ 1926.3 Inspections - right of entry.

(a) It shall be a condition of each contract which is subject to section 107 of the Contract Work Hours and Safety Standards Act that the Secretary of Labor or any authorized representative shall have a right of entry to any site of contract performance for the following purposes:

　(1) To inspect or investigate the matter of compliance with the safety and health standards contained in subpart C of this part and following subparts; and

　(2) To carry out the duties of the Secretary under section 107(b) of the Act.

(b) For the purpose of carrying out his investigative duties under the Act, the Secretary of Labor may, by agreement, use with or without reimbursement the services, personnel, and facilities of any State or Federal agency. Any agreements with States under this section shall be similar to those provided for under the Walsh-Healey Public Contracts Act under 41 CFR part 50-205.

§ 1926.4 Rules of practice for administrative adjudications for enforcement of safety and healthstandards.

(a) The rules of practice for administrative adjudications for the enforcement of the safety and health standards contained in subpart C of this part and the following subparts shall be the same as those published in part 6 of this title with respect to safety and health violations of the Service Contract Act of 1965 (69 Stat. 1035), except as provided in paragraph (b) of this section.

(b) In the case of debarment, the findings required by section 107(d) of the Act shall be made by the hearing examiner or the Assistant Secretary of Labor for Occupational Safety and Health, as the case may be.
Whenever, as provided in section 107(d)(2), a contractor requests termination of debarment before the end of the 3-year period prescribed in that section, the request shall be filed in writing with the Assistant Secretary of Labor for Occupational Safety and Health who shall publish a notice in the FEDERAL REGISTER that the request has been received and afford interested persons an opportunity to be heard upon the request, and thereafter the provisions of part 6 of this title shall apply with respect to prehearing conferences, hearings and related matters, and decisions and orders.

§ 1926.5 OMB control numbers under the Paperwork Reduction Act.

The following sections or paragraphs each contain a collection of information requirement which has been approved by the Office of Management and Budget under the control number listed.

29 CFR citation	OMB control No.
1926.33	1218-0065
1926.50	1218-0093
1926.52	1218-0048
1926.53	1218-0103
1926.59	1218-0072
1926.60	1218-0183
1926.62	1218-0189
1926.64	1218-0200
1926.65	1218-0202
1926.103	1218-0099
1926.200	1218-0132
1926.250	1218-0093
1926.251	1218-0233
1926.403	1218-0130
1926.404	1218-0130
1926.405	1218-0130
1926.407	1218-0130
1926.408	1218-0130
1926.453(a)(2)	1218-0216
1926.502	1218-0197
1926.503	1218-0197
1926.550(a)(1)	1218-0115

29 CFR citation	OMB control No.
1926.550(a)(2)	1218-0115
1926.550(a)(4)	1218-0115
1926.550(a)(6)	1218-0113
1926.550(a)(11)	1218-0054
1926.550(a)(16)	1218-0115
1926.550(b)(2)	1218-0232
1926.550(g)	1218-0151
1926.552	1218-0231
1926.652	1218-0137
1926.703	1218-0095
1926.800	1218-0067
1926.803	1218-0067
1926.900	1218-0217
1926.903	1218-0227
1926.1080	1218-0069
1926.1081	1218-0069
1926.1083	1218-0069
1926.1090	1218-0069
1926.1091	1218-0069
1926.1101	1218-0134
1926.1103	1218-0085
1926.1104	1218-0084
1926.1106	1218-0086
1926.1107	1218-0083
1926.1108	1218-0087
1926.1109	1218-0089
1926.1110	1218-0082
1926.1111	1218-0090
1926.1112	1218-0080
1926.1113	1218-0079
1926.1114	1218-0088
1926.1115	1218-0044
1926.1116	1218-0081
1926.1117	1218-0010
1926.1118	1218-0104
1926.1124	1218-0267
1926.1126	1218-0252
1926.1127	1218-0186
1926.1128	1218-0129
1926.1129	1218-0128

29 CFR citation	OMB control No.
1926.1144	1218-0101
1926.1145	1218-0126
1926.1147	1218-0108
1926.1148	1218-0145
1926.1153	1218-0266
1926.1203	1218-0258
1926.1204	1218-0258
1926.1205	1218-0258
1926.1206	1218-0258
1926.1207	1218-0258
1926.1208	1218-0258
1926.1209	1218-0258
1926.1210	1218-0258
1926.1211	1218-0258
1926.1212	1218-0258
1926.1213	1218-0258
1926.1402	1218-0261
1926.1403	1218-0261
1926.1404	1218-0261
1926.1406	1218-0261
1926.1407	1218-0261
1926.1408	1218-0261
1926.1409	1218-0261
1926.1410	1218-0261
1926.1411	1218-0261
1926.1412	1218-0261
1926.1413	1218-0261
1926.1414	1218-0261
1926.1417	1218-0261
1926.1423	1218-0261
1926.1424	1218-0261
1926.1427	1218-0270
1926.1428	1218-0261
1926.1431	1218-0261
1926.1433	1218-0261
1926.1434	1218-0261
1926.1435	1218-0261
1926.1436	1218-0261
1926.1437	1218-0261
1926.1441	1218-0261

[61 FR 5509, Feb. 13, 1996, as amended at 63 FR 3814, Jan. 27, 1998; 63 FR 13340, Mar. 19, 1998; 63 FR 17094, Apr. 8, 1998; 64 FR 18810, Apr. 16, 1999; 71 FR 38086, July 5, 2006; 75 FR 68430, Nov. 8, 2010; 81 FR 48710, July 26, 2016; 81 FR 53268, Aug. 12, 2016; 83 FR 9703, Mar. 7, 2018; 84 FR 34785, July 19, 2019]

§ 1926.6 Incorporation by reference.

(a) The standards of agencies of the U.S. Government, and organizations which are not agencies of the U.S. Government which are incorporated by reference in this part, have the same force and effect as other standards in this part. Only the mandatory provisions (*i.e.,* provisions containing the word "shall" or other mandatory language) of standards incorporated by reference are adopted as standards under the Occupational Safety and Health Act.

(b) The standards listed in this section are incorporated by reference into this part with the approval of the Director of the Federal Register in accordance with 5 U.S.C. 552(a) and 1 CFR part 51. To enforce any edition other than that specified in this section, OSHA must publish a document in the FEDERAL REGISTER and the material must be available to the public.

(c) Copies of standards listed in this section and issued by private standards organizations are available for purchase from the issuing organizations at the addresses or through the other contact information listed below for these private standards organizations. In addition, the standards are available for inspection at any Regional Office of the Occupational Safety and Health Administration (OSHA), or at the OSHA Docket Office, U.S. Department of Labor, 200 Constitution Avenue NW, Room N-3508, Washington, DC 20210; telephone: 202-693-2350 (TTY number: 877-889-5627). These standards are also available for inspection at the National Archives and Records Administration (NARA). For information on the availability of these standards at NARA, telephone: 202-741-6030, or go to *www.archives.gov/federal-register/cfr/ibr- locations.html*.

(d) The following material is available for purchase from the American Conference of Governmental Industrial Hygienists (ACGIH), 1330 Kemper Meadow Drive, Cincinnati, OH 45240; telephone: 513-742-6163; fax: 513-742-3355; e-mail: *mail@acgih.org*, Web site: *http://www.acgih.org*.

(1) Threshold Limit Values of Airborne Contaminants for 1970, 1970, IBR approved for § 1926.55(a) and appendix A of § 1926.55.

(2) [Reserved]

(e) The following material is available for purchase from the American National Standards Institute (ANSI), 25 West 43rd Street, Fourth Floor, New York, NY 10036; telephone: 212-642-4900; fax: 212-302-1286; e-mail: *info@ansi.org;* Web site: *http://www.ansi.org/*.

(1) ANSI A10.3-1970, Safety Requirements for Explosive-Actuated Fastening Tools, IBR approved for § 1926.302(e).

(2) ANSI A10.4-1963, Safety Requirements for Workmen's Hoists, IBR approved for §

(3) 1926.552(c).ANSI A10.5-1969, Safety Requirements for Material Hoists, IBR approved for §

(4) 1926.552(b).

ANSI A11.1-1965 (R1970), Practice for Industrial Lighting, IBR approved for § 1926.56(b).

(5) ANSI A17.1-1965, Elevators, Dumbwaiters, Escalators, and Moving Walks, IBR approved for § 1926.552(d).

(6) ANSI A17.1a-1967, Elevators, Dumbwaiters, Escalators, and Moving Walks Supplement, IBR approved for § 1926.552(d).

(7) ANSI A17.1b-1968, Elevators, Dumbwaiters, Escalators, and Moving Walks Supplement, IBR approved for § 1926.552(d).

(8) ANSI A17.1c-1969, Elevators, Dumbwaiters, Escalators, and Moving Walks Supplement, IBR approved for § 1926.552(d).

(9) ANSI A17.1d-1970, Elevators, Dumbwaiters, Escalators, and Moving Walks Supplement, IBR approved for § 1926.552(d).

(10) ANSI A17.2-1960, Practice for the Inspection of Elevators (Inspector's Manual), IBR approved for §1926.552(d).

(11) ANSI A17.2a-1965, Practice for the Inspection of Elevators (Inspector's Manual) Supplement, IBR approved for § 1926.552(d).

(12) ANSI A17.2b-1967, Practice for the Inspection of Elevators (Inspector's Manual) Supplement, IBR approved for § 1926.552(d).

(13) ANSI A92.2-1969, Vehicle Mounted Elevating and Rotating Work Platforms, IBR approved for §§ 1926.453(a) and 1926.453(b).

(14) ANSI B7.1-1970, Safety Code for the Use, Care, and Protection of Abrasive Wheels, IBR approved for §§ 1926.57(g), 1926.303(b), 1926.303(c), and 1926.303(d).

(15) ANSI B20.1-1957, Safety Code for Conveyors, Cableways, and Related Equipment, IBR approved for §1926.555(a).

(16) ANSI B56.1-1969, Safety Standards for Powered Industrial Trucks, IBR approved for §

(17)-(22) 1926.602(c).[Reserved]

(23) ANSI O1.1-1961, Safety Code for Woodworking Machinery, IBR approved for § 1926.304(f).

(24) ANSI Z35.1-1968, Specifications for Accident Prevention Signs; IBR approved for § 1926.200(b), (c), and 1 (i). Copies available for purchase from the IHS Standards Store, 15 Inverness Way East, Englewood, CO 80112; telephone: 1-877-413-5184; Web site: *www.global.ihs.com*.

(25) ANSI Z35.2-1968, Specifications for Accident Prevention Tags, IBR approved for §

(26) 1926.200(i).ANSI Z49.1-1967, Safety in Welding and Cutting, IBR approved for §

(27) 1926.350(j).

USA Z53.1-1967 (also referred to as ANSI Z53.1-1967), Safety Color Code for Marking Physical Hazards, ANSI approved October 9, 1967; IBR approved for § 1926.200(c). Copies available for purchase from the IHS Standards Store, 15 Inverness Way East, Englewood, CO 80112; telephone:1-877-413-5184; Web site: *www.global.ihs.com*.

(28) ANSI Z535.1-2006 (R2011), Safety Colors, reaffirmed July 19, 2011; IBR approved for § 1926.200(c).Copies available for purchase from the:

  (i) American National Standards Institute's e-Standards Store, 25 W 43rd Street, 4th Floor, NewYork, NY 10036; telephone: 212-642-4980; Web site: *http://webstore.ansi.org/;*

(ii) IHS Standards Store, 15 Inverness Way East, Englewood, CO 80112; telephone: 877-413-5184; Web site: *www.global.ihs.com;* or

(iii) TechStreet Store, 3916 Ranchero Dr., Ann Arbor, MI 48108; telephone: 877-699-9277; Web site: *www.techstreet.com.*

(29) ANSI Z535.2-2011, Environmental and Facility Safety Signs, published September 15, 2011; IBR approved for § 1926.200(b), (c), and (i). Copies available for purchase from the:

(i) American National Standards Institute's e-Standards Store, 25 W 43rd Street, 4th Floor, NewYork, NY 10036; telephone: 212-642-4980; Web site: *http://webstore.ansi.org/;*

(ii) IHS Standards Store, 15 Inverness Way East, Englewood, CO 80112; telephone: 877-413-5184; Web site: *www.global.ihs.com;* or

(iii) TechStreet Store, 3916 Ranchero Dr., Ann Arbor, MI 48108; telephone: 877-699-9277; Web site: *www.techstreet.com.*

(30) ANSI Z535.5-2011, Safety Tags and Barricade Tapes (for Temporary Hazards), published September15, 2011, including Errata, November 14, 2011; IBR approved for § 1926.200(h) and (i). Copies available for purchase from the:

(i) American National Standards Institute's e-Standards Store, 25 W 43rd Street, 4th Floor, NewYork, NY 10036; telephone: 212-642-4980; Web site: *http://webstore.ansi.org/;*

(ii) IHS Standards Store, 15 Inverness Way East, Englewood, CO 80112; telephone: 877-413-5184; Web site: *www.global.ihs.com;* or

(iii) TechStreet Store, 3916 Ranchero Dr., Ann Arbor, MI 48108; telephone: 877-699-9277; Web site: *www.techstreet.com.*

(31) ANSI/ISEA Z87.1-2010, Occupational and Educational Personal Eye and Face Protection Devices, Approved April 3, 2010; IBR approved for § 1926.102(b). Copies are available for purchase from:

(i) American National Standards Institute's e-Standards Store, 25 W 43rd Street, 4th Floor, NewYork, NY 10036; telephone: (212) 642-4980; Web site: *http://webstore.ansi.org/;*

(ii) IHS Standards Store, 15 Inverness Way East, Englewood, CO 80112; telephone: (877) 413-5184;Web site: *http://global.ihs.com;* or

(iii) TechStreet Store, 3916 Ranchero Dr., Ann Arbor, MI 48108; telephone: (877) 699-9277; Web site: *http://techstreet.com.*

(32) ANSI Z87.1-2003, Occupational and Educational Personal Eye and Face Protection Devices, Approved June 19, 2003; IBR approved for § 1926.102(b). Copies available for purchase from the:

(i) American National Standards Institute's e-Standards Store, 25 W 43rd Street, 4th Floor, NewYork, NY 10036; telephone: (212) 642-4980; Web site: *http://webstore.ansi.org/;*

(ii) IHS Standards Store, 15 Inverness Way East, Englewood, CO 80112; telephone: (877) 413-5184;Web site: *http://global.ihs.com;* or

(iii) TechStreet Store, 3916 Ranchero Dr., Ann Arbor, MI 48108; telephone: (877) 699-9277; Web site: *http://techstreet.com*.

(33) ANSI Z87.1-1989 (R-1998), Practice for Occupational and Educational Eye and Face Protection, Reaffirmation approved January 4, 1999; IBR approved for § 1926.102(b). Copies are available forpurchase from:

(i) American National Standards Institute's e-Standards Store, 25 W 43rd Street, 4th Floor, NewYork, NY 10036; telephone: (212) 642-4980; Web site: *http://webstore.ansi.org/;*

(ii) IHS Standards Store, 15 Inverness Way East, Englewood, CO 80112; telephone: (877) 413-5184;Web site: *http://global.ihs.com;* or

(iii) TechStreet Store, 3916 Ranchero Dr., Ann Arbor, MI 48108; telephone: (877) 699-9277; Web site: *http://techstreet.com*.

(34) American National Standards Institute (ANSI) Z89.1-2009, American National Standard for Industrial Head Protection, approved January 26, 2009; IBR approved for § 1926.100(b)(1)(i). Copies of ANSI Z89.1-2009 are available for purchase only from the International Safety Equipment Association, 1901 North Moore Street, Arlington, VA 22209-1762; telephone: 703-525-1695; fax: 703-528-2148; Web site: *www.safetyequipment.org*.

(35) American National Standards Institute (ANSI) Z89.1-2003, American National Standard for Industrial Head Protection; IBR approved for § 1926.100(b)(1)(ii). Copies of ANSI Z89.1-2003 are available for purchase only from the International Safety Equipment Association, 1901 North Moore Street, Arlington, VA 22209-1762; telephone: 703-525-1695; fax: 703-528-2148; Web site: *www.safetyequipment.org*.

(36) American National Standards Institute (ANSI) Z89.1-1997, American National Standard for Personnel Protection - Protective Headwear for Industrial Workers - Requirements; IBR approved for § 1926.100(b)(1)(iii). Copies of ANSI Z89.1-1997 are available for purchase only from the International Safety Equipment Association, 1901 North Moore Street, Arlington, VA 22209-1762; telephone:
703-525-1695; fax: 703-528-2148; Web site: *www.safetyequipment.org*.

(f) The following material is available for purchase from standards resellers such as the Document Center Inc., 111 Industrial Road, Suite 9, Belmont, CA 94002; telephone: 650-591-7600; fax: 650-591-7617; e-mail: *info@document-center.com*; Web site: *http://www.document-center.com/*:

(1) ANSI B15.1-1953 (R1958), Safety Code for Mechanical Power-Transmission Apparatus, revised 1958, IBR approved for § 1926.300(b)(2).

(2) ANSI B30.5-1968, Crawler, Locomotive, and Truck Cranes, approved Dec. 16, 1968, IBR approved for
§ 1926.1433(a).

(g) The following material is available for purchase from the American Society for Testing and Materials (ASTM), ASTM International, 100 Barr Harbor Drive, PO Box C700, West Conshohocken, PA, 19428-2959;telephone: 610-832-9585; fax: 610-832-9555; e-mail: *service@astm.org;* Web site: *http://www.astm.org/*:

(1) ASTM A370-1968, Methods and Definitions for Mechanical Testing and Steel Products, IBR approved for § 1926.1001(f).

(2) [Reserved]

(3) ASTM D56-1969, Standard Method of Test for Flash Point by the Tag Closed Tester, IBR approved for § 1926.155(i).

(4) ASTM D93-1969, Standard Method of Test for Flash Point by the Pensky Martens Closed Tester, IBR approved for § 1926.155(i).

(5) ASTM D323-1958 (R1968), Standard Method of Test for Vapor Pressure of Petroleum Products (Reid Method), IBR approved for § 1926.155(m).

(h) The following material is available for purchase from the American Society of Mechanical Engineers (ASME), Three Park Avenue, New York, NY 10016; telephone: 1-800-843-2763; fax: 973-882-1717; e-mail: *infocentral@asme.org;* Web site: *http://www.asme.org/:*

(1) ASME B30.2-2005, Overhead and Gantry Cranes (Top Running Bridge, Single or Multiple Girder, Top Running Trolley Hoist), issued Dec. 30, 2005 ("ASME B30.2-2005"), IBR approved for § 1926.1438(b).

(2) ASME B30.5-2004, Mobile and Locomotive Cranes, issued Sept. 27, 2004 ("ASME B30.5-2004"), IBR approved for §§ 1926.1414(b); 1926.1414(e); 1926.1433(b).

(3) ASME B30.7-2001, Base-Mounted Drum Hoists, issued Jan. 21, 2002 ("ASME B30.7-2001"), IBR approved for § 1926.1436(e).

(4) ASME B30.14-2004, Side Boom Tractors, issued Sept. 20, 2004 ("ASME B30.14-2004"), IBR approved for § 1926.1440(c).

(5) ASME Boiler and Pressure Vessel Code, Section VIII, 1968, IBR approved for §§ 1926.152(i), 1926.306(a), and 1926.603(a).

(6) ASME Power Boilers, Section I, 1968, IBR approved for § 1926.603(a).

(i) The following material is available for purchase from the American Society of Agricultural and Biological Engineers (ASABE), 2950 Niles Road, St. Joseph, MI 49085; telephone: 269-429-0300; fax: 269-429-3852; e-mail: *hq@asabe.org;* Web site: *http://www.asabe.org/:*

(1) ASAE R313.1-1971, Soil Cone Penetrometer, reaffirmed 1975, IBR approved for §

(2) 1926.1002(e).[Reserved]

(j) The following material is available for purchase from the American Welding Society (AWS), 550 N.W. LeJeune Road, Miami, Florida 33126; telephone: 1-800-443-9353; Web site: *http://www.aws.org/:*

(1) AWS D1.1/D1.1M:2002, Structural Welding Code - Steel, 18th ed., ANSI approved Aug. 31, 2001 ("AWS D1.1/D1.1M:2002"), IBR approved for § 1926.1436(c).

(2) ANSI/AWS D14.3-94, Specification for Welding Earthmoving and Construction Equipment, ANSI approved Jun. 11, 1993 ("ANSI/AWS D14.3-94"), IBR approved for § 1926.1436(c).

(k) The following material is available for purchase from the British Standards Institution (BSI), 389 Chiswick High Road, London, W4 4AL, United Kingdom; telephone: + 44 20 8996 9001; fax: + 44 20 8996 7001; e-mail: *cservices@bsigroup.com;* Web site: *http://www.bsigroup.com/:*

(1) BS EN 13000:2004, Cranes - Mobile Cranes, published Jan. 4, 2006 ("BS EN 13000:2004"), IBR

approved for § 1926.1433(c).

(2) BS EN 14439:2006, Cranes - Safety - Tower Cranes, published Jan. 31, 2007 ("BS EN 14439:2006"),IBR approved for § 1926.1433(c).

(l) The following material is available for purchase from the Bureau of Reclamation, United States Department of the Interior, 1849 C Street, NW., Washington DC 20240; telephone: 202-208-4501; Web site:*http://www.usbr.gov/*:

(1) Safety and Health Regulations for Construction, Part II, Sept. 1971, IBR approved for §

(2) 1926.1000(f).[Reserved]

(m) The following material is available for purchase from the California Department of Industrial Relations, 455 Golden Gate Avenue, San Francisco CA 94102; telephone: (415) 703-5070; e-mail: *info@dir.ca.gov*; Web site: *http://www.dir.ca.gov/*:

(1) Construction Safety Orders, IBR approved for §

(2) 1926.1000(f).[Reserved]

(n) The following material is available from the Federal Highway Administration, United States Department of Transportation, 1200 New Jersey Avenue SE, Washington, DC 20590; telephone: 202-366-4000; website:*www.fhwa.dot.gov/*:

(1) Manual on Uniform Traffic Control Devices for Streets and Highways, 2009 Edition, December 2009 (including Revision 1 dated May 2012 and Revision 2 dated May 2012), ("MUTCD") IBR approved for
§§ 1926.200(g) and 1926.201(a).

(2) [Reserved]

(o) The following material is available for purchase from the General Services Administration (GSA), 1800 F Street, NW., Washington, DC 20405; telephone: (202) 501-0800; Web site: *http://www.gsa.gov/*:

(1) QQ-P-416, Federal Specification Plating Cadmium (Electrodeposited), IBR approved for § 1926.104(e).

(2) [Reserved]

(p) The following material is available for purchase from the Institute of Makers of Explosives (IME), 1120 19th Street, NW., Suite 310, Washington, DC 20036; telephone: 202-429-9280; fax: 202-429-9280; e-mail: *info@ime.org*; Web site: *http://www.ime.org/*:

(1) IME Pub. No. 2, American Table of Distances for Storage of Explosives, Jun. 5, 1964, IBR approvedfor § 1926.914(a).

(2) IME Pub. No. 20, Radio Frequency Energy - A Potential Hazard in the Use of Electric Blasting Caps,Mar. 1968, IBR approved for § 1926.900(k).

(q) The following material is available from the International Labour Organization (ILO), 4 route des Morillons,CH-1211 Genève 22, Switzerland; telephone: +41 (0) 22 799 6111; fax: +41 (0) 22 798 8685; website:*//www.ilo.org/*:

(1) Guidelines for the Use of the ILO International Classification of Radiographs of Pneumoconioses, Revised Edition 2011, Occupational safety and health series; 22 (Rev.2011), IBR approved for §

1926.1101.

(2) [Reserved]

(r) The following material is available for purchase from the International Organization for Standardization(ISO), 1, ch. de la Voie-Creuse, Case postale 56, CH-1211 Geneva 20, Switzerland; telephone: + 41 22 749
01 11; fax: + 41 22 733 34 30; Web site: *http://www.iso.org/:*

(1) ISO 3471:2008(E), Earth-moving machinery - Roll-over protective structures - Laboratory tests and performance requirements, Fourth Edition, Aug. 8, 2008 ("ISO 3471:2008"), IBR approved for §§ 1926.1001(c) and 1926.1002(c).

(2) ISO 5700:2013(E), Tractors for agriculture and forestry - Roll-over protective structures - Static test method and acceptance conditions, Fifth Edition, May 1, 2013 ("ISO 5700:2013"), IBR approved for §1926.1002(c).

(3) ISO 27850:2013(E), Tractors for agriculture and forestry - Falling object protective structures - Test procedures and performance requirements, First Edition, May.01, 2013 ("ISO 27850:2013"), IBR approved for § 1926.1003(c).

(4) ISO 11660-1:2008(E), Cranes - Access, guards and restraints - Part 1: General, 2d ed., Feb. 15, 2008("ISO 11660-1:2008(E)"), IBR approved for § 1926.1423(c).

(5) ISO 11660-2:1994(E), Cranes - Access, guards and restraints - Part 2: Mobile cranes, 1994 ("ISO 11660-2:1994(E)"), IBR approved for § 1926.1423(c).

(6) ISO 11660-3:2008(E), Cranes - Access, guards and restraints - Part 3: Tower cranes, 2d ed., Feb. 15,2008 ("ISO 11660-3:2008(E)"), IBR approved for § 1926.1423(c).

(s) The following material is available for purchase from the National Fire Protection Association (NFPA), 1 Batterymarch Park, Quincy, MA 02169; telephone: 617-770-3000; fax: 617-770-0700; Web site: *http://www.nfpa.org/:*

(1) NFPA 10A-1970, Maintenance and Use of Portable Fire Extinguishers, IBR approved for § 1926.150(c).

(2) NFPA 13-1969, Standard for the Installation of Sprinkler Systems, IBR approved for §

(3) 1926.152(d).NFPA 30-1969, The Flammable and Combustible Liquids Code, IBR approved for §

(4) 1926.152(c).

NFPA 80-1970, Standard for Fire Doors and Windows, Class E or F Openings, IBR approved for § 1926.152(b).

(5) NFPA 251-1969, Standard Methods of Fire Test of Building Construction and Material, IBR approvedfor §§ 1926.152(b) and 1926.155(f).

(6) NFPA 385-1966, Standard for Tank Vehicles for Flammable and Combustible Liquids, IBR approvedfor § 1926.152(g).

(t) The following material is available for purchase from the Power Crane and Shovel Association (PCSA),6737 W. Washington Street, Suite 2400, Milwaukee, WI 53214; telephone: 1-800-369-2310; fax:
414-272-1170; Web site: *http://www.aem.org/CBC/ProdSpec/PCSA/:*

(1) PCSA Std. No. 1, Mobile Crane and Excavator Standards, 1968, IBR approved for § 1926.602(b).

(2) PCSA Std. No. 2, Mobile Hydraulic Crane Standards, 1968 ("PCSA Std. No. 2 (1968)"), IBR approved for §§ 1926.602(b) and 1926.1433(a).

(3) PCSA Std. No. 3, Mobile Hydraulic Excavator Standards, 1969, IBR approved for § 1926.602(b).

(u) The following material is available from the Society of Automotive Engineers (SAE), 400 Commonwealth Drive, Warrendale, PA 15096; telephone: 1-877-606-7323; fax: 724-776-0790; website: *www.sae.org/*:

(1) SAE 1970 Handbook, IBR approved for § 1926.602(b).

(2) SAE J166-1971, Trucks and Wagons, IBR approved for § 1926.602(a).

(3) SAE J167, Protective Frame with Overhead Protection-Test Procedures and Performance Requirements, approved July 1970, IBR approved for § 1926.1003(b).

(4) SAE J168, Protective Enclosures-Test Procedures and Performance Requirements, approved July1970, IBR approved for § 1926.1002(b).

(5) SAE J185 (reaf. May 2003), Access Systems for Off-Road Machines, reaffirmed May 2003 ("SAEJ185 (May 1993)"), IBR approved for § 1926.1423(c).

(6) SAE J236-1971, Self-Propelled Graders, IBR approved for § 1926.602(a).

(7) SAE J237-1971, Front End Loaders and Dozers, IBR approved for §

(8) 1926.602(a).SAE J319b-1971, Self-Propelled Scrapers, IBR approved for §

(9) 1926.602(a).

SAE J320a, Minimum Performance Criteria for Roll-Over Protective Structure for Rubber-Tired, Self- Propelled Scrapers, revised July 1969 (editorial change July 1970), IBR approved for § 1926.1001(b).

(10) SAE J321a-1970, Fenders for Pneumatic-Tired Earthmoving Haulage Equipment, IBR approved for §1926.602(a).

(11) SAE J333a-1970, Operator Protection for Agricultural and Light Industrial Tractors, IBR approved for
§ 1926.602(a).

(12) SAE J334a, Protective Frame Test Procedures and Performance Requirements, revised July 1970,IBR approved for § 1926.1002(b).

(13) SAE J386-1969, Seat Belts for Construction Equipment, IBR approved for § 1926.602(a).

(14) SAE J394, Minimum Performance Criteria for Roll-Over Protective Structure for Rubber-Tired Front End Loaders and Rubber-Tired Dozers, approved July 1969 (editorial change July 1970), IBR approved for § 1926.1001(b).

(15) SAE J395, Minimum Performance Criteria for Roll-Over Protective Structure for Crawler Tractors and Crawler-Type Loaders, approved July 1969 (editorial change July 1970), IBR approved for § 1926.1001(b).

(16) SAE J396, Minimum Performance Criteria for Roll-Over Protective Structure for Motor Graders, approved July 1969 (editorial change July 1970), IBR approved for § 1926.1001(b).

(17) SAE J397, Critical Zone Characteristics and Dimensions for Operators of Construction and Industrial Machinery, approved July 1969, IBR approved for § 1926.1001(b).

(18) SAE J987 (rev. Jun. 2003), Lattice Boom Cranes - Method of Test, revised Jun. 2003 ("SAE J987(Jun. 2003)"), IBR approved for § 1926.1433(c).

(19) SAE J1063 (rev. Nov. 1993), Cantilevered Boom Crane Structures - Method of Test, revised Nov. 1993("SAE J1063 (Nov. 1993)"), IBR approved for § 1926.1433(c).

(v) The following material is available for purchase from the United States Army Corps of Engineers, 441 G Street, NW., Washington, DC 20314; telephone: 202-761-0011; e-mail: *hq-publicaffairs@usace.army.mil;* Web site: *http://www.usace.army.mil/:*

(1) EM-385-1-1, General Safety Requirements, Mar. 1967, IBR approved for §

(2) 1926.1000(f). [Reserved]

[75 FR 48130, Aug. 9, 2010, as amended at 77 FR 37600, June 22, 2012; 78 FR 35566, June 13, 2013; 78 FR 66641, Nov. 6, 2013;
79 FR 20692, Apr. 11, 2014; 81 FR 16092, Mar. 25, 2016; 84 FR 21574, May 14, 2019]

Subpart B - General Interpretations

Authority: Sec. 107, Contract Work Hours and Safety Standards Act (Construction Safety Act) (40 U.S.C. 333).

§ 1926.10 Scope of subpart.

(a) This subpart contains the general rules of the Secretary of Labor interpreting and applying the construction safety and health provisions of section 107 of the Contract Work Hours and Safety Standards Act (83 Stat. 96). Section 107 requires as a condition of each contract which is entered into under legislation subject to Reorganization Plan Number 14 of 1950 (64 Stat. 1267), and which is for construction, alteration, and/or repair, including painting and decorating, that no contractor or subcontractor contracting for any part of the contract work shall require any laborer or mechanic employed in the performance of the contract to work in surroundings or under working conditions which are unsanitary, hazardous, or dangerous to his health or safety, as determined under construction safety and health standards promulgated by the Secretary by regulation.

§ 1926.11 Coverage under section 103 of the act distinguished.

(a) *Coverage under section 103.* It is important to note that the coverage of section 107 differs from that for the overtime requirements of the Contract Work Hours and Safety Standards Act. The application of the overtime requirements is governed by section 103, which subject to specific exemptions, includes:

(1) Federal contracts requiring or involving the employment of laborers or mechanics (thus including, but not limited to, contracts for construction), and

(2) contracts assisted in whole or in part by Federal loans, grants, or guarantees under any statute "providing wage standards for such work." The statutes "providing wage standards for such work" include statutes for construction which require the payment of minimum wages in accordance with prevailing wage findings by the Secretary of Labor in accordance with the Davis-Bacon Act. A provision to section 103 excludes from the overtime requirements work where the Federal assistance is only in the form of a loan guarantee or

insurance.

(b) **Coverage under section 107.** To be covered by section 107 of the Contract Work Hours and Safety Standards Act, a contract must be one which

(1) is entered into under a statute that is subject to Reorganization Plan No. 14 of 1950 (64 Stat. 1267); and

(2) is for "construction, alteration, and/or repair, including painting and decorating."

§ 1926.12 Reorganization Plan No. 14 of 1950.

(a) **General provisions.** Reorganization Plan No. 14 of 1950 relates to the prescribing by the Secretary of Labor of "appropriate standards, regulations, and procedures" with respect to the enforcement of labor standards under Federal and federally assisted contracts which are subject to various statutes subject to the Plan. The rules of the Secretary of Labor implementing the Plan are published in part 5 of this title.
Briefly, the statutes subject to the Plan include the Davis-Bacon Act, including its extension to Federal-aid highway legislation subject to 23 U.S.C. 113, and other statutes subject to the Plan by its original terms, statutes by which the Plan is expressly applied, such as the Contract Work Hours Standards Act by virtue of section 104(d) thereof.

(b) **The Plan.**

(1) The statutes subject to Reorganization Plan No. 14 of 1950 are cited and briefly described in the remaining paragraphs of this section. These descriptions are general in nature and not intended to convey the full scope of the work to be performed under each statute. The individual statutes should be resorted to for a more detailed scope of the work.

(2) **Federal-Aid Highway Acts.** The provisions codified in 23 U.S.C. 113 apply to the initial construction, reconstruction, or improvement work performed by contractors or subcontractors on highway projects on the Federal-aid systems, the primary and secondary, as well as their extensions in urban areas, and the Interstate System, authorized under the highway laws providing for the expenditure of Federal funds upon the Federal-aid system. As cited in 41 Op. A.G. 488, 496, the Attorney General ruled that the Federal-Aid Highway Acts are subject to Reorganization Plan No. 14 of 1950.

(3) **National Housing Act (12 U.S.C. 1713, 1715a, 1715e, 1715k, 1715l(d)(3) and (4), 1715v, 1715w, 1715x, 1743, 1747, 1748, 1748h-2, 1750g, 1715l(h)(1), 1715z(j)(1), 1715z-1, 1715y(d), Subchapter 1x-A and 1x-B, 1715z-7).** This act covers construction which is financed with assistance by the Federal Government through programs of loan and mortgage insurance for the following purposes:

(i) Rental Housing - Section 1713 provides mortgage and insurance on rental housing of eight or more units and on mobile-home courts.

(ii) Section 1715a - Repealed.

(iii) Cooperative Housing - Section 1715e authorizes mortgage insurance on cooperative housing of five or more units as well as supplementary loans for improvement of repair or resale of memberships.

(iv) Urban Renewal Housing - Section 1715k provides mortgage insurance on single family or multifamily housing in approved urban renewal areas.

(v) Low or Moderate Income Housing - Section 1715L(d) (3) and (4) insures mortgages on low-costsingle family or multifamily housing.

(vi) Housing for Elderly - Section 1715v provides mortgage insurance on rental housing for elderlyor handicapped persons.

(vii) Nursing Homes - Section 1715w authorizes mortgage insurance on nursing home facilities andmajor equipment.

(viii) Experimental Housing - Section 1715x provides mortgage insurance on single family or multifamily housing with experimental design of materials.

(ix) War Housing Insurance - Section 1743 not active.

(x) Yield Insurance - Section 1747 insures investment returns on multifamily housing.

(xi) Armed Services Housing - Section 1748b to assist in relieving acute shortage and urgent needfor family housing at or in areas adjacent to military installations.

(xii) Defense Housing for Impacted Areas - Section 1748h-2 provides mortgage insurance on singlefamily or multifamily housing for sale or rent primarily to military or civilian personnel of the Armed Services, National Aeronautics and Space Administration, or Atomic Energy Commission.

(xiii) Defense Rental Housing - Section 1750g provides for mortgage insurance in critical defense housing areas.

(xiv) Rehabilitation - Section 1715L (h)(1) provides mortgage insurance for nonprofit organizations to finance the purchase and rehabilitation of deteriorating or substandard housing for subsequent resale to low-income home purchasers. There must be located on the property five or more single family dwellings of detached, semidetached, or row construction.

(xv) Homeowner Assistance - Section 1715Z(j)(1) authorizes mortgage insurance to nonprofit organizations or public bodies or agencies executed to finance sale of individual dwellings tolower income individuals or families. Also includes the rehabilitation of such housing if it is deteriorating or substandard for subsequent resale to lower income home purchasers.

(xvi) Rental Housing Assistance - Section 1715Z-1 authorizes mortgage insurance and interest reduction payments on behalf of owners of rental housing projects designed for occupancy by lower income families. Payments are also authorized for certain State or locally aided projects.

(xvii) Condominium Housing - Section 1715y(d) provides mortgage insurance on property purchased for the development of building sites. This includes waterlines and water supply installations, sewer lines and sewage disposal installations, steam, gas, and electrical lines and installations,roads, streets, curbs, gutters, sidewalks, storm drainage facilities, and other installations or work.

(xviii) Group Medical Practice Facilities - Subchapter LX-B authorizes mortgage insurance for the financing of construction and equipment, of facilities for group practice of medicine, optometry,or dentistry.

(xix) Nonprofit Hospitals - 1715z-7 authorizes mortgage insurance to cover new and rehabilitated hospitals, including initial equipment.

(4) ***Hospital Survey and Construction Act, as amended by the Hospital and Medical Facilities Amendments of 1964 (42 U.S.C. 291e).*** The provisions of this Act cover construction contracts madeby State or local authorities or private institutions under Federal grant-in-aid programs for the construction of hospitals and other medical facilities.

(5) ***Federal Airport Act (49 U.S.C. 1114(b)).*** The act provides grant-in-aid funds for airport construction limited to general site preparation runways, taxiways, aprons, lighting appurtenant thereto, and fire,rescue, and maintenance buildings. The act excludes construction intended for use as a public parking facility for passenger automobiles and the cost of construction of any part of an airport building except such of those buildings or parts of buildings to house facilities or activities directly related to the safety of persons at the airport.

(6) ***Housing Act of 1949 (42 U.S.C. 1459).*** Construction contracts awarded by local authorities financedwith the assistance of loans and grants from the Federal Government. The construction programsare for slum clearance and urban renewal which includes rehabilitation grants, neighborhood development programs, neighborhood renewal plans, community renewal, demolition projects, and assistance for blighted areas. See the Housing Act of 1964, paragraph (b)(21) of this section, concerning financial assistance for low-rent housing for domestic farm labor.

(7) ***School Survey and Construction Act of 1950 (20 U.S.C. 636).*** This act provides for a Federal grant-in- aid program to assist in the construction of schools in federally affected areas.

(8) ***Defense Housing & Community Facilities & Services Act of 1951 (42 U.S.C. 1592i).*** Inactive Program.
public housing and slum clearance projects awarded by local authorities. These projects are financed with the assistance of loans and grants from the Federal Government. The slum clearanceis the demolition and removal of buildings from any slum area to be used for a low-rent housing project.

(10) ***Federal Civil Defense Act of 1950 (50 U.S.C. App. 2281).*** This act provides for Federal assistance tothe several States and their political subdivisions in the field of civil defense which includes procurement, construction, leasing, or renovating of materials and facilities.

(11) ***Delaware River Basin Compact (sec. 15.1, 75 Stat. 714).*** This joint resolution creates, by intergovernmental compact between the United States, Delaware, New Jersey, New York, and Pennsylvania, a regional agency for planning, conservation, utilization, development, management and control of the water and related sources of the Delaware River.

(12) ***Cooperative Research Act (20 U.S.C. 332a(c)).*** This act provides Federal grants to a university, college, or other appropriate public or nonprofit private agency or institution for part or all of the costof constructing a facility for research or for research and related purposes. Research and related purposes means research, research training, surveys, or demonstrations in the field of education, or the dissemination of information derived therefrom, or all of such activities, including (but without limitation) experimental schools, except that such term does not include research, research training, surveys, or demonstrations in the field of sectarian instruction or the dissemination of information derived therefrom. Construction includes new buildings, and the acquisition, expansion, remodeling, replacement, and alteration of existing buildings and the equipping of new buildings and existing buildings.

(13) ***Health Professions Educational Assistance Act of 1963 (42 U.S.C. 292d (c)(4), 293a(c)(5)).*** The provisions of this act provide for grants to assist public and nonprofit medical, dental, and similarschools for the construction, expansion, or renovation of teaching facilities.

(14) ***Mental Retardation Facilities Construction Act (42 U.S.C. 295(a)(2)(D), 2662(5), 2675(a)(5)).*** This act authorizes Federal financial assistance in the construction of centers for research on mental retardation and related aspects of human development, of university-affiliated facilities for the mentally retarded and of facilities for the mentally retarded.

(15) ***Community Mental Health Centers Act (42 U.S.C. 2685(a)(5)).*** This act authorizes Federal grants for the construction of public and other nonprofit community mental health centers.

(16) ***Higher Education Facilities Act of 1963 (20 U.S.C. 753).*** This act authorizes the grant or loan of Federal funds to assist public and other nonprofit institutions of higher education in financing the construction, rehabilitation, or improvement of academic and related facilities in undergraduate and graduate schools.

(17) ***Vocational Educational Act of 1963 (20 U.S.C. 35f).*** This act provides for Federal grants to the various States for construction of area vocational education school facilities.

(18) ***Library Services and Construction Act (20 U.S.C. 355e(a)(4)).*** This act provides for Federal assistance to the various States for the construction of public libraries.
assist States and local public bodies and agencies thereof in financing the acquisition, construction, reconstruction, and improvement of facilities and equipment for use, by operation or lease or otherwise, in mass transportation service in urban areas and in coordinating such service with highway and other transportation in such areas.

(20) ***Economic Opportunity Act of 1964 (42 U.S.C. 2947).*** This act covers construction which is financed with assistance of the Federal Government for the following purposes:

   (i) Authorizes Federal assistance for construction of projects, buildings and works which will provide young men and women in rural and urban residential centers with education, vocational training, and useful work experience (Title I).

   (ii) Authorizes financial assistance for construction work planned and carried out at the community level for antipoverty programs (Title II):

      (*a*) Authorizes loans to low income rural families by assisting them to acquire or improve real estate or reduce encumbrances or erect improvements thereon, and to participate in cooperative associations and/or to finance nonagricultural enterprises which will enable such families to supplement their income (Title III);

      (*b*) Authorizes loans to local cooperative associations furnishing essential processing, purchasing, or marketing services, supplies, or facilities predominantly to low-income rural families (Title III);

      (*c*) Authorizes financial assistance to States, political subdivisions of States, public and nonprofit agencies, institutions, organizations, farm associations, or individuals in establishing housing, sanitation, education, and child day-care programs for migrants and other seasonally employed agricultural employees and their families (Title III).

   (iii) Authorizes loans or guarantees loans to small businesses for construction work (Title IV).

   (iv) Authorizes the payment of the cost of experimental, pilot, or demonstration projects to foster State programs providing construction work experience or training for unemployed fathers and needy people (Title V).

(21) ***Housing Act of 1964 (42 U.S.C. 1486(f); 42 U.S.C. 1452b(e)).*** Provides financial assistance for low-rent housing for domestic farm labor. The Act further provides for loans, through

public or private agencies, where feasible, to owners or tenants of property in urban renewal areas to finance rehabilitation required to conform the property to applicable code requirements or carry out the objectives of the urban renewal plan for the area.

(22) ***The Commercial Fisheries Research and Development Act of 1964 (16 U.S.C. 779e(b))***. This Act authorizes financial assistance to State agencies for construction projects designed for the research and development of the commercial fisheries resources of the Nation.

(23) ***The Nurse Training Act of 1964 (42 U.S.C. 296a(b)(5))***. This act provides for grants to assist in the construction of new facilities for collegiate, associate degree, and diploma schools of nursing, or replacement or rehabilitation of existing facilities of such schools.

(24) ***Elementary and Secondary Education Act of 1965 (20 U.S.C. 241i, 848)***. The purpose of the act is to provide financial assistance to local educational agencies serving areas with concentrations of children from low-income families for construction in connection with the expansion or improvement of their educational programs.

(25) ***Federal Water Pollution Control Act, as amended by the Water Quality Act of 1965 (3 U.S.C. 466e(g))***. Provides for financial assistance to States or municipalities for construction of facilities in connection with the prevention and control of water pollution. This includes projects that will control the discharge into any waters of untreated or inadequately treated sewage.

(26) ***Appalachian Regional Development Act of 1965 (40 U.S.C. App. 402)***. Authorizes Federal assistance in the construction of an Appalachian development highway system; construction of multicounty demonstration health facilities, hospitals, regional health, diagnostic and treatment centers, and other facilities for health; seal and fill voids in abandoned mines and to rehabilitate strip mine areas; construction of school facilities for vocational education; and to assist in construction of sewage treatment works.

(27) ***National Technical Institute for the Deaf Act (20 U.S.C. 684(b)(5))***. Provides for financial assistance for institutions of higher education for the establishment, construction, including equipment and operation, of a National Institution for the Deaf.

(28) ***Housing Act of 1959 (12 U.S.C. 1701(q)(c)(3))***. This act authorizes loans to nonprofit corporations to be used for the construction of housing and related facilities for elderly families. Also, the provisions of the act provide for rehabilitation, alteration, conversion or improvement of existing structures which are otherwise inadequate for proposed dwellings used by such families.

(29) ***College Housing Act of 1950, as amended (12 U.S.C. 1749a(f))***. This act provides for Federal loans to assist educational institutions in providing housing and other educational facilities for students and faculties.

(30) ***Housing and Urban Development Act of 1965 (42 U.S.C. 1500c-3, 3107)***. This act provides for Federal assistance for the following purposes:

   (i) Grants to States and local public bodies to assist in any construction work to be carried out under the open-space land and urban beautification provisions contained therein. It provides for parks and recreation areas, conservation of land and other natural resources, and historical and scenic purposes.

   (ii) Grants to local public bodies and agencies to finance specific projects for basic public water facilities (including works for the storage, treatment, purification, and distribution of water), and for basic public sewer facilities (other than "treatment works" as defined in the Federal Water Pollution Control Act).

(iii) Grants to any local public body or agency to assist in financing neighborhood facilities. These facilities must be necessary for carrying out a program of health, recreational, social, or similar community service and located so as to be available for the use of the area's low or moderate income residents.

(31) ***National Foundation on the Arts and the Humanities Act of 1965 (20 U.S.C. 954(k)).*** The act establishes the "National Foundation on the Arts and the Humanities" which may provide matching grants to groups (nonprofit organizations and State and other public organizations) and to individuals engaged in creative and performing arts for the entire range of artistic activity, including construction of necessary facilities.

(32) ***Public Works and Economic Development Act of 1965 (42 U.S.C. 3222).*** This act provides for Federal assistance for the following purposes:

(i) Grants for the acquisition or development of land or improvements for public works or development facility usage in redevelopment areas. It authorizes loans to assist in financing the purchase or development of land for public works which will assist in the creation of long-term employment opportunities in the area.

(ii) Loans for the purchase or development of land and facilities (including machinery and equipment) for industrial or commercial usage within redevelopment areas; guarantee of loans for working capital made to private borrowers by private lending institutions in connection with direct loan projects; and to contract to pay to, or on behalf of, business entities locating in redevelopment areas, a portion of the interest costs which they incur in financing their expansions from private sources.

(iii) Loans and grants to create economic development centers within designated county economic development districts.

(33) ***High-Speed Ground Transportation Study (40 U.S.C. 1636(b)).*** This act provides for financial assistance for construction activities in connection with research and development of different forms of high-speed ground transportation and demonstration projects relating to intercity rail passenger service.

(34) ***Heart Disease, Cancer and Stroke Amendments of 1965 (42 U.S.C. 299(b)(4)).*** This act provides for grants to public or nonprofit private universities, medical schools, research, institutions, hospitals, and other public and nonprofit agencies and institutions, or associations thereof to assist in construction and equipment of facilities in connection with research, training, demonstration of patient care, diagnostic and treatment related to heart disease, cancer, stroke, and other major diseases.

(35) ***Mental Retardation Facilities and Community Mental Health Centers Construction Act Amendments of 1965 (20 U.S.C. 618(g)).*** These provisions provide for grants to institutions of higher education for construction of facilities for research or for research and related purposes relating to education for mentally retarded, hard of hearing, deaf, speech impaired, visually handicapped, seriously emotionally disturbed, crippled, or other health impaired children who by reason thereof require special education.

(36) ***Vocational Rehabilitation Act Amendments of 1965 (29 U.S.C. 41a(b)(4)).*** This act authorizes grants to assist in meeting the costs of construction of public or other nonprofit workshops and rehabilitation facilities.

(37) ***Clean Air and Solid Waste Disposal Acts (42 U.S.C. 3256).*** This act provides for financial assistance to public (Federal, State, interstate, or local) authorities, agencies, and institutions, private agencies and institutions, and individuals in the construction of facilities for solid-waste disposal. The term construction includes the installation of initial equipment.

(38) ***Medical Library Assistance Act of 1965 (42 U.S.C. 280b-3(b)(3)).*** This act provides for grants to public or private non-profit agencies or institutions for the cost of construction of medical library facilities.

(39) ***Veterans Nursing Home Care Act (38 U.S.C. 5035(a)(8)).*** The construction industry health and safety standards do not apply to this act since it is not subject to Reorganization Plan No. 14 of 1950.

(40) ***National Capital Transportation Act of 1965 (40 U.S.C. 682(b)(4)).*** This act provides for Federal assistance to the National Capital Transportation Agency for construction of a rail rapid transit system and related facilities for the Nation's Capital.

(41) ***Alaska Centennial - 1967 (80 Stat. 82).*** The program under this legislation has expired.

(42) ***Model Secondary School for the Deaf Act (80 Stat. 1028).*** This act provides for funds to establish and operate, including construction and initial equipment of new buildings, expansion, remodeling, and alteration of existing buildings and equipment thereof, a model secondary school for the deaf to serve the residents of the District of Columbia and nearby States.

(43) ***Allied Health Professions Personnel Training Act of 1966 (42 U.S.C. 295h(b)(2)(E)).*** This act provides for grants to assist in the construction of new facilities for training centers for allied health professions, or replacement or rehabilitation of existing facilities for such centers.

(44) ***Demonstration Cities and Metropolitan Development Act of 1966 (42 U.S.C. 3310; 12 U.S.C. 1715c; 42 U.S.C. 1416).*** This act provides for Federal assistance for the following purposes:

   (i) Grants to assist in the construction, rehabilitation, alteration, or repair of residential property only if such residential property is designed for residential use for eight or more families to enable city demonstration agencies to carry out comprehensive city demonstration programs (42 U.S.C. 3310).

   (ii) Amends the National Housing Act (12 U.S.C. 1715c) and the Housing Act of 1937 (42 U.S.C. 1416). See these acts for coverage.

(45) ***Air Quality Act of 1967 (42 U.S.C. 1857j-3).*** This act provides for Federal assistance to public or nonprofit agencies, institutions, and organizations and to individuals, and contracts with public or private agencies, institutions, or persons for construction of research and development facilities and demonstration plants relating to the application of preventing or controlling discharges into the air of various types of pollutants.

(46) ***Elementary and Secondary Education Amendments of 1967 (Title VII - Bilingual Education Act) (20 U.S.C. 880b-6).*** This act provides for Federal assistance to local educational agencies or to an institution of higher education applying jointly with a local educational agency for minor remodeling projects in connection with bilingual education programs to meet the special needs of children with limited English-speaking ability in the United States.

(47) ***Vocational Rehabilitation Amendments of 1967 (29 U.S.C. 42a(c)(3)).*** This act authorizes Federal assistance to any public or nonprofit private agency or organization for the construction of a center for vocational rehabilitation of handicapped individuals who are both deaf and blind which shall beknown as the National Center for Deaf-Blind Youths and Adults. Construction includes new buildingsand expansion, remodeling, alteration and renovation of existing buildings, and initial equipment ofsuch new, newly acquired, expanded, remodeled, altered, or renovated buildings.

(48) ***National Visitor Center Facilities Act of 1968 (40 U.S.C. 808).*** This act authorizes agreements and leases with the owner of property in the District of Columbia known as Union Station for the use of all or a part of such property for a national visitor center to be known as the National Visitor Center.The agreements and leases shall provide for such alterations of the Union Station Building as necessary to provide adequate facilities for visitors. They also provide for the construction of a parking facility, including necessary approaches and ramps.

(49) ***Juvenile Delinquency Prevention and Control Act of 1968 (42 U.S.C. 3843).*** This act provides for Federal grants to State, county, municipal, or other public agency or combination thereof for the construction of facilities to be used in connection with rehabilitation services for the diagnosis, treatment, and rehabilitation of delinquent youths and youths in danger of becoming delinquent.

(50) ***Housing and Urban Development Act of 1968 (including New Communities Act of 1968) (42 U.S.C.3909).*** This act provides for Federal assistance for the following purposes:

   (i) Guarantees, and commitments to guarantee, the bonds, debentures, notes, and other obligations issued by new community developers to help finance new community developmentprojects.

   (ii) Amends section 212(a) of the National Housing Act, adding section 236 for "Rental Housing forLower Income Families" and section 242 "Mortgage Insurance for Nonprofit Hospitals" thereto.

(51) ***Public Health Service Act Amendment (Alcoholic and Narcotic Addict Rehabilitation Amendments of 1968) (42 U.S.C. 2681, et seq.).*** This act provides for grants to a public and nonprofit private agency or organization for construction projects consisting of any facilities (including post-hospitalization treatment facilities for the prevention and treatment of alcoholism or treatment of narcotic addicts.)

(52) ***Vocational Education Amendments of 1968 (20 U.S.C. 1246).*** This act provides for grants to States for the construction of area vocational education school facilities. The act further provides grants to public educational agencies, organizations, or institutions for construction of residential schools toprovide vocational education for the purpose of demonstrating the feasibility and desirability of suchschools. The act still further provides grants to State boards, to colleges and universities, to public educational agencies, organizations or institutions to reduce the cost of borrowing funds for the construction of residential schools and dormitories.

(53) ***Postal Reorganization Act (39 U.S.C. 410(d)(2)).*** This Act provides for construction, modification, alteration, repair, and other improvements of postal facilities located in leased buildings.

(54) ***Airport and Airway Development Act of 1970 (Pub. L. 91-258, section 52(b)(7)).*** This Act provides for Federal financial assistance to States and localities for the construction, improvement, or repair ofpublic airports.

(55) ***Public Law 91-230.*** This Act provides for federal financial assistance to institutions of higher learning for the construction of a National Center on Educational Media and Materials for theHandicapped. The program under this statute expires on July 1, 1971. Public Law 91-230, section 662(1).

(ii) ***Education of the Handicapped Act (20 U.S.C. 12326, 1404(a))***. This Act provides for financial assistance to States for construction, expansion, remodeling, or alteration of facilities for theeducation of handicapped children at the preschool, elementary school, and secondary schoollevels.

(56) ***Housing and Urban Development Act of 1970 (Pub. L. 91-609, section 707(b))***. This Act provides forgrants to States and local public agencies to help finance the development of open-space or other land in urban areas for open-space uses. This Act becomes effective on July 1, 1971.

(57) ***Developmental Disabilities Services and Facilities Construction Amendments of 1970 (Pub. L. 91-517, section 135(a)(5))***. This Act authorizes grants to States for construction of facilities for the provision of services to persons with developmental disabilities who are unable to pay for such services.

(58) ***Rail Passenger Service Act of 1970 (Pub. L. 91-518, section 405(d))***. This statute provides that the National Railroad Passenger Corporation may construct physical facilities necessary to intercity railpassenger operations within the basic national rail passenger system designated by the Secretary of Transportation.

(c) **VA and FHA housing.** In the course of the legislative development of section 107, it was recognized that section 107 would not apply to housing construction for which insurance was issued by the Federal Housing Authority and Veterans' Administration for individual home ownership. Concerning construction under the National Housing Act, Reorganization Plan No. 14 of 1950 applies to construction which is subject to the minimum wage requirements of section 212(a) thereof (12 U.S.C. 1715c).

§ 1926.13 Interpretation of statutory terms.

(a) The terms *construction, alteration,* and *repair* used in section 107 of the Act are also used in section 1 of the Davis-Bacon Act (40 U.S.C. 276a), providing minimum wage protection on Federal construction contracts, and section 1 of the Miller Act (40 U.S.C. 270a), providing performance and payment bond protection on Federal construction contracts. Similarly, the terms *contractor* and *subcontractor* are used in those statutes, as well as in Copeland (Anti-Kickback) Act (40 U.S.C. 276c) and the Contract Work Hoursand Safety Standards Act itself, which apply concurrently with the Miller Act and the Davis-Bacon Act onFederal construction contracts and also apply to most federally assisted construction contracts. The use of the same or identical terms in these statutes which apply concurrently with section 107 of the Act have considerable precedential value in ascertaining the coverage of section 107.

(b) It should be noted that section 1 of the Davis-Bacon Act limits minimum wage protection to laborers and mechanics "employed directly" upon the "site of the work." There is no comparable limitation in section 107 of the Act. Section 107 expressly requires as a self-executing condition of each covered contract that no contractor or subcontractor shall require "any laborer or mechanic employed in the performance of thecontract to work in surroundings or under working conditions which are unsanitary, hazardous, or dangerous to his health or safety" as these health and safety standards are applied in the rules of the Secretary of Labor.

(c) The term *subcontractor* under section 107 is considered to mean a person who agrees to perform any part of the labor or material requirements of a contract for construction, alteration or repair. Cf. MacEvoy Co. v. United States, 322 U.S. 102, 108-9 (1944). A person who undertakes to perform a portion of a contract involving the furnishing of supplies or materials will be considered a "subcontractor" under this part and section 107 if the work in question involves the performance of construction work and is to be performed:

(1) Directly on or near the construction site, or

(2) by the employer for the specific project on a customized basis. Thus, a supplier of materials which will become an integral part of the construction is a "subcontractor" if the supplier fabricates or assembles the goods or materials in question specifically for the construction project and the work involved may be said to be construction activity. If the goods or materials in question are ordinarily sold to other customers from regular inventory, the supplier is not a "subcontractor." Generally, the

furnishing of prestressed concrete beams and prestressed structural steel would be considered manufacturing; therefore a supplier of such materials would not be considered a "subcontractor." An example of material supplied "for the specific project on a customized basis" as that phrase is used in this section would be ventilating ducts, fabricated in a shop away from the construction job site and specifically cut for the project according to design specifications. On the other hand, if a contractor buys standard size nails from a foundry, the foundry would not be a covered "subcontractor." Ordinarily a contract for the supplying of construction equipment to a contractor would not, in and of itself, be considered a "subcontractor" for purposes of this part.

§ 1926.14 Federal contract for "mixed" types of performance.

(a) It is the intent of the Congress to provide safety and health protection of Federal, federally financed, or federally assisted construction. See, for example, H. Report No. 91-241, 91st Cong., first session, p. 1 (1969). Thus, it is clear that when a Federal contract calls for mixed types of performance, such as both manufacturing and construction, section 107 would apply to the construction. By its express terms, section 107 applies to a contract which is "for construction, alteration, and/or repair." Such a contract is not required to be exclusively for such services. The application of the section is not limited to contracts which permit an overall characterization as "construction contracts." The text of section 107 is not so limited.

(b) When the mixed types of performances include both construction and manufacturing, see also § 1926.15(b) concerning the relationship between the Walsh-Healey Public Contracts Act and section 107.

§ 1926.15 Relationship to the Service Contract Act; Walsh-Healey Public Contracts Act.

(a) A contract for "construction" is one for nonpersonal service. See, e.g., 41 CFR 1-1.208. Section 2(e) of the Service Contract Act of 1965 requires as a condition of every Federal contract (and bid specification therefor) exceeding $2,500, the "principal purpose" of which is to furnish services to the United States through the use of "service employees," that certain safety and health standards be met. See 29 CFR part 1925, which contains the Department rules concerning these standards. Section 7 of the Service Contract Act provides that the Act shall not apply to "any contract of the United States or District of Columbia for construction, alteration, and/or repair, including painting and

(b)   decorating of public buildings or public works." It is clear from the legislative history of section 107 that no gaps in coverage between the two statutes are intended.

(b)   The Walsh-Healey Public Contracts Act requires that contracts entered into by any Federal agency for themanufacture or furnishing of materials, supplies, articles, and equipment in any amount exceeding $10,000 must contain, among other provisions, a requirement that "no part of such contract will be performed nor will any of the materials, supplies, articles or equipment to be manufactured or furnished under said contract be manufactured or fabricated in any plants, factories, buildings, or surroundings or under working conditions which are unsanitary or hazardous or dangerous to the health and safety of employees engaged in the performance of said contract." The rules of the Secretary concerning these standards are published in 41 CFR part 50-204, and express the Secretary of Labor's interpretation and application of section 1(e) of the Walsh-Healey Public Contracts Act to certain particular working conditions. None of the described working conditions are intended to deal with construction activities, although such activities may conceivably be a part of a contract which is subject to the Walsh-Healey Public Contracts Act. Nevertheless, such activities remain subject to the general statutory duty prescribedby section 1(e). Section 103(b) of the Contract Work Hours and Safety Standards Act provides, among other things, that the Act shall not apply to any work required to be done in accordance with the provisions of the Walsh-Healey Public Contracts Act.

§ 1926.16 Rules of construction.

(a)   The prime contractor and any subcontractors may make their own arrangements with respect to obligations which might be more appropriately treated on a jobsite basis rather than individually. Thus, for example, the prime contractor and his subcontractors may wish to make an express agreement that the prime contractor or one of the subcontractors will provide all required first-aid or toilet facilities, thus relieving the subcontractors from the actual, but not any legal, responsibility (or, as the case may be, relieving the other subcontractors from this responsibility). In no case shall the prime contractor be relieved of overall responsibility for compliance with the requirements of this part for all work to be performed under the contract.

(b)   By contracting for full performance of a contract subject to section 107 of the Act, the prime contractor assumes all obligations prescribed as employer obligations under the standards contained in this part, whether or not he subcontracts any part of the work.

(c)   To the extent that a subcontractor of any tier agrees to perform any part of the contract, he also assumesresponsibility for complying with the standards in this part with respect to that part. Thus, the prime contractor assumes the entire responsibility under the contract and the subcontractor assumes responsibility with respect to his portion of the work. With respect to subcontracted work, the prime contractor and any subcontractor or subcontractors shall be deemed to have joint responsibility.

(d)   Where joint responsibility exists, both the prime contractor and his subcontractor or subcontractors, regardless of tier, shall be considered subject to the enforcement provisions of the Act.

Subpart C - General Safety and Health Provisions

Authority: 40 U.S.C. 3701 *et seq.;* 29 U.S.C. 653, 655, 657; Secretary of Labor's Order No. 12-71 (36 FR 8754), 8-76

(41 FR 25059), 9-83 (48 FR 35736), 6-96 (62 FR 111), 5-2007 (72 FR 31160), or 1-2012 (77 FR 3912) as applicable;

and 29 CFR part 1911.

§ 1926.20 General safety and health provisions.

(a) *Contractor requirements.*

    (1) Section 107 of the Act requires that it shall be a condition of each contract which is entered into under legislation subject to Reorganization Plan Number 14 of 1950 (64 Stat. 1267), as defined in §1926.12, and is for construction, alteration, and/or repair, including painting and decorating, that no contractor or subcontractor for any part of the contract work shall require any laborer or mechanic employed in the performance of the contract to work in surroundings or under working conditions which are unsanitary, hazardous, or dangerous to his health or safety.

(b) *Accident prevention responsibilities.*

    (1) It shall be the responsibility of the employer to initiate and maintain such programs as may be necessary to comply with this part.

    (2) Such programs shall provide for frequent and regular inspections of the job sites, materials, and equipment to be made by competent persons designated by the employers.

    (3) The use of any machinery, tool, material, or equipment which is not in compliance with any applicable requirement of this part is prohibited. Such machine, tool, material, or equipment shall either be identified as unsafe by tagging or locking the controls to render them inoperable or shall be physically removed from its place of operation.

    (4) The employer shall permit only those employees qualified by training or experience to operate equipment and machinery.

(c) The standards contained in this part shall apply with respect to employments performed in a workplace in a State, the District of Columbia, the Commonwealth of Puerto Rico, the Virgin Islands, American Samoa, Guam, the Commonwealth of the Northern Mariana Islands, Wake Island, Outer Continental Shelf lands defined in the Outer Continental Shelf Lands Act, and Johnston Island.

(d)

    (1) If a particular standard is specifically applicable to a condition, practice, means, method, operation, or process, it shall prevail over any different general standard which might otherwise be applicable to the same condition, practice, means, method, operation, or process.

    (2) On the other hand, any standard shall apply according to its terms to any employment and place of employment in any industry, even though particular standards are also prescribed for the industry to the extent that none of such particular standards applies.

(e) In the event a standard protects on its face a class of persons larger than employees, the standard shall be applicable under this part only to employees and their employment and places of employment.

(f) *Compliance duties owed to each employee -*

    (1) *Personal protective equipment.* Standards in this part requiring the employer to provide personal protective equipment (PPE), including respirators and other types of PPE, because of hazards to employees impose a separate compliance duty with respect to each employee covered by the requirement. The employer must provide PPE to each employee required to use

the PPE, and each failure to provide PPE to an employee may be considered a separate violation.

(2) **Training.** Standards in this part requiring training on hazards and related matters, such as standards requiring that employees receive training or that the employer train employees, provide training to employees, or institute or implement a training program, impose a separate compliance duty with respect to each employee covered by the requirement. The employer must train each affected employee in the manner required by the standard, and each failure to train an employee may be considered a separate violation.

[44 FR 8577, Feb. 9, 1979; 44 FR 20940, Apr. 6, 1979, as amended at 58 FR 35078, June 30, 1993; 73 FR 75588, Dec. 12, 2008; 85 FR 8735, Feb. 18, 2020; 85 FR 8735, Feb. 18, 2020]

§ 1926.21 Safety training and education.

(a) **General requirements.** The Secretary shall, pursuant to section 107(f) of the Act, establish and supervise programs for the education and training of employers and employees in the recognition, avoidance and prevention of unsafe conditions in employments covered by the act.

(b) **Employer responsibility.**

(1) The employer should avail himself of the safety and health training programs the Secretary provides.

(2) The employer shall instruct each employee in the recognition and avoidance of unsafe conditions and the regulations applicable to his work environment to control or eliminate any hazards or other exposure to illness or injury.

(3) Employees required to handle or use poisons, caustics, and other harmful substances shall be instructed regarding the safe handling and use, and be made aware of the potential hazards, personal hygiene, and personal protective measures required.

(4) In job site areas where harmful plants or animals are present, employees who may be exposed shall be instructed regarding the potential hazards, and how to avoid injury, and the first aid procedures to be used in the event of injury.

(5) Employees required to handle or use flammable liquids, gases, or toxic materials shall be instructed in the safe handling and use of these materials and made aware of the specific requirements contained in subparts D, F, and other applicable subparts of this part.

[44 FR 8577, Feb. 9, 1979; 44 FR 20940, Apr. 6, 1979, as amended at 80 FR 25518, May 4, 2015]

§ 1926.22 Recording and reporting of injuries. [Reserved]

§ 1926.23 First aid and medical attention.

First aid services and provisions for medical care shall be made available by the employer for every employee covered by these regulations. Regulations prescribing specific requirements for first aid, medical attention, and emergency facilities are contained in subpart D of this part.

§ 1926.24 Fire protection and prevention.

The employer shall be responsible for the development and maintenance of an effective fire protection and prevention program at the job site throughout all phases of the construction, repair, alteration, or demolition work. The employer shall ensure the availability of the fire protection and suppression equipment required by subpart F of this part.

§ 1926.25 Housekeeping.

(a) During the course of construction, alteration, or repairs, form and scrap lumber with protruding nails, and all other debris, shall be kept cleared from work areas, passageways, and stairs, in and around buildings or other structures.

(b) Combustible scrap and debris shall be removed at regular intervals during the course of construction. Safe means shall be provided to facilitate such removal.

(c) Containers shall be provided for the collection and separation of waste, trash, oily and used rags, andother refuse. Containers used for garbage and other oily, flammable, or hazardous wastes, such as caustics, acids, harmful dusts, etc. shall be equipped with covers. Garbage and other waste shall be disposed of at frequent and regular intervals.

§ 1926.26 Illumination.

Construction areas, aisles, stairs, ramps, runways, corridors, offices, shops, and storage areas where work is in progress shall be lighted with either natural or artificial illumination. The minimum illumination requirements for work areas are contained in subpart D of this part.

§ 1926.27 Sanitation.

Health and sanitation requirements for drinking water are contained in subpart D of this part.

§ 1926.28 Personal protective equipment.

(a) The employer is responsible for requiring the wearing of appropriate personal protective equipment in all operations where there is an exposure to hazardous conditions or where this part indicates the need for using such equipment to reduce the hazards to the employees.

(b) Regulations governing the use, selection, and maintenance of personal protective and lifesaving equipment are described under subpart E of this part.

§ 1926.29 Acceptable certifications.

(a) *Pressure vessels.* Current and valid certification by an insurance company or regulatory authority shall bedeemed as acceptable evidence of safe installation, inspection, and testing of pressure vessels provided by the employer.

(b) *Boilers.* Boilers provided by the employer shall be deemed to be in compliance with the requirements of this part when evidence of current and valid certification by an insurance company or regulatory authority attesting to the safe installation, inspection, and testing is presented.

(c) *Other requirements.* Regulations prescribing specific requirements for other types of pressure vessels and similar equipment are contained in subparts F and O of this part.

§ 1926.30 Shipbuilding and ship repairing.

(a) *General.* Shipbuilding, ship repairing, alterations, and maintenance performed on ships under Government contract, except naval ship construction, is work subject to the Act.

(b) **Applicable safety and health standards.** For the purpose of work carried out under this section, the safety and health regulations in part 1915 of this title, Shipyard Employment, shall apply.

[44 FR 8577, Feb. 9, 1979; 44 FR 20940, Apr. 6, 1979, as amended at 61 FR 9249, Mar. 7, 1996]

§ 1926.32 Definitions.

The following definitions shall apply in the application of the regulations in this part:

(a) **Act** means section 107 of the Contract Work Hours and Safety Standards Act, commonly known as theConstruction Safety Act (86 Stat. 96; 40 U.S.C. 333).

(b) **ANSI** means American National Standards Institute.

(c) **Approved** means sanctioned, endorsed, accredited, certified, or accepted as satisfactory by a duly constituted and nationally recognized authority or agency.

(d) **Authorized person** means a person approved or assigned by the employer to perform a specific type ofduty or duties or to be at a specific location or locations at the jobsite.

(e) **Administration** means the Occupational Safety and Health Administration.

(f) **Competent person** means one who is capable of identifying existing and predictable hazards in the surroundings or working conditions which are unsanitary, hazardous, or dangerous to employees, and who has authorization to take prompt corrective measures to eliminate them.

(g) **Construction work.** For purposes of this section, *Construction work* means work for construction, alteration, and/or repair, including painting and decorating.

(h) **Defect** means any characteristic or condition which tends to weaken or reduce the strength of the tool, object, or structure of which it is a part.

(i) **Designated person** means "authorized person" as defined in paragraph (d) of this section.

(j) **Employee** means every laborer or mechanic under the Act regardless of the contractual relationship which may be alleged to exist between the laborer and mechanic and the contractor or subcontractor who engaged him. "Laborer and mechanic" are not defined in the Act, but the identical terms are used in the Davis-Bacon Act (40 U.S.C. 276a), which provides for minimum wage protection on Federal and federallyassisted construction contracts. The use of the same term in a statute which often applies concurrently with section 107 of the Act has considerable precedential value in ascertaining the meaning of "laborer and mechanic" as used in the Act. *Laborer* generally means one who performs manual labor or who laborsat an occupation requiring physical strength; *mechanic* generally means a worker skilled with tools. See 18 Comp. Gen. 341.

(k) **Employer** means contractor or subcontractor within the meaning of the Act and of this part.

(l) **Hazardous substance** means a substance which, by reason of being explosive, flammable, poisonous,corrosive, oxidizing, irritating, or otherwise harmful, is likely to cause death or injury.

(m) **Qualified** means one who, by possession of a recognized degree, certificate, or professional standing, or who by extensive knowledge, training, and experience, has successfully demonstrated his ability to solveor resolve problems relating to the subject matter, the work, or the project.

(n) **Safety factor** means the ratio of the ultimate breaking strength of a member or piece of material or equipment to the actual working stress or safe load when in use.

(o) **Secretary** means the Secretary of Labor.

(p) **SAE** means Society of Automotive Engineers.

(q) **Shall** means mandatory.

(r) **Should** means recommended.

(s) **Suitable** means that which fits, and has the qualities or qualifications to meet a given purpose, occasion, condition, function, or circumstance.

[44 FR 8577, Feb. 9, 1979; 44 FR 20940, Apr. 6, 1979, as amended at 58 FR 35078, June 30, 1993]

§ 1926.33 Access to employee exposure and medical records.

Note: The requirements applicable to construction work under this section are identical to those set forth at § 1910.1020 of this chapter.

[61 FR 31431, June 20, 1996]

§ 1926.34 Means of egress.

(a) **General.** In every building or structure exits shall be so arranged and maintained as to provide free and unobstructed egress from all parts of the building or structure at all times when it is occupied. No lock or fastening to prevent free escape from the inside of any building shall be installed except in mental, penal, or corrective institutions where supervisory personnel is continually on duty and effective provisions are made to remove occupants in case of fire or other emergency.

(b) **Exit marking.** Exits shall be marked by a readily visible sign. Access to exits shall be marked by readily visible signs in all cases where the exit or way to reach it is not immediately visible to the occupants.

(c) **Maintenance and workmanship.** Means of egress shall be continually maintained free of all obstructions or impediments to full instant use in the case of fire or other emergency.

[58 FR 35083, June 30, 1993]

§ 1926.35 Employee emergency action plans.

(a) **Scope and application.** This section applies to all emergency action plans required by a particular OSHA standard. The emergency action plan shall be in writing (except as provided in the last sentence of paragraph (e)(3) of this section) and shall cover those designated actions employers and employees must take to ensure employee safety from fire and other emergencies.

(b) **Elements.** The following elements, at a minimum, shall be included in the

(1) plan:Emergency escape procedures and emergency escape route

(2) assignments;

Procedures to be followed by employees who remain to operate critical plant operations before they evacuate;

(3) Procedures to account for all employees after emergency evacuation has been completed;

(4)

(5)

Rescue and medical duties for those employees who are to perform them;

The preferred means of reporting fires and other emergencies; and

(6) Names or regular job titles of persons or departments who can be contacted for further information or explanation of duties under the plan.

(c) *Alarm system.*

(1) The employer shall establish an employee alarm system which complies with § 1926.159.

(2) If the employee alarm system is used for alerting fire brigade members, or for other purposes, a distinctive signal for each purpose shall be used.

(d) *Evacuation.* The employer shall establish in the emergency action plan the types of evacuation to be used in emergency circumstances.

(e) *Training.*

(1) Before implementing the emergency action plan, the employer shall designate and train a sufficient number of persons to assist in the safe and orderly emergency evacuation of employees.

(2) The employer shall review the plan with each employee covered by the plan at the following times:

(i) Initially when the plan is developed,

(ii) Whenever the employee's responsibilities or designated actions under the plan change, and

(iii) Whenever the plan is changed.

(3) The employer shall review with each employee upon initial assignment those parts of the plan which the employee must know to protect the employee in the event of an emergency. The written plan shall be kept at the workplace and made available for employee review. For those employers with 10 or fewer employees the plan may be communicated orally to employees and the employer need not maintain a written plan.

[58 FR 35083, June 30, 1993]

Subpart D - Occupational Health and Environmental Controls

Authority: 40 U.S.C. 3704; 29 U.S.C. 653, 655, and 657; and Secretary of Labor's Order No. 12-71 (36 FR 8754), 8-76 (41 FR 25059), 9-83 (48 FR 35736), 1-90 (55 FR 9033), 6-96 (62 FR 111), 3-2000 (65 FR 50017), 5-2002 (67 FR

65008), 5-2007 (72 FR 31159), 4-2010 (75 FR 55355), 1-2012 (77 FR 3912), or 8-2020 (85 FR 58393), as applicable;

and 29 CFR part 1911.

Sections 1926.59, 1926.60, and 1926.65 also issued under 5 U.S.C. 553 and 29 CFR part 1911.

Section 1926.61 also issued under 49 U.S.C. 1801-1819 and 5 U.S.C. 553.

Section 1926.62 also issued under sec. 1031, Public Law 102-550, 106 Stat. 3672 (42 U.S.C. 4853).

Section 1926.65 also issued under sec. 126, Public Law 99-499, 100 Stat. 1614 (reprinted at 29 U.S.C.A. 655 Note)

and 5 U.S.C. 553.

§ 1926.50 Medical services and first aid.

(a) The employer shall insure the availability of medical personnel for advice and consultation on matters of occupational health.

(b) Provisions shall be made prior to commencement of the project for prompt medical attention in case of serious injury.

(c) In the absence of an infirmary, clinic, hospital, or physician, that is reasonably accessible in terms of time and distance to the worksite, which is available for the treatment of injured employees, a person who has a valid certificate in first-aid training from the U.S. Bureau of Mines, the American Red Cross, or equivalent training that can be verified by documentary evidence, shall be available at the worksite to render first aid.

(d)

   (1) First aid supplies shall be easily accessible when required.

   (2) The contents of the first aid kit shall be placed in a weatherproof container with individual sealed packages for each type of item, and shall be checked by the employer before being sent out on each job and at least weekly on each job to ensure that the expended items are replaced.

(e) Proper equipment for prompt transportation of the injured person to a physician or hospital, or a communication system for contacting necessary ambulance service, shall be provided.

(f)

   (1) In areas where 911 emergency dispatch services are not available, the telephone numbers of the physicians, hospitals, or ambulances shall be conspicuously posted.

   (2) In areas where 911 emergency dispatch services are available and an employer uses a communication system for contacting necessary emergency-medical service, the employer must:

      (i) Ensure that the communication system is effective in contacting the emergency-medical service; and

      (ii)

         (A) When using a communication system in an area that does not automatically supply the caller's latitude and longitude information to the 911 emergency dispatcher, the employer must post in a conspicuous location at the worksite either:

            (1) The latitude and longitude of the worksite; or

            (2) Other location-identification information that communicates effectively to employees the location of the worksite.

         (B) The requirement specified in paragraph (f)(2)(ii)(A) of this section does not apply to worksites with readily available telephone land lines that have 911 emergency service that automatically identifies the location of the caller.

(g) Where the eyes or body of any person may be exposed to injurious corrosive materials, suitable facilities for quick drenching or flushing of the eyes and body shall be provided within the work area for immediate emergency use.

Appendix A to § 1926.50 - First aid Kits (Non-Mandatory)

First aid supplies are required to be easily accessible under paragraph § 1926.50(d)(1). An example of the minimal contents of a generic first aid kit is described in American National Standard (ANSI) Z308.1-1978 "Minimum Requirements for Industrial Unit-Type First-aid Kits". The contents of the kit listed in the ANSI standard should be adequate for small work sites. When larger operations or multiple operations are beingconducted at the same location, employers should determine the need for additional first aid kits at the worksite, additional types of first aid equipment and supplies and additional quantities and types of supplies and equipment in the first aid kits.

In a similar fashion, employers who have unique or changing first-aid needs in their workplace may need to enhance their first-aid kits. The employer can use the OSHA 300 log, OSHA 301 log, or other reports to identify these unique problems. Consultation from the local fire/rescue department, appropriate medical professional, or local emergency room may be helpful to employers in these circumstances. By assessing the specific needs of their workplace, employers can ensure that reasonably anticipated supplies are available. Employers should assess the specific needs of their worksite periodically and augment the first aid kit appropriately.

If it is reasonably anticipated employees will be exposed to blood or other potentially infectious materials while using first-aid supplies, employers should provide personal protective equipment (PPE). Appropriate PPE includes gloves, gowns, face shields, masks and eye protection (see "Occupational Exposure to Blood borne Pathogens", 29 CFR 1910.1030(d)(3)) (56 FR 64175).

[44 FR 8577, Feb. 9, 1979; 44 FR 20940, Apr. 6, 1979, as amended at 49 FR 18295, Apr. 30, 1984; 58 FR 35084, June 30, 1993; 61 FR 5510, Feb. 13, 1996; 63 FR 33469, June 18, 1998; 76 FR 80740, Dec. 27, 2011; 84 FR 21575, May 14, 2019]

§ 1926.51 Sanitation.

(a) **Potable water.**

    (1) An adequate supply of potable water shall be provided in all places of employment.

    (2) Portable containers used to dispense drinking water shall be capable of being tightly closed, and equipped with a tap. Water shall not be dipped from containers.

    (3) Any container used to distribute drinking water shall be clearly marked as to the nature of its contents and not used for any other purpose.

    (4) The common drinking cup is prohibited.

    (5) Where single service cups (to be used but once) are supplied, both a sanitary container for the unused cups and a receptacle for disposing of the used cups shall be provided.

    (6) **Potable water** means water that meets the standards for drinking purposes of the State or local authority having jurisdiction, or water that meets the quality standards prescribed by the U.S. Environmental Protection Agency's National Primary Drinking Water Regulations (40 CFR part 141).

(b) **Nonpotable water.**

(1) Outlets for nonpotable water, such as water for industrial or firefighting purposes only, shall beidentified by signs meeting the requirements of subpart G of this part, to indicate clearly that thewater is unsafe and is not to be used for drinking, washing, or cooking purposes.

(2) There shall be no cross-connection, open or potential, between a system furnishing potable water and a system furnishing nonpotable water.

(c) **Toilets at construction jobsites.**

(1) Toilets shall be provided for employees according to the following table:

Table D-1

Number of employees	Minimum number of facilities
20 or less	1.
20 or more	1 toilet seat and 1 urinal per 40 workers.
200 or more	1 toilet seat and 1 urinal per 50 workers.

(2) Under temporary field conditions, provisions shall be made to assure not less than one toilet facilityis available.

(3) Job sites, not provided with a sanitary sewer, shall be provided with one of the following toilet facilities unless prohibited by local codes:

    (i) Privies (where their use will not contaminate ground or surface

    (ii) water);Chemical toilets;

    (iii) Recirculating

    (iv) toilets;

(4) Combustion

    toilets.

The requirements of this paragraph (c) for sanitation facilities shall not apply to mobile crews havingtransportation readily available to nearby toilet facilities.

(d) **Food handling.**

(1) All employees' food service facilities and operations shall meet the applicable laws, ordinances, andregulations of the jurisdictions in which they are located.

(2) All employee food service facilities and operations shall be carried out in accordance with sound hygienic principles. In all places of employment where all or part of the food service is provided, the food dispensed shall be wholesome, free from spoilage, and shall be processed, prepared, handled, and stored in such a manner as to be protected against contamination.

(e) **Temporary sleeping quarters.** When temporary sleeping quarters are provided, they shall be heated, ventilated, and lighted.

(f) **Washing facilities.**

(1) The employer shall provide adequate washing facilities for employees engaged in the

application of paints, coating, herbicides, or insecticides, or in other operations where contaminants may be harmful to the employees. Such facilities shall be in near proximity to the worksite and shall be so equipped as to enable employees to remove such substances.

(2) *General.* Washing facilities shall be maintained in a sanitary condition.

(3) *Lavatories.*

   (i) Lavatories shall be made available in all places of employment. The requirements of this subdivision do not apply to mobile crews or to normally unattended work locations if employees working at these locations have transportation readily available to nearby washing facilities which meet the other requirements of this paragraph.

   (ii) Each lavatory shall be provided with hot and cold running water, or tepid running

   (iii) water. Hand soap or similar cleansing agents shall be provided.

   (iv) Individual hand towels or sections thereof, of cloth or paper, air blowers or clean individual sections of continuous cloth toweling, convenient to the lavatories, shall be provided.

(4) *Showers.*

   (i) Whenever showers are required by a particular standard, the showers shall be provided in accordance with paragraphs (f)(4) (ii) through (v) of this section.

   (ii) One shower shall be provided for each 10 employees of each sex, or numerical fraction thereof, who are required to shower during the same shift.

   (iii) Body soap or other appropriate cleansing agents convenient to the showers shall be provided as specified in paragraph (f)(3)(iii) of this section.

   (iv) Showers shall be provided with hot and cold water feeding a common discharge line.

   (v) Employees who use showers shall be provided with individual clean towels.

(g) *Eating and drinking areas.* No employee shall be allowed to consume food or beverages in a toilet room nor in any area exposed to a toxic material.

(h) *Vermin control.* Every enclosed workplace shall be so constructed, equipped, and maintained, so far as reasonably practicable, as to prevent the entrance or harborage of rodents, insects, and other vermin. A continuing and effective extermination program shall be instituted where their presence is detected.

(i) *Change rooms.* Whenever employees are required by a particular standard to wear protective clothing because of the possibility of contamination with toxic materials, change rooms equipped with storage facilities for street clothes and separate storage facilities for the protective clothing shall be provided.

[44 FR 8577, Feb. 9, 1979; 44 FR 20940, Apr. 6, 1979, as amended at 58 FR 35084, June 30, 1993; 76 FR 33611, June 8, 2011]

§ 1926.52 Occupational noise exposure.

(a) Protection against the effects of noise exposure shall be provided when the sound levels exceed those shown in Table D-2 of this section when measured on the A-scale of a standard sound level meter at slow response.

(b) When employees are subjected to sound levels exceeding those listed in Table D-2 of this section, feasibleadministrative or engineering controls shall be utilized. If such controls fail to reduce sound levels withinthe levels of the table, personal protective equipment as required in subpart E, shall be provided and used to reduce sound levels within the levels of the table.

(c) If the variations in noise level involve maxima at intervals of 1 second or less, it is to be considered continuous.

(d)

(1) In all cases where the sound levels exceed the values shown herein, a continuing, effective hearingconservation program shall be administered.

Table D-2 - Permissible Noise Exposures

Duration per day, hours	Sound level dBA slow response
8	90
6	92
4	95
3	97
2	100
1½	102
1	105
½	110
¼ or less	115

(2)

(i) When the daily noise exposure is composed of two or more periods of noise exposure of different levels, their combined effect should be considered, rather than the individual effect ofeach. Exposure to different levels for various periods of time shall be computed according to the formula set forth in paragraph (d)(2)(ii) of this section.

(ii) $F_e = (T_1/L_1) + (T_2/L_2) + \cdots + (T_n/L_n)$ Where:

F_e = The equivalent noise exposure factor.

T = The period of noise exposure at any essentially constant level.

L = The duration of the permissible noise exposure at the constant level (from Table D-2).

If the value of F_e exceeds unity (1) the exposure exceeds permissible levels.

(iii) A sample computation showing an application of the formula in paragraph (d)(2)(ii) of thissection is as follows. An employee is exposed at these levels for these periods:

110 db A ¼

hour.   100

db A ½

hour. 90 db

A 1½ hours.

$F_e = (¼/½) + (½/2) + (1½/8)$

$F_e = 0.500 + 0.25 + 0.188$

$F_e = 0.938$

Since the value of F_e does not exceed unity, the exposure is within permissible limits.

(e) Exposure to impulsive or impact noise should not exceed 140 dB peak sound pressure level.

§ 1926.53 Ionizing radiation.

(a) In construction and related activities involving the use of sources of ionizing radiation, the pertinent provisions of the Nuclear Regulatory Commission's Standards for Protection Against Radiation (10 CFR part 20), relating to protection against occupational radiation exposure, shall apply.

(b) Any activity which involves the use of radioactive materials or X-rays, whether or not under license from the Nuclear Regulatory Commission, shall be performed by competent persons specially trained in the proper and safe operation of such equipment. In the case of materials used under Commission license, only persons actually licensed, or competent persons under direction and supervision of the licensee, shall perform such work.

(c)-(r) [Reserved]

Note: The requirements applicable to construction work under paragraphs (c) through (r) of this section are identical to those set forth at paragraphs (a) through (p) of § 1910.1096 of this chapter.

[44 FR 8577, Feb. 9, 1979; 44 FR 20940, Apr. 6, 1979, as amended at 61 FR 5510, Feb. 13, 1996; 61 FR 31431, June 20, 1996]

§ 1926.54 Nonionizing radiation.

(a) Only qualified and trained employees shall be assigned to install, adjust, and operate laser equipment.

(b) Proof of qualification of the laser equipment operator shall be available and in possession of the operator at all times.

(c) Employees, when working in areas in which a potential exposure to direct or reflected laser light greater than 0.005 watts (5 milliwatts) exists, shall be provided with antilaser eye protection devices as specified in subpart E of this part.

(d) Areas in which lasers are used shall be posted with standard laser warning placards.

(e) Beam shutters or caps shall be utilized, or the laser turned off, when laser transmission is not actually required. When the laser is left unattended for a substantial period of time, such as during lunch hour, overnight, or at change of shifts, the laser shall be turned off.

(f) Only mechanical or electronic means shall be used as a detector for guiding the internal alignment of the laser.

(g) The laser beam shall not be directed at employees.

(h) When it is raining or snowing, or when there is dust or fog in the air, the operation of laser systems shall be prohibited where practicable; in any event, employees shall be kept out of range of the area of source and target during such weather conditions.

(i) Laser equipment shall bear a label to indicate maximum output.

(j) Employees shall not be exposed to light intensities above:

  (1) Direct staring: 1 micro-watt per square centimeter;

  (2) Incidental observing: 1 milliwatt per square

  (3) centimeter; Diffused reflected light: $2^1/_2$ watts per

(k) square centimeter.

Laser unit in operation should be set up above the heads of the employees, when possible.

(l) Employees shall not be exposed to microwave power densities in excess of 10 milliwatts per square centimeter.

§ 1926.55 Gases, vapors, fumes, dusts, and mists.

(a) Employers must limit an employee's exposure to any substance listed in Table 1 or 2 of this section inaccordance with the following:

(1) **Substances with limits preceded by (C) - Ceiling Values.** An employee's exposure, as determined from breathing-zone air samples, to any substance in Table 1 of this section with a permissible exposure limit preceded by (C) must at no time exceed the exposure limit specified for that substance. If instantaneous monitoring is not feasible, then the employer must assess the ceiling as a 15-minute time-weighted average exposure that the employer cannot exceed at any time during the working day.

(2) **Other substances - 8-hour Time Weighted Averages.** An employee's exposure, as determined from breathing-zone air samples, to any substance in Table 1 or 2 of this section with a permissible exposure limit not preceded by (C) must not exceed the limit specified for that substance measured as an 8-hour time-weighted average in any work shift.

(b) To achieve compliance with paragraph (a) of this section, administrative or engineering controls must first be implemented whenever feasible. When such controls are not feasible to achieve full compliance, protective equipment or other protective measures shall be used to keep the exposure of employees to air contaminants within the limits prescribed in this section. Any equipment and technical measures used for this purpose must first be approved for each particular use by a competent industrial hygienist or other technically qualified person. Whenever respirators are used, their use shall comply with § 1926.103.

(c) Paragraphs (a) and (b) of this section do not apply to the exposure of employees to airborne asbestos, tremolite, anthophyllite, or actinolite dust. Whenever any employee is exposed to airborne asbestos, tremolite, anthophyllite, or actinolite dust, the requirements of § 1926.1101 shall apply.

(d) Paragraphs (a) and (b) of this section do not apply to the exposure of employees to formaldehyde. Whenever any employee is exposed to formaldehyde, the requirements of § 1910.1048 of this title shall apply.

Table 1 to § 1926.55 - Permissible Exposure Limits for AirborneContaminants

Substance	CAS No.[d]	ppm[a]	mg/m^3 [b]	Skin designation *
Abate; see Temephos				
Acetaldehyde	75-07-0	200	360	-
Acetic acid	64-19-7	10	25	-
Acetic anhydride	108-24-7	5	20	-
Acetone	67-64-1	1000	2400	-
Acetonitrile	75-05-8	40	70	-
2-Acetylaminofluorine; see § 1926.1114	53-96-3			
Acetylene	74-86-2	E		
Acetylene dichloride; see 1,2-Dichloroethylene				
Acetylene tetrabromide	79-27-6	1	14	-
Acrolein	107-02-8	0.1	0.25	-
Acrylamide	79-06-1	-	0.3	X
Acrylonitrile; see § 1926.1145	107-13-1			
Aldrin	309-00-2	-	0.25	X

Substance	CAS No.[d]	ppm[a]	mg/m³ [b]	Skin designation *
Allyl alcohol	107-18-6	2	5	X
Allyl chloride	107-05-1	1	3	-
Allyl glycidyl ether (AGE)	106-92-3	(C)10	(C)45	-
Allyl propyl disulfide	2179-59-1	2	12	-
alpha-Alumina	1344-28-1			
Total dust			-	-
Respirable fraction			-	-
Alundum; see alpha-Alumina				
4-Aminodiphenyl; see § 1926.1111	92-67-1			
2-Aminoethanol; see Ethanolamine				
2-Aminopyridine	504-29-0	0.5	2	-
Ammonia	7664-41-7	50	35	-
Ammonium sulfamate	7773-06-0			
Total dust		-	15	-
Respirable fraction		-	5	
n-Amyl acetate	628-63-7	100	525	-
sec-Amyl acetate	626-38-0	125	650	-
Aniline and homologs	62-53-3	5	19	X
Anisidine (o-, p-isomers)	29191-52-4	-	0.5	X
Antimony and compounds (as Sb)	7440-36-0	-	0.5	-
ANTU (alpha Naphthylthiourea)	86-88-4	-	0.3	-
Argon	7440-37-1	E		
Arsenic, inorganic compounds (as As); see § 1926.1118	7440-38-2	-	-	-
Arsenic, organic compounds (as As)	7440-38-2	-	0.5	
Arsine	7784-42-1	0.05	0.2	-
Asbestos; see § 1926.1101				
Azinphos-methyl	86-50-0	-	0.2	X
Barium, soluble compounds (as Ba)	7440-39-3	-	0.5	-
Benzene[g]; see § 1926.1128	71-43-2			
Benzidine; see § 1926.1110	92-87-5			
p-Benzoquinone; see Quinone				
Benzo(a)pyrene; see Coal tar pitch volatiles				
Benzoyl peroxide	94-36-0	-	5	-
Benzyl chloride	100-44-7	1	5	-
Beryllium and beryllium compounds (as Be); see 1926.1124[(q)]	7440-41-7	-	0.002	-
Biphenyl; see Diphenyl				
Bisphenol A; see Diglycidyl ether				
Boron oxide	1303-86-2			

Substance	CAS No.[d]	ppm[a]	mg/m³ [b]	Skin designation *
Total dust		-	15	-
Boron tribromide	10294-33-4	1	10	-
Boron trifluoride	7637-07-2	(C)1	(C)3	-
Bromine	7726-95-6	0.1	0.7	-
Bromine pentafluoride	7789-30-2	0.1	0.7	-
Bromoform	75-25-2	0.5	5	X
Butadiene (1,3-Butadiene); see 29 CFR 1910.1051; 29 CFR 1910.19(l)	106-99-0	STEL 1 ppm/5 ppm		-
Butanethiol; see Butyl mercaptan				
2-Butanone (Methyl ethyl ketone)	78-93-3	200	590	-
2-Butoxyethanol	111-76-2	50	240	X
n-Butyl-acetate	123-86-4	150	710	-
sec-Butyl acetate	105-46-4	200	950	-
tert-Butyl acetate	540-88-5	200	950	-
n-Butyl alcohol	71-36-3	100	300	-
sec-Butyl alcohol	78-92-2	150	450	-
tert-Butyl alcohol	75-65-0	100	300	-
Butylamine	109-73-9	(C)5	(C)15	X
tert-Butyl chromate (as CrO_3); see 1926.1126[n]	1189-85-1			
n-Butyl glycidyl ether (BGE)	2426-08-6	50	270	-
Butyl mercaptan	109-79-5	0.5	1.5	-
p-tert-Butyltoluene	98-51-1	10	60	-
Cadmium (as Cd); see 1926.1127	7440-43-9			
Calcium carbonate	1317-65-3			
Total dust			-	-
Respirable fraction			-	
Calcium oxide	1305-78-8	-	5	-
Calcium sulfate	7778-18-9			
Total dust		-	15	-
Respirable fraction		-	5	-
Camphor, synthetic	76-22-2	-	2	-
Carbaryl (Sevin)	63-25-2	-	5	-
Carbon black	1333-86-4	-	3.5	-
Carbon dioxide	124-38-9	5000	9000	-
Carbon disulfide	75-15-0	20	60	X
Carbon monoxide	630-08-0	50	55	-
Carbon tetrachloride	56-23-5	10	65	X
Cellulose	9004-34-6			

Substance	CAS No.[d]	ppm[a]	mg/m³ [b]	Skin designation *
Total dust			-	-
Respirable fraction			-	-
Chlordane	57-74-9	-	0.5	X
Chlorinated camphene	8001-35-2	-	0.5	X
Chlorinated diphenyl oxide	55720-99-5	-	0.5	-
Chlorine	7782-50-5	1	3	-
Chlorine dioxide	10049-04-4	0.1	0.3	
Chlorine trifluoride	7790-91-2	(C)0.1	(C)0.4	-
Chloroacetaldehyde	107-20-0	(C)1	(C)3	-
a-Chloroacetophenone (Phenacyl chloride)	532-27-4	0.05	0.3	-
Chlorobenzene	108-90-7	75	350	-
o-Chlorobenzylidene malononitrile	2698-41-1	0.05	0.4	-
Chlorobromomethane	74-97-5	200	1050	-
2-Chloro-1,3-butadiene; see beta-Chloroprene				
Chlorodiphenyl (42% Chlorine) (PCB)	53469-21-9	-	1	X
Chlorodiphenyl (54% Chlorine) (PCB)	11097-69-1	-	0.5	X
1-Chloro,2,3-epoxypropane; see Epichlorohydrin				
2-Chloroethanol; see Ethylene chlorohydrin				
Chloroethylene; see Vinyl chloride				
Chloroform (Trichloromethane)	67-66-3	(C)50	(C)240	-
bis(Chloromethyl) ether; see § 1926.1108	542-88-1			
Chloromethyl methyl ether; see § 1926.1106	107-30-2			
1-Chloro-1-nitropropane	600-25-9	20	100	-
Chloropicrin	76-06-2	0.1	0.7	-
beta-Chloroprene	126-99-8	25	90	X
Chromium (II) compounds				
(as Cr)	7440-47-3	-	0.5	-
Chromium (III) compounds				
(as Cr)	7440-47-3	-	0.5	-
Chromium (VI) compounds; See 1926.1126°				
Chromium metal and insol. salts (as Cr)	7440-47-3	-	1	-
Chrysene; see Coal tar pitch volatiles				
Coal tar pitch volatiles (benzene soluble fraction), anthracene, BaP, phenanthrene, acridine, chrysene, pyrene	65996-93-2	-	0.2	-
Cobalt metal, dust, and fume (as Co)	7440-48-4	-	0.1	-
Copper	7440-50-8			
Fume (as Cu)		-	0.1	-
Dusts and mists (as Cu)		-	1	-

Substance	CAS No.[d]	ppm[a]	mg/m³ [b]	Skin designation *
Corundum; see Emery				
Cotton dust (raw)		-	1	
Crag herbicide (Sesone)	136-78-7			
Total dust			-	-
Respirable fraction			-	-
Cresol, all isomers	1319-77-3	5	22	X
Crotonaldehyde	123-73-9; 4170-30-3	2	6	
Cumene	98-82-8	50	245	X
Cyanides (as CN)	Varies with Compound	-	5	X
Cyanogen	460-19-5	10	-	-
Cyclohexane	110-82-7	300	1050	-
Cyclohexanol	108-93-0	50	200	-
Cyclohexanone	108-94-1	50	200	-
Cyclohexene	110-83-8	300	1015	-
Cyclonite	121-82-4	-	1.5	X
Cyclopentadiene	542-92-7	75	200	
DDT, see Dichlorodiphenyltrichloroethane				
DDVP, see Dichlorvos				
2,4-D (Dichlorophenoxyacetic acid)	94-75-7	-	10	-
Decaborane	17702-41-9	0.05	0.3	X
Demeton (Systox)	8065-48-3	-	0.1	X
Diacetone alcohol (4-Hydroxy-4-methyl-2-pentanone)	123-42-2	50	240	-
1,2-Diaminoethane; see Ethylenediamine				
Diazomethane	334-88-3	0.2	0.4	-
Diborane	19287-45-7	0.1	0.1	-
1,2-Dibromo-3-chloropropane (DBCP); see § 1926.1144	96-12-8			-
1,2-Dibromoethane; see Ethylene dibromide				
Dibutyl phosphate	107-66-4	1	5	-
Dibutyl phthalate	84-74-2	-	5	-
Dichloroacetylene	7572-29-4	(C)0.1	(C)0.4	
o-Dichlorobenzene	95-50-1	(C)50	(C)300	-
p-Dichlorobenzene	106-46-7	75	450	-
3,3'-Dichlorobenzidine; see § 1926.1107	91-94-1			
Dichlorodifluoromethane	75-71-8	1000	4950	-
1,3-Dichloro-5,5-dimethyl hydantoin	118-52-5	-	0.2	-
Dichlorodiphenyltrichloroethane (DDT)	50-29-3	-	1	X
1,1-Dichloroethane	75-34-3	100	400	-

Substance	CAS No.[d]	ppm[a]	mg/m³ [b]	Skin designation *
1,2-Dichloroethane; see Ethylene dichloride				
1,2-Dichloroethylene	540-59-0	200	790	-
Dichloroethyl ether	111-44-4	(C)15	(C)90	X
Dichloromethane; see Methylene chloride				
Dichloromonofluoromethane	75-43-4	1000	4200	-
1,1-Dichloro-1-nitroethane	594-72-9	(C)10	(C)60	-
1,2-Dichloropropane; see Propylene dichloride				
Dichlorotetrafluoroethane	76-14-2	1000	7000	-
Dichlorvos (DDVP)	62-73-7	-	1	X
Dieldrin	60-57-1	-	0.25	X
Diethylamine	109-89-7	25	75	-
2-Diethylaminoethanol	100-37-8	10	50	X
Diethylene triamine	111-40-0	(C)10	(C)42	X
Diethyl ether; see Ethyl ether				
Difluorodibromomethane	75-61-6	100	860	-
Diglycidyl ether (DGE)	2238-07-5	(C)0.5	(C)2.8	-
Dihydroxybenzene; see Hydroquinone				
Diisobutyl ketone	108-83-8	50	290	-
Diisopropylamine	108-18-9	5	20	X
4-Dimethylaminoazobenzene; see § 1926.1115	60-11-7			
Dimethoxymethane; see Methylal				
Dimethyl acetamide	127-19-5	10	35	X
Dimethylamine	124-40-3	10	18	-
Dimethylaminobenzene; see Xylidine				
Dimethylaniline (N,N-Dimethylaniline)	121-69-7	5	25	X
Dimethylbenzene; see Xylene				
Dimethyl-1,2-dibromo- 2,2-dichloroethyl phosphate	300-76-5	-	3	-
Dimethylformamide	68-12-2	10	30	X
2,6-Dimethyl-4-heptanone; see Diisobutyl ketone				
1,1-Dimethylhydrazine	57-14-7	0.5	1	X
Dimethylphthalate	131-11-3	-	5	-
Dimethyl sulfate	77-78-3	1	5	X
Dinitrobenzene (all isomers)			1	X
(ortho)	528-29-0			
(meta)	99-65-0			
(para)	100-25-4			
Dinitro-o-cresol	534-52-1	-	0.2	X
Dinitrotoluene	25321-14-6	-	1.5	X

Substance	CAS No.[d]	ppm[a]	mg/m³ [b]	Skin designation *
Dioxane (Diethylene dioxide)	123-91-1	100	360	X
Diphenyl (Biphenyl)	92-52-4	0.2	1	-
Diphenylamine	122-39-4	-	10	-
Diphenylmethane diisocyanate; see Methylenebisphenyl isocyanate				
Dipropylene glycol methyl ether	34590-94-8	100	600	X
Di-sec octyl phthalate (Di-(2-ethylhexyl) phthalate)	117-81-7	-	5	-
Emery	12415-34-8			
Total dust			-	-
Respirable fraction			-	-
Endosulfan	115-29-7	-	0.1	X
Endrin	72-20-8	-	0.1	X
Epichlorohydrin	106-89-8	5	19	X
EPN	2104-64-5	-	0.5	X
1,2-Epoxypropane; see Propylene oxide				
2,3-Epoxy-1-propanol; see Glycidol				
Ethane	74-84-0	E		
Ethanethiol; see Ethyl mercaptan				
Ethanolamine	141-43-5	3	6	-
2-Ethoxyethanol (Cellosolve)	110-80-5	200	740	X
2-Ethoxyethyl acetate (Cellosolve acetate)	111-15-9	100	540	X
Ethyl acetate	141-78-6	400	1400	-
Ethyl acrylate	140-88-5	25	100	X
Ethyl alcohol (Ethanol)	64-17-5	1000	1900	-
Ethylamine	75-04-7	10	18	-
Ethyl amyl ketone (5-Methyl-3-heptanone)	541-85-5	25	130	-
Ethyl benzene	100-41-4	100	435	-
Ethyl bromide	74-96-4	200	890	-
Ethyl butyl ketone (3-Heptanone)	106-35-4	50	230	-
Ethyl chloride	75-00-3	1000	2600	-
Ethyl ether	60-29-7	400	1200	-
Ethyl formate	109-94-4	100	300	-
Ethyl mercaptan	75-08-1	0.5	1	-
Ethyl silicate	78-10-4	100	850	-
Ethylene	74-85-1	E		
Ethylene chlorohydrin	107-07-3	5	16	X
Ethylenediamine	107-15-3	10	25	-
Ethylene dibromide	106-93-4	(C)25	(C)190	X
Ethylene dichloride (1,2-Dichloroethane)	107-06-2	50	200	-

Substance	CAS No.[d]	ppm[a]	mg/m³ [b]	Skin designation *
Ethylene glycol dinitrate	628-96-6	(C)0.2	(C)1	X
Ethylene glycol methyl acetate; see Methyl cellosolve acetate				
Ethyleneimine; see § 1926.1112	151-56-4			
Ethylene oxide; see § 1926.1147	75-21-8			
Ethylidene chloride; see 1,1-Dichloroethane				
N-Ethylmorpholine	100-74-3	20	94	X
Ferbam	14484-64-1			
Total dust		·	15	·
Ferrovanadium dust	12604-58-9	·	1	·
Fibrous Glass				
Total dust				·
Respirable fraction		·		·
Fluorides (as F)	Varies with compound	·	2.5	·
Fluorine	7782-41-4	0.1	0.2	
Fluorotrichloromethane (Trichlorofluoromethane)	75-69-4	1000	5600	
Formaldehyde; see § 1926.1148	50-00-0			
Formic acid	64-18-6	5	9	·
Furfural	98-01-1	5	20	X
Furfuryl alcohol	98-00-0	50	200	·
Gasoline	8006-61-9		A³	·
Glycerin (mist)	56-81-5			
Total dust		·		·
Respirable fraction		·		·
Glycidol	556-52-5	50	150	·
Glycol monoethyl ether; see 2-Ethoxyethanol				
Graphite, natural, respirable dust	7782-42-5	(²)	(²)	(²)
Graphite, synthetic				
Total dust		·		·
Respirable fraction		·		·
Guthion; see Azinphos methyl				
Gypsum	13397-24-5			
Total dust		·		·
Respirable fraction		·		·
Hafnium	7440-58-6	·	0.5	·
Helium	7440-59-7	E		
Heptachlor	76-44-8	·	0.5	X
Heptane (n-Heptane)	142-82-5	500	2000	·

Substance	CAS No.[d]	ppm[a]	mg/m³ [b]	Skin designation *
Hexachloroethane	67-72-1	1	10	X
Hexachloronaphthalene	1335-87-1	-	0.2	X
n-Hexane	110-54-3	500	1800	-
2-Hexanone (Methyl n-butyl ketone)	591-78-6	100	410	-
Hexone (Methyl isobutyl ketone)	108-10-1	100	410	-
sec-Hexyl acetate	108-84-9	50	300	-
Hydrazine	302-01-2	1	1.3	X
Hydrogen	1333-74-0	E		
Hydrogen bromide	10035-10-6	3	10	
Hydrogen chloride	7647-01-0	(C)5	(C)7	-
Hydrogen cyanide	74-90-8	10	11	X
Hydrogen fluoride (as F)	7664-39-3	3	2	-
Hydrogen peroxide	7722-84-1	1	1.4	-
Hydrogen selenide (as Se)	7783-07-5	0.05	.02	-
Hydrogen sulfide	7783-06-4	10	15	-
Hydroquinone	123-31-9	-	2	-
Indene	95-13-6	10	45	-
Indium and compounds (as In)	7440-74-6	-	0.1	-
Iodine	7553-56-2	(C)0.1	(C)1	-
Iron oxide fume	1309-37-1	-	10	-
Iron salts (soluble) (as Fe)	Varies with compound	-	1	-
Isoamyl acetate	123-92-2	100	525	-
Isoamyl alcohol (primary and secondary)	123-51-3	100	360	-
Isobutyl acetate	110-19-0	150	700	-
Isobutyl alcohol	78-83-1	100	300	-
Isophorone	78-59-1	25	140	-
Isopropyl acetate	108-21-4	250	950	-
Isopropyl alcohol	67-63-0	400	980	-
Isopropylamine	75-31-0	5	12	-
Isopropyl ether	108-20-3	500	2100	-
Isopropyl glycidyl ether (IGE)	4016-14-2	50	240	-
Kaolin	1332-58-7			
Total dust			-	-
Respirable fraction			-	-
Ketene	463-51-4	0.5	0.9	-
Lead, inorganic (as Pb); see 1926.62	7439-92-1			
Limestone	1317-65-3			
Total dust			-	-

Substance	CAS No.[d]	ppm[a]	mg/m³ [b]	Skin designation *
Respirable fraction		·		·
Lindane	58-89-9	·	0.5	X
Lithium hydride	7580-67-8	·	0.025	·
L.P.G. (Liquefied petroleum gas)	68476-85-7	1000	1800	
Magnesite	546-93-0			
Total dust			·	·
Respirable fraction			·	·
Magnesium oxide fume	1309-48-4			
Total particulate		15	·	·
Malathion	121-75-5			
Total dust		·	15	X
Maleic anhydride	108-31-6	0.25		
Manganese compounds (as Mn)	7439-96-5	·	(C)5	·
Manganese fume (as Mn)	7439-96-5	·	(C)5	·
Marble	1317-65-3			
Total dust		·		·
Respirable fraction		·		·
Mercury (aryl and inorganic)(as Hg)	7439-97-6		0.1	X
Mercury (organo) alkyl compounds (as Hg)	7439-97-6	·	0.01	X
Mercury (vapor) (as Hg)	7439-97-6	·	0.1	X
Mesityl oxide	141-79-7	25	100	·
Methane	74-82-8	E		
Methanethiol; see Methyl mercaptan				
Methoxychlor	72-43-5			
Total dust		·	15	·
2-Methoxyethanol (Methyl cellosolve)	109-86-4	25	80	X
2-Methoxyethyl acetate (Methyl cellosolve acetate)	110-49-6	25	120	X
Methyl acetate	79-20-9	200	610	·
Methyl acetylene (Propyne)	74-99-7	1000	1650	·
Methyl acetylene-propadiene mixture (MAPP)		1000	1800	·
Methyl acrylate	96-33-3	10	35	X
Methylal (Dimethoxy-methane)	109-87-5	1000	3100	·
Methyl alcohol	67-56-1	200	260	·
Methylamine	74-89-5	10	12	·
Methyl amyl alcohol; see Methyl isobutyl carbinol				
Methyl n-amyl ketone	110-43-0	100	465	·
Methyl bromide	74-83-9	(C)20	(C)80	X
Methyl butyl ketone; see 2-Hexanone				

Substance	CAS No.[d]	ppm[a]	mg/m³ [b]	Skin designation *
Methyl cellosolve; see 2-Methoxyethanol				
Methyl cellosolve acetate; see 2-Methoxyethyl acetate				
Methylene chloride; see § 1910.1052				
Methyl chloroform (1,1,1-Trichloroethane)	71-55-6	350	1900	-
Methylcyclohexane	108-87-2	500	2000	-
Methylcyclohexanol	25639-42-3	100	470	-
o-Methylcyclohexanone	583-60-8	100	460	X
Methylene chloride	75-09-2	500	1740	-
Methylenedianiline (MDA)	101-77-9			
Methyl ethyl ketone (MEK); see 2-Butanone				
Methyl formate	107-31-3	100	250	-
Methyl hydrazine (Monomethyl hydrazine)	60-34-4	(C)0.2	(C)0.35	X
Methyl iodide	74-88-4	5	28	X
Methyl isoamyl ketone	110-12-3	100	475	-
Methyl isobutyl carbinol	108-11-2	25	100	X
Methyl isobutyl ketone; see Hexone				
Methyl isocyanate	624-83-9	0.02	0.05	X
Methyl mercaptan	74-93-1	0.5	1	-
Methyl methacrylate	80-62-6	100	410	-
Methyl propyl ketone; see 2-Pentanone				
Methyl silicate	681-84-5	(C)5	(C)30	-
alpha-Methyl styrene	98-83-9	(C)100	(C)480	-
Methylene bisphenyl isocyanate (MDI)	101-68-8	(C)0.02	(C)0.2	-
Mica; see Silicates				
Molybdenum (as Mo)	7439-98-7			
Soluble compounds		-	5	-
Insoluble compounds				
Total dust		-	15	-
Monomethyl aniline	100-61-8	2	9	X
Monomethyl hydrazine; see Methyl hydrazine				
Morpholine	110-91-8	20	70	X
Naphtha (Coal tar)	8030-30-6	100	400	-
Naphthalene	91-20-3	10	50	-
alpha-Naphthylamine; see § 1926.1104	134-32-7			
beta-Naphthylamine; see § 1926.1109	91-59-8			-
Neon	7440-01-9	E		
Nickel carbonyl (as Ni)	13463-39-3	0.001	0.007	-
Nickel, metal and insoluble compounds (as Ni)	7440-02-0	-	1	-

Substance	CAS No.[d]	ppm[a]	mg/m³ [b]	Skin designation *
Nickel, soluble compounds (as Ni)	7440-02-0	-	1	-
Nicotine	54-11-5	-	0.5	X
Nitric acid	7697-37-2	2	5	-
Nitric oxide	10102-43-9	25	30	-
p-Nitroaniline	100-01-6	1	6	X
Nitrobenzene	98-95-3	1	5	X
p-Nitrochlorobenzene	100-00-5	-	1	X
4-Nitrodiphenyl; see § 1926.1103	92-93-3			
Nitroethane	79-24-3	100	310	-
Nitrogen	7727-37-9	E		
Nitrogen dioxide	10102-44-0	(C)5	(C)9	-
Nitrogen trifluoride	7783-54-2	10	29	-
Nitroglycerin	55-63-0	(C)0.2	(C)2	X
Nitromethane	75-52-5	100	250	-
1-Nitropropane	108-03-2	25	90	-
2-Nitropropane	79-46-9	25	90	-
N-Nitrosodimethylamine; see § 1926.1116	62-79-9			-
Nitrotoluene (all isomers)		5	30	X
o-isomer	88-72-2;			
m-isomer	99-08-1;			
p-isomer	99-99-0			
Nitrotrichloromethane; see Chloropicrin				
Nitrous oxide	10024-97-2	E		
Octachloronaphthalene	2234-13-1	-	0.1	X
Octane	111-65-9	400	1900	-
Oil mist, mineral	8012-95-1	-	5	-
Osmium tetroxide (as Os)	20816-12-0	-	0.002	-
Oxalic acid	144-62-7	-	1	-
Oxygen difluoride	7783-41-7	0.05	0.1	-
Ozone	10028-15-6	0.1	0.2	-
Paraquat, respirable dust	4685-14-7;	-	0.5	X
	1910-42-5;			
	2074-50-2			
Parathion	56-38-2	-	0.1	X
Particulates not otherwise regulated				
Total dust organic and inorganic		-	15	-
PCB; see Chlorodiphenyl (42% and 54% chlorine)				
Pentaborane	19624-22-7	0.005	0.01	-

Substance	CAS No.[d]	ppm[a]	mg/m³ [b]	Skin designation *
Pentachloronaphthalene	1321-64-8	-	0.5	X
Pentachlorophenol	87-86-5	-	0.5	X
Pentaerythritol	115-77-5			
Total dust		-		-
Respirable fraction		-		
Pentane	109-66-0	500	1500	-
2-Pentanone (Methyl propyl ketone)	107-87-9	200	700	-
Perchloroethylene (Tetrachloroethylene)	127-18-4	100	670	-
Perchloromethyl mercaptan	594-42-3	0.1	0.8	-
Perchloryl fluoride	7616-94-6	3	13.5	-
Petroleum distillates (Naphtha)(Rubber Solvent)			A³	-
Phenol	108-95-2	5	19	X
p-Phenylene diamine	106-50-3	-	0.1	X
Phenyl ether, vapor	101-84-8	1	7	-
Phenyl ether-biphenyl mixture, vapor		1	7	-
Phenylethylene; see Styrene				
Phenyl glycidyl ether (PGE)	122-60-1	10	60	-
Phenylhydrazine	100-63-0	5	22	X
Phosdrin (Mevinphos)	7786-34-7	-	0.1	X
Phosgene (Carbonyl chloride)	75-44-5	0.1	0.4	-
Phosphine	7803-51-2	0.3	0.4	-
Phosphoric acid	7664-38-2	-	1	-
Phosphorus (yellow)	7723-14-0	-	0.1	-
Phosphorus pentachloride	10026-13-8	-	1	-
Phosphorus pentasulfide	1314-80-3	-	1	-
Phosphorus trichloride	7719-12-2	0.5	3	-
Phthalic anhydride	85-44-9	2	12	-
Picric acid	88-89-1	-	0.1	X
Pindone (2-Pivalyl-1,3-indandione)	83-26-1	-	0.1	-
Plaster of Paris	26499-65-0			
Total dust		-		-
Respirable fraction		-		-
Platinum (as Pt)	7440-06-4			
Metal		-	-	-
Soluble salts		-	0.002	
Polytetrafluoroethylene decomposition products			A²	
Portland cement	65997-15-1			
Total dust		-	15	-

Substance	CAS No.[d]	ppm[a]	mg/m³[b]	Skin designation *
Respirable fraction		5		-
Propane	74-98-6	E		
Propargyl alcohol	107-19-7	1	-	X
beta-Propriolactone; see § 1926.1113	57-57-8			
n-Propyl acetate	109-60-4	200	840	-
n-Propyl alcohol	71-23-8	200	500	-
n-Propyl nitrate	627-13-4	25	110	-
Propylene dichloride	78-87-5	75	350	-
Propylene imine	75-55-8	2	5	X
Propylene oxide	75-56-9	100	240	-
Propyne; see Methyl acetylene				
Pyrethrum	8003-34-7	-	5	-
Pyridine	110-86-1	5	15	-
Quinone	106-51-4	0.1	0.4	-
RDX; see Cyclonite				
Rhodium (as Rh), metal fume and insoluble compounds	7440-16-6	-	0.1	-
Rhodium (as Rh), soluble compounds	7440-16-6	-	0.001	-
Ronnel	299-84-3	-	10	-
Rotenone	83-79-4	-	5	-
Rouge				
Total dust			-	-
Respirable fraction			-	-
Selenium compounds (as Se)	7782-49-2	-	0.2	-
Selenium hexafluoride (as Se)	7783-79-1	0.05	0.4	-
Silica, amorphous, precipitated and gel	112926-00-8	(^{2})	(^{2})	(^{2})
Silica, amorphous, diatomaceous earth, containing less than 1% crystalline silica	61790-53-2	(^{2})	(^{2})	(^{2})
Silica, crystalline, respirable dust				
Cristobalite; see 1926.1153	14464-46-1			
Quartz; see 1926.1153[5]	14808-60-7			
Tripoli (as quartz); see 1926.1153[5]	1317-95-9			
Tridymite; see 1926.1153	15468-32-3			
Silica, fused, respirable dust	60676-86-0	(^{2})	(^{2})	(^{2})
Silicates (less than 1% crystalline silica)				
Mica (respirable dust)	12001-26-2	(^{2})	(^{2})	(^{2})
Soapstone, total dust		(^{2})	(^{2})	(^{2})
Soapstone, respirable dust		(^{2})	(^{2})	(^{2})
Talc (containing asbestos); use asbestos limit; see § 1926.1101				

Substance	CAS No.[d]	ppm[a]	mg/m³ [b]	Skin designation *
Talc (containing no asbestos), respirable dust	14807-96-6	([2])	([2])	([2])
Tremolite, asbestiform; see § 1926.1101				
Silicon carbide	409-21-2			
Total dust			-	-
Respirable fraction			-	-
Silver, metal and soluble compounds (as Ag)	7440-22-4	-	0.01	-
Soapstone; see Silicates				
Sodium fluoroacetate	62-74-8	-	0.05	X
Sodium hydroxide	1310-73-2	-	2	-
Starch	9005-25-8			
Total dust			-	-
Respirable fraction			-	-
Stibine	7803-52-3	0.1	0.5	-
Stoddard solvent	8052-41-3	200	1150	-
Strychnine	57-24-9	-	0.15	-
Styrene	100-42-5	(C)100	(C)420	-
Sucrose	57-50-1			
Total dust			-	-
Respirable fraction			-	-
Sulfur dioxide	7446-09-5	5	13	-
Sulfur hexafluoride	2551-62-4	1000	6000	-
Sulfuric acid	7664-93-9	-	1	-
Sulfur monochloride	10025-67-9	1	6	-
Sulfur pentafluoride	5714-22-7	0.025	0.25	-
Sulfuryl fluoride	2699-79-8	5	20	-
Systox, see Demeton				
2,4,5-T (2,4,5-trichlorophenoxyacetic acid)	93-76-5	-	10	-
Talc; see Silicates -				
Tantalum, metal and oxide dust	7440-25-7	-	5	-
TEDP (Sulfotep)	3689-24-5	-	0.2	X
Teflon decomposition products			A2	
Tellurium and compounds (as Te)	13494-80-9	-	0.1	-
Tellurium hexafluoride (as Te)	7783-80-4	0.02	0.2	-
Temephos	3383-96-8			
Total dust			-	-
Respirable fraction			-	-
TEPP (Tetraethyl pyrophosphate)	107-49-3	-	0.05	X
Terphenyls	26140-60-3	(C)1	(C)9	-

Substance	CAS No.[d]	ppm[a]	mg/m³ [b]	Skin designation *
1,1,1,2-Tetrachloro-2,2-difluoroethane	76-11-9	500	4170	-
1,1,2,2-Tetrachloro-1,2-difluoroethane	76-12-0	500	4170	-
1,1,2,2-Tetrachloroethane	79-34-5	5	35	X
Tetrachloroethylene; see Perchloroethylene				
Tetrachloromethane; see Carbon tetrachloride				
Tetrachloronaphthalene	1335-88-2	-	2	X
Tetraethyl lead (as Pb)	78-00-2	-	0.1	X
Tetrahydrofuran	109-99-9	200	590	-
Tetramethyl lead, (as Pb)	75-74-1	-	0.15	X
Tetramethyl succinonitrile	3333-52-6	0.5	3	X
Tetranitromethane	509-14-8	1	8	-
Tetryl (2,4,6-Trinitrophenylmethylnitramine)	479-45-8	-	1.5	X
Thallium, soluble compounds (as Tl)	7440-28-0	-	0.1	X
Thiram	137-26-8	-	5	-
Tin, inorganic compounds (except oxides) (as Sn)	7440-31-5	-	2	-
Tin, organic compounds (as Sn)	7440-31-5	-	0.1	-
Tin oxide (as Sn)	21651-19-4	-	-	-
Total dust			-	
Respirable fraction			-	-
Titanium dioxide	13463-67-7			
Total dust			-	-
Toluene	108-88-3	200	750	-
Toluene-2,4-diisocyanate (TDI)	584-84-9	(C)0.02	(C)0.14	-
o-Toluidine	95-53-4	5	22	X
Toxaphene; see Chlorinated camphene				
Tremolite; see Silicates				
Tributyl phosphate	126-73-8	-	5	-
1,1,1-Trichloroethane; see Methyl chloroform				
1,1,2-Trichloroethane	79-00-5	10	45	X
Trichloroethylene	79-01-6	100	535	-
Trichloromethane; see Chloroform				
Trichloronaphthalene	1321-65-9	-	5	X
1,2,3-Trichloropropane	96-18-4	50	300	-
1,1,2-Trichloro-1,2,2-trifluoroethane	76-13-1	1000	7600	-
Triethylamine	121-44-8	25	100	-
Trifluorobromomethane	75-63-8	1000	6100	-
Trimethyl benzene	25551-13-7	25	120	-
2,4,6-Trinitrophenol; see Picric acid				

Substance	CAS No.[d]	ppm[a]	mg/m³ [b]	Skin designation *
2,4,6-Trinitrophenylmethylnitramine; see Tetryl				
2,4,6-Trinitrotoluene (TNT)	118-96-7	-	1.5	X
Triorthocresyl phosphate	78-30-8	-	0.1	-
Triphenyl phosphate	115-86-6	-	3	-
Tungsten (as W)	7440-33-7			
Insoluble compounds		-	5	-
Soluble compounds		-	1	-
Turpentine	8006-64-2	100	560	-
Uranium (as U)	7440-61-1			
Soluble compounds		-	0.2	-
Insoluble compounds		-	0.2	-
Vanadium	1314-62-1			
Respirable dust (as V_2O_5)		-	(C)0.5	-
Fume (as V_2O_5)		-	(C)0.1	-
Vegetable oil mist				
Total dust			-	-
Respirable fraction			-	-
Vinyl benzene; see Styrene				
Vinyl chloride; see § 1926.1117	75-01-4			
Vinyl cyanide; see Acrylonitrile				
Vinyl toluene	25013-15-4	100	480	-
Warfarin	81-81-2	-	0.1	-
Xylenes (o-, m-, p-isomers)	1330-20-7	100	435	-
Xylidine	1300-73-8	5	25	X
Yttrium	7440-65-5	-	1	-
Zinc chloride fume	7646-85-7	-	1	-
Zinc oxide fume	1314-13-2	-	5	-
Zinc oxide	1314-13-2			
Total dust		-	15	-
Respirable fraction		-	5	-
Zirconium compounds (as Zr)	7440-67-7	-	5	

Table 2 to § 1926.55 - Mineral Dusts

Substance	mppcf[j]
SILICA:	
Crystalline	250[k]
Quartz. Threshold Limit calculated from the formula[p]	% SiO_2 + 5

Substance	mppcf[j]
Cristobalite	
Amorphous, including natural diatomaceous earth	20
SILICATES (less than 1% crystalline silica)	
Mica	20
Portland cement	50
Soapstone	20
Talc (non-asbestiform)	20
Talc (fibrous), use asbestos limit	-
Graphite (natural)	15
Inert or Nuisance Particulates:[m]	50 (or 15 mg/m^3 whichever is the smaller) of total dust <1% SiO_2
[Inert or Nuisance Dusts includes all mineral, inorganic, and organic dusts as indicated by examples in TLV's appendix D]	
Conversion factors	
mppcf × 35.3 = million particles per cubic meter = particles per c.c.	

Footnotes to Tables 1 and 2 of this section:

[1] [Reserved]

[2] See Table 2 of this section.

[3] Use Asbestos Limit § 1926.1101.

[4] [Reserved]

[5] See Table 2 of this section for the exposure limit for any operations or sectors where theexposure limit in § 1926.1153 is stayed or is otherwise not in effect.

* An "X" designation in the "Skin Designation" column indicates that the substance is a dermalhazard.

[a] Parts of vapor or gas per million parts of contaminated air by volume at 25 °C and 760 torr.

[b] Milligrams of substance per cubic meter of air. When entry is in this column only, the value isexact; when listed with a ppm entry, it is approximate.

[c] [Reserved]

[d] The CAS number is for information only. Enforcement is based on the substance name. For anentry covering more than one metal compound, measured as the metal, the CAS number for the metal is given - not CAS numbers for the individual compounds.

[e-f] [Reserved]

g For sectors excluded from § 1926.1128 the limit is 10 ppm TWA.

h-i [Reserved]

j Millions of particles per cubic foot of air, based on impinger samples counted by light-fieldtechniques.

k The percentage of crystalline silica in the formula is the amount determined from airborne samples, except in those instances in which other methods have been shown to be applicable.

l [Reserved]

m Covers all organic and inorganic particulates not otherwise regulated. Same as Particulates NotOtherwise Regulated.

n If the exposure limit in § 1926.1126 is stayed or is otherwise not in effect, the exposure limit is a ceiling of 0.1 mg/m^3.

o If the exposure limit in § 1926.1126 is stayed or is otherwise not in effect, the exposure limit is 0.1 mg/m^3 (as CrO_3) as an 8-hour TWA.

p This standard applies to any operations or sectors for which the respirable crystalline silica standard, 1926.1153, is stayed or otherwise is not in effect.

q This standard applies to any operations or sectors for which the beryllium standard, 1926.1124, is stayed or otherwise is not in effect.

The 1970 TLV uses letter designations instead of a numerical value as follows:

A^1 [Reserved]

A^2 Polytetrafluoroethylene decomposition products. Because these products decompose in part byhydrolysis in alkaline solution, they can be quantitatively determined in air as fluoride to provide an index of exposure. No TLV is recommended pending determination of the toxicity of the products, but air concentrations should be minimal.

A^3 Gasoline and/or Petroleum Distillates. The composition of these materials varies greatly and thus a single TLV for all types of these materials is no longer applicable. The content of benzene,other aromatics and additives should be determined to arrive at the appropriate TLV.

E Simple asphyxiants. The limiting factor is the available oxygen which shall be at least 19.5% andbe within the requirements addressing explosion in part 1926.

[39 FR 22801, June 24, 1974, as amended at 51 FR 37007, Oct. 17, 1986; 52 FR 46312, Dec. 4, 1987; 58 FR 35089, June 30, 1993;

61 FR 9249, 9250, Mar. 7, 1996; 61 FR 56856, Nov. 4, 1996; 62 FR 1619, Jan. 10, 1997; 71 FR 10381, Feb. 28, 2006; 71 FR 36009,
June 23, 2006; 81 FR 16875, Mar. 25, 2016; 81 FR 31168, May 18, 2016; 81 FR 60273, Sept. 1, 2016; 82 FR 2750, Jan. 9, 2017; 84
FR 21576, May 14, 2019]

§ 1926.56 Illumination.

(a) *General.* Construction areas, ramps, runways, corridors, offices, shops, and storage areas shall be lighted to not less than the minimum illumination intensities listed in Table D-3 while any work is in progress:

Table D-3 - Minimum Illumination Intensities in Foot-Candles

Foot-candles	Area or operation
5	General construction area lighting.
3	General construction areas, concrete placement, excavation and waste areas, accessways, active storage areas, loading platforms, refueling, and field maintenance areas.
5	Indoors: warehouses, corridors, hallways, and exitways.
5	Tunnels, shafts, and general underground work areas: (Exception: minimum of 10 foot-candles is required at tunnel and shaft heading during drilling, mucking, and scaling. Bureau of Mines approved cap lights shall be acceptable for use in the tunnel heading.)
10	General construction plant and shops (e.g., batch plants, screening plants, mechanical and electrical equipment rooms, carpenter shops, rigging lofts and active storerooms, barracks or living quarters, locker or dressing rooms, mess halls, and indoor toilets and workrooms).
30	First aid stations, infirmaries, and offices.

(b) *Other areas.* For areas or operations not covered above, refer to the American National Standard A11.1-1965, R1970, Practice for Industrial Lighting, for recommended values of illumination.

§ 1926.57 Ventilation.

(a) *General.* Whenever hazardous substances such as dusts, fumes, mists, vapors, or gases exist or are produced in the course of construction work, their concentrations shall not exceed the limits specified in
§ 1926.55(a). When ventilation is used as an engineering control method, the system shall be installed and operated according to the requirements of this section.

(b) *Local exhaust ventilation.* Local exhaust ventilation when used as described in (a) shall be designed to prevent dispersion into the air of dusts, fumes, mists, vapors, and gases in concentrations causing harmful exposure. Such exhaust systems shall be so designed that dusts, fumes, mists, vapors, or gases are not drawn through the work area of employees.

(c) *Design and operation.* Exhaust fans, jets, ducts, hoods, separators, and all necessary appurtenances, including refuse receptacles, shall be so designed, constructed, maintained and operated as to ensure the required protection by maintaining a volume and velocity of exhaust air sufficient to gather dusts, fumes, vapors, or gases from said equipment or process, and to convey them to suitable points of safe disposal, thereby preventing their dispersion in harmful quantities into the atmosphere where employees work.

(d) **Duration of operations.**

(1) The exhaust system shall be in operation continually during all operations which it is designed to serve. If the employee remains in the contaminated zone, the system shall continue to operate afterthe cessation of said operations, the length of time to depend upon the individual circumstances andeffectiveness of the general ventilation system.

(2) Since dust capable of causing disability is, according to the best medical opinion, of microscopic size, tending to remain for hours in suspension in still air, it is essential that the exhaust system becontinued in operation for a time after the work process or equipment served by the same shall haveceased, in order to ensure the removal of the harmful elements to the required extent. For the samereason, employees wearing respiratory equipment should not remove same immediately until the atmosphere seems clear.

(e) **Disposal of exhaust materials.** The air outlet from every dust separator, and the dusts, fumes, mists, vapors, or gases collected by an exhaust or ventilating system shall discharge to the outside atmosphere.Collecting systems which return air to work area may be used if concentrations which accumulate in the work area air do not result in harmful exposure to employees. Dust and refuse discharged from an exhaust system shall be disposed of in such a manner that it will not result in harmful exposure to employees.

(f) **Abrasive blasting** -

(1) **Definitions applicable to this paragraph** -

(i) **Abrasive.** A solid substance used in an abrasive blasting operation.

(ii) **Abrasive-blasting respirator.** A respirator constructed so that it covers the wearer's head, neck, and shoulders to protect the wearer from rebounding abrasive.

(iii) **Blast cleaning barrel.** A complete enclosure which rotates on an axis, or which has an internal moving tread to tumble the parts, in order to expose various surfaces of the parts to the actionof an automatic blast spray.

(iv) **Blast cleaning room.** A complete enclosure in which blasting operations are performed and where the operator works inside of the room to operate the blasting nozzle and direct the flowof the abrasive material.

(v) **Blasting cabinet.** An enclosure where the operator stands outside and operates the blastingnozzle through an opening or openings in the enclosure.

(vi) **Clean air.** Air of such purity that it will not cause harm or discomfort to an individual if it is inhaled for extended periods of time.

(vii) **Dust collector.** A device or combination of devices for separating dust from the air handled byan exhaust ventilation system.

(viii) **Exhaust ventilation system.** A system for removing contaminated air from a space, comprising two or more of the following elements

(A) enclosure

(B) or hood,

(C) duct work,

- dust collecting equipment,
- (D) exhauster, and
- (E) discharge stack.

- (ix) **Particulate-filter respirator.** An air purifying respirator, commonly referred to as a dust or a fume respirator, which removes most of the dust or fume from the air passing through the device.

- (x) **Respirable dust.** Airborne dust in sizes capable of passing through the upper respiratory system to reach the lower lung passages.

- (xi) **Rotary blast cleaning table.** An enclosure where the pieces to be cleaned are positioned on a rotating table and are passed automatically through a series of blast sprays.

- (xii) **Abrasive blasting.** The forcible application of an abrasive to a surface by pneumatic pressure, hydraulic pressure, or centrifugal force.

(2) *Dust hazards from abrasive blasting.*

- (i) Abrasives and the surface coatings on the materials blasted are shattered and pulverized during blasting operations and the dust formed will contain particles of respirable size. The composition and toxicity of the dust from these sources shall be considered in making an evaluation of the potential health hazards.

- (ii) The concentration of respirable dust or fume in the breathing zone of the abrasive-blasting operator or any other worker shall be kept below the levels specified in § 1926.55 or other pertinent sections of this part.

- (iii) Organic abrasives which are combustible shall be used only in automatic systems. Where flammable or explosive dust mixtures may be present, the construction of the equipment, including the exhaust system and all electric wiring, shall conform to the requirements of American National Standard Installation of Blower and Exhaust Systems for Dust, Stock, and Vapor Removal or Conveying, Z33.1-1961 (NFPA 91-1961), and subpart S of this part. The blast nozzle shall be bonded and grounded to prevent the build up of static charges. Where flammable or explosive dust mixtures may be present, the abrasive blasting enclosure, the ducts, and the dust collector shall be constructed with loose panels or explosion venting areas, located on sides away from any occupied area, to provide for pressure relief in case of explosion, following the principles set forth in the National Fire Protection Association Explosion Venting Guide. NFPA 68-1954.

(3) *Blast-cleaning enclosures.*

- (i) Blast-cleaning enclosures shall be exhaust ventilated in such a way that a continuous inward flow of air will be maintained at all openings in the enclosure during the blasting operation.

    - (A) All air inlets and access openings shall be baffled or so arranged that by the combination of inward air flow and baffling the escape of abrasive or dust particles into an adjacent work area will be minimized and visible spurts of dust will not be observed.

    - (B) The rate of exhaust shall be sufficient to provide prompt clearance of the dust-laden air within the enclosure after the cessation of blasting.

(C) Before the enclosure is opened, the blast shall be turned off and the exhaust system shallbe run for a sufficient period of time to remove the dusty air within the enclosure.

(D) Safety glass protected by screening shall be used in observation windows, where harddeep-cutting abrasives are used.

(E) Slit abrasive-resistant baffles shall be installed in multiple sets at all small access openings where dust might escape, and shall be inspected regularly and replaced whenneeded.

(1) Doors shall be flanged and tight when closed.

(2) Doors on blast-cleaning rooms shall be operable from both inside and outside, exceptthat where there is a small operator access door, the large work access door may be closed or opened from the outside only.

(4) **Exhaust ventilation systems.**

(i) The construction, installation, inspection, and maintenance of exhaust systems shall conformto the principles and requirements set forth in American National Standard Fundamentals Governing the Design and Operation of Local Exhaust Systems, Z9.2-1960, and ANSI Z33.1-1961.

(a) When dust leaks are noted, repairs shall be made as soon as possible.

(b) The static pressure drop at the exhaust ducts leading from the equipment shall be checkedwhen the installation is completed and periodically thereafter to assure continued satisfactory operation. Whenever an appreciable change in the pressure drop indicates a partial blockage,the system shall be cleaned and returned to normal operating condition.

(ii) In installations where the abrasive is recirculated, the exhaust ventilation system for the blasting enclosure shall not be relied upon for the removal of fines from the spent abrasiveinstead of an abrasive separator. An abrasive separator shall be provided for the purpose.

(iii) The air exhausted from blast-cleaning equipment shall be discharged through dust collecting equipment. Dust collectors shall be set up so that the accumulated dust can be emptied and removed without contaminating other working areas.

(5) **Personal protective equipment.**

(i) Employers must use only respirators approved by NIOSH under 42 CFR part 84 for protectingemployees from dusts produced during abrasive-blasting operations.

(ii) Abrasive-blasting respirators shall be worn by all abrasive-blasting

(A) operators: When working inside of blast-cleaning rooms, or

(B) When using silica sand in manual blasting operations where the nozzle and blast are notphysically separated from the operator in an exhaust ventilated enclosure, or

(C) Where concentrations of toxic dust dispersed by the abrasive blasting may exceed the limits set in § 1926.55 or other pertinent sections of this part and the nozzle and blast arenot physically separated from the operator in an exhaust-ventilated enclosure.

 (iii) Properly fitted particulate-filter respirators, commonly referred to as dust-filter respirators, may be used for short, intermittent, or occasional dust exposures such as cleanup, dumping of dust collectors, or unloading shipments of sand at a receiving point when it is not feasible to controlthe dust by enclosure, exhaust ventilation, or other means. The respirators used must be approved by NIOSH under 42 CFR part 84 for protection against the specific type of dust encountered.

 (iv) A respiratory protection program as defined and described in § 1926.103, shall be establishedwherever it is necessary to use respiratory protective equipment.

 (v) Operators shall be equipped with heavy canvas or leather gloves and aprons or equivalent protection to protect them from the impact of abrasives. Safety shoes shall be worn to protectagainst foot injury where heavy pieces of work are handled.

  (A) Safety shoes shall conform to the requirements of American National Standard for Men'sSafety-Toe Footwear, Z41.1-1967.

  (B) Equipment for protection of the eyes and face shall be supplied to the operator when the respirator design does not provide such protection and to any other personnel working inthe vicinity of abrasive blasting operations. This equipment shall conform to the requirements of § 1926.102.

(6) ***Air supply and air compressors.*** Air for abrasive-blasting respirators must be free of harmful quantities of dusts, mists, or noxious gases, and must meet the requirements for supplied-air quality and use specified in 29 CFR 1910.134(i).

(7) ***Operational procedures and general safety.*** Dust shall not be permitted to accumulate on the floor oron ledges outside of an abrasive-blasting enclosure, and dust spills shall be cleaned up promptly.
Aisles and walkways shall be kept clear of steel shot or similar abrasive which may create a slippinghazard.

(8) ***Scope.*** This paragraph applies to all operations where an abrasive is forcibly applied to a surface bypneumatic or hydraulic pressure, or by centrifugal force. It does not apply to steam blasting, or steam cleaning, or hydraulic cleaning methods where work is done without the aid of abrasives.

(g) ***Grinding, polishing, and bufing***

 (1) ***operations*** - Definitions applicable to

  (i) his paragraph -

   Abrasive cutting-off wheels. Organic-bonded wheels, the thickness of which is not more than one forty-eighth of their diameter for those up to, and including, 20 inches (50.8 cm) in diameter, and not more than one-sixtieth of their diameter for those larger than 20 inches (50.8cm) in diameter, used for a multitude of operations variously known as cutting, cutting off, grooving, slotting, coping, and jointing, and the like. The wheels may be "solid" consisting of organic-bonded abrasive material throughout, "steel centered" consisting of a steel disc with a rim of organic-bonded material moulded around the periphery, or of the "inserted tooth" type consisting of a steel disc with

(ii) **Belts.** All power-driven, flexible, coated bands used for grinding, polishing, or buffing purposes.

(iii) **Branch pipe.** The part of an exhaust system piping that is connected directly to the hood or enclosure.

(iv) **Cradle.** A movable fixture, upon which the part to be ground or polished is placed.

(v) **Disc wheels.** All power-driven rotatable discs faced with abrasive materials, artificial or natural, and used for grinding or polishing on the side of the assembled disc.

(vi) **Entry loss.** The loss in static pressure caused by air flowing into a duct or hood. It is usually expressed in inches of water gauge.

(vii) **Exhaust system.** A system consisting of branch pipes connected to hoods or enclosures, one or more header pipes, an exhaust fan, means for separating solid contaminants from the air flowing in the system, and a discharge stack to outside.

(viii) **Grinding wheels.** All power-driven rotatable grinding or abrasive wheels, except disc wheels as defined in this standard, consisting of abrasive particles held together by artificial or natural bonds and used for peripheral grinding.

(ix) **Header pipe (main pipe).** A pipe into which one or more branch pipes enter and which connects such branch pipes to the remainder of the exhaust system.

(x) **Hoods and enclosures.** The partial or complete enclosure around the wheel or disc through which air enters an exhaust system during operation.

(xi) **Horizontal double-spindle disc grinder.** A grinding machine carrying two power-driven, rotatable, coaxial, horizontal spindles upon the inside ends of which are mounted abrasive disc wheels used for grinding two surfaces simultaneously.

(xii) Horizontal single-spindle disc grinder. A grinding machine carrying an abrasive disc wheel upon one or both ends of a power-driven, rotatable single horizontal spindle.

(xiii) **Polishing and bufing wheels.** All power-driven rotatable wheels composed all or in part of textile fabrics, wood, felt, leather, paper, and may be coated with abrasives on the periphery of the wheel for purposes of polishing, buffing, and light grinding.

(xiv) **Portable grinder.** Any power-driven rotatable grinding, polishing, or buffing wheel mounted in such manner that it may be manually manipulated.

(xv) **Scratch brush wheels.** All power-driven rotatable wheels made from wire or bristles, and used for scratch cleaning and brushing purposes.

(xvi) **Swing-frame grinder.** Any power-driven rotatable grinding, polishing, or buffing wheel mounted in such a manner that the wheel with its supporting framework can be manipulated over stationary objects.

(xvii) **Velocity pressure (vp).** The kinetic pressure in the direction of flow necessary to cause a fluid atrest to flow at a given velocity. It is usually expressed in inches of water gauge.

(xviii) **Vertical spindle disc grinder.** A grinding machine having a vertical, rotatable power-driven spindle carrying a horizontal abrasive disc wheel.

(2) **Application.** Wherever dry grinding, dry polishing or buffing is performed, and employee exposure, without regard to the use of respirators, exceeds the permissible exposure limits prescribed in § 1926.55 or other pertinent sections of this part, a local exhaust ventilation system shall be providedand used to maintain employee exposures within the prescribed limits.

(3) **Hood and branch pipe requirements.**

(i) Hoods connected to exhaust systems shall be used, and such hoods shall be designed, located,and placed so that the dust or dirt particles shall fall or be projected into the hoods in the direction of the air flow. No wheels, discs, straps, or belts shall be operated in such manner and in such direction as to cause the dust and dirt particles to be thrown into the operator's breathing zone.

(ii) Grinding wheels on floor stands, pedestals, benches, and special-purpose grinding machines and abrasive cutting-off wheels shall have not less than the minimum exhaust volumes shownin Table D-57.1 with a recommended minimum duct velocity of 4,500 feet per minute in the branch and 3,500 feet per minute in the main. The entry losses from all hoods except the vertical-spindle disc grinder hood, shall equal 0.65 velocity pressure for a straight takeoff and 0.45 velocity pressure for a tapered takeoff. The entry loss for the vertical-spindle disc grinderhood is shown in figure D-57.1 (following paragraph (g) of this section).

Table D-57.1 - Grinding and Abrasive Cutting-Off Wheels

Wheel diameter, inches (cm)	Wheel width, inches (cm)	Minimum exhaust volume (feet^3/min.)
To 9 (22.86)	1$^1/_2$ (3.81)	220
Over 9 to 16 (22.86 to 40.64)	2 (5.08)	390
Over 16 to 19 (40.64 to 48.26)	3 (7.62)	500
Over 19 to 24 (48.26 to 60.96)	4 (10.16)	610
Over 24 to 30 (60.96 to 76.2)	5 (12.7)	880
Over 30 to 36 (76.2 to 91.44)	6 (15.24)	1,200

For any wheel wider than wheel diameters shown in Table D-57.1, increase the exhaust volumeby the ratio of the new width to the width shown.

Example: If wheel width = 4$^1/_2$ inches (11.43 cm),then 4.5 ÷ 4 × 610 = 686 (rounded to 690).

(iii) Scratch-brush wheels and all buffing and polishing wheels mounted on floor stands, pedestals, benches, or special-purpose machines shall have not less than the minimum exhaust volumeshown in Table D-57.2.

Table D-57.2 - Buffing and Polishing Wheels

Wheel diameter, inches (cm)	Wheel width, inches cm)	Minimum exhaust volume (feet^3/min.)

To 9 (22.86)	2 (5.08)	300
Over 9 to 16 (22.86 to 40.64)	3 (7.62)	500
Over 16 to 19 (40.64 to 48.26)	4 (10.16)	610
Over 19 to 24 (48.26 to 60.96)	5 (12.7)	740
Over 24 to 30 (60.96 to 76.2)	6 (15.24)	1,040
Over 30 to 36 (76.2 to 91.44)	6 (15.24)	1,200

(iv) Grinding wheels or discs for horizontal single-spindle disc grinders shall be hooded to collect the dust or dirt generated by the grinding operation and the hoods shall be connected to branchpipes having exhaust volumes as shown in Table D-57.3.

Table D-57.3 - Horizontal Single-Spindle Disc Grinder

Disc diameter, inches (cm)	Exhaust volume (ft.^{3}/min.)
Up to 12 (30.48)	220
Over 12 to 19 (30.48 to 48.26)	390
Over 19 to 30 (48.26 to 76.2)	610
Over 30 to 36 (76.2 to 91.44)	880

(v) Grinding wheels or discs for horizontal double-spindle disc grinders shall have a hood enclosingthe grinding chamber and the hood shall be connected to one or more branch pipes having exhaust volumes as shown in Table D-57.4.

Table D-57.4 - Horizontal Double-Spindle Disc Grinder

Disc diameter, inches (cm)	Exhaust volume (ft.^{3}/min.)
Up to 19 (48.26)	610
Over 19 to 25 (48.26 to 63.5)	880
Over 25 to 30 (63.5 to 76.2)	1,200
Over 30 to 53 (76.2 to 134.62)	1,770
Over 53 to 72 (134.62 to 182.88)	6,280

(vi) Grinding wheels or discs for vertical single-spindle disc grinders shall be encircled with hoodsto remove the dust generated in the operation. The hoods shall be connected to one or morebranch pipes having exhaust volumes as shown in Table D-57.5.

Table D-57.5 - Vertical Spindle Disc Grinder

Disc diameter, inches (cm)	One-half or more of disc covered		Disc not covered	
	Number[1]	Exhaust foot^3/min.	Number[1]	Exhaust foot^3/min.
Up to 20 (50.8)	1	500	2	780
Over 20 to 30 (50.8 to 76.2)	2	780	2	1,480
Over 30 to 53 (76.2 to 134.62)	2	1,770	4	3,530
Over 53 to 72 (134.62 to 182.88)	2	3,140	5	6,010

[1] Number of exhaust outlets around periphery of hood, or equal distribution provided by othermeans.

(vii) Grinding and polishing belts shall be provided with hoods to remove dust and dirt generated inthe operations and the hoods shall be connected to branch pipes having exhaust volumes as shown in Table D-57.6.

Table D-57.6 - Grinding and Polishing Belts

Belts width, inches (cm)	Exhaust volume (ft.^{3}/min.)
Up to 3 (7.62)	220
Over 3 to 5 (7.62 to 12.7)	300
Over 5 to 7 (12.7 to 17.78)	390
Over 7 to 9 (17.78 to 22.86)	500
Over 9 to 11 (22.86 to 27.94)	610
Over 11 to 13 (27.94 to 33.02)	740

(viii) Cradles and swing-frame grinders. Where cradles are used for handling the parts to be ground, polished, or buffed, requiring large partial enclosures to house the complete operation, a minimum average air velocity of 150 feet per minute shall be maintained over the entire opening of the enclosure. Swing-frame grinders shall also be exhausted in the same manner asprovided for cradles. (See fig. D-57.3)

(ix) Where the work is outside the hood, air volumes must be increased as shown in AmericanStandard Fundamentals Governing the Design and Operation of Local Exhaust Systems, Z9.2-1960 (section 4, exhaust hoods).

(4) *Exhaust systems.*

(i) Exhaust systems for grinding, polishing, and buffing operations should be designed in accordance with American Standard Fundamentals Governing the Design and Operation ofLocal Exhaust Systems, Z9.2-1960.

(ii) Exhaust systems for grinding, polishing, and buffing operations shall be tested in the manner described in American Standard Fundamentals Governing the Design and Operation of LocalExhaust Systems, Z9.2-1960.

(iii) All exhaust systems shall be provided with suitable dust collectors.

(5) *Hood and enclosure design.*

(i)

(A) It is the dual function of grinding and abrasive cutting-off wheel hoods to protect the operator from the hazards of bursting wheels as well as to provide a means for the removal of dust and dirt generated. All hoods shall be not less in structural strength than specified in the American National Standard Safety Code for the Use, Care, and Protectionof Abrasive Wheels, B7.1-1970.

(B) Due to the variety of work and types of grinding machines employed, it is necessary to develop hoods adaptable to the particular machine in question, and such hoods shall belocated as close as possible to the operation.

(ii) Exhaust hoods for floor stands, pedestals, and bench grinders shall be designed in accordance with figure D-57.2. The adjustable tongue shown in the figure shall be kept in working order and shall be adjusted within one-fourth inch (0.635 cm) of the wheel periphery at all times.

(iii) Swing-frame grinders shall be provided with exhaust booths as indicated in figure D-57.3.

(iv) Portable grinding operations, whenever the nature of the work permits, shall be conducted within a partial enclosure. The opening in the enclosure shall be no larger than is actually required in the operation and an average face air velocity of not less than 200 feet per minuteshall be maintained.

(v) Hoods for polishing and buffing and scratch-brush wheels shall be constructed to conform asclosely to figure D-57.4 as the nature of the work will permit.

(vi) Cradle grinding and polishing operations shall be performed within a partial enclosure similar tofigure D-57.5. The operator shall be positioned outside the working face of the opening of the enclosure. The face opening of the enclosure should not be any greater in area than that actually required for the performance of the operation and the average air velocity into the working face of the enclosure shall not be less than 150 feet per minute.

(vii) Hoods for horizontal single-spindle disc grinders shall be constructed to conform as closely as possible to the hood shown in figure D-57.6. It is essential that there be a space between the back of the wheel and the hood, and a space around the periphery of the wheel of at least 1 inch (2.54 cm) in order to permit the suction to act around the wheel periphery. The opening on the side of the disc shall be no larger than is required for the grinding operation, but must neverbe less than twice the area of the branch outlet.

(viii) Horizontal double-spindle disc grinders shall have a hood encircling the wheels and grindingchamber similar to that illustrated in figure D-57.7. The openings for passing the work into thegrinding chamber should be kept as small as possible, but must never be less than twice thearea of the branch outlets.

(ix) Vertical-spindle disc grinders shall be encircled with a hood so constructed that the heavy dustis drawn off a surface of the disc and the lighter dust exhausted through a continuous slot at the top of the hood as shown in figure D-57.1.

(x) Grinding and polishing belt hoods shall be constructed as close to the operation as possible.The hood should extend almost to the belt, and 1-inch (2.54 cm) wide openings should be provided on either side. Figure D-57.8 shows a typical hood for a belt operation.

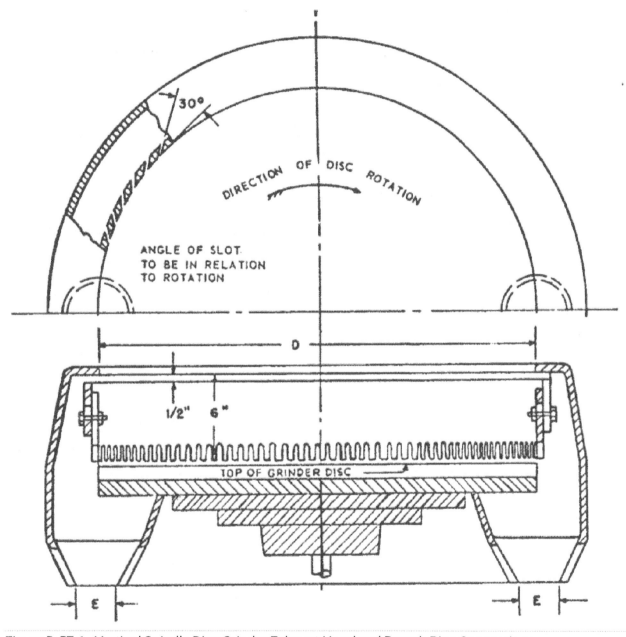

Figure D-57.1 - Vertical Spindle Disc Grinder Exhaust Hood and Branch Pipe Connections

Dia. *D* inches (cm)		Exhaust *E*		Volume Exhausted at 4,500 ft/min ft³/min	Note
Min.	Max.	No Pipes	Dia.		
	20 (50.8)	1	4¼ (10.795)	500	When one-half or more of the disc can be hooded, use exhaust ducts as shown at the left.
Over 20 (50.8)	30 (76.2)	2	4 (10.16)	780	

Dia. *D* inches (cm)		Exhaust *E*		Volume Exhausted at 4,500 ft/min ft³/min	Note
Min.	Max.	No Pipes	Dia.		
Over 30 (76.2)	72 (182.88)	2	6 (15.24)	1,770	
Over 53 (134.62)	72 (182.88)	2	8 (20.32)	3,140	
	20 (50.8)	2	4 (10.16)	780	When no hood can be used over disc, use exhaust ducts as shown at left.
Over 20 (50.8)	20 (50.8)	2	4 (10.16)	780	
Over 30 (76.2)	30 (76.2)	2	5$^1/_2$ (13.97)	1,480	
Over 53 (134.62)	53 (134.62)	4	6 (15.24)	3,530	
	72 (182.88)	5	7 (17.78)	6,010	

Entry loss = 1.0 slot velocity pressure + 0.5 branch velocity pressure.

Minimum slot velocity = 2,000 ft/min - $^1/_2$-inch (1.27 cm) slot width.

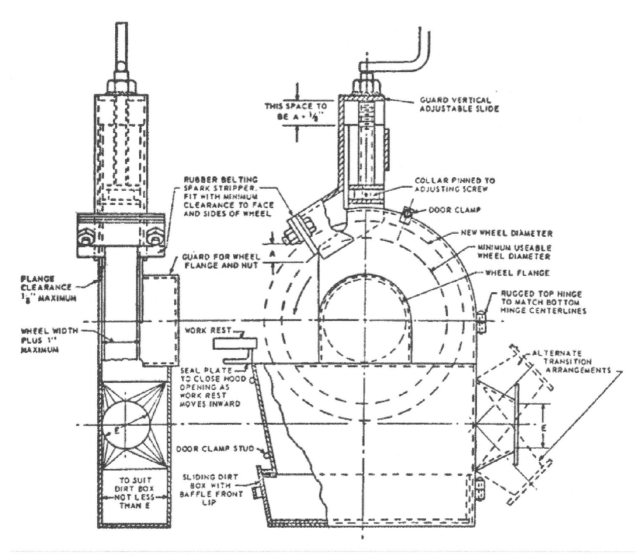

Figure D-57.2 - Standard Grinder Hood

Wheel dimension, inches (centimeters)			Exhaust outlet, inches (centimeters) E	Volume of air at 4,500 ft/min
Diameter		Width, Max		
Min= d	Max= D			
	9 (22.86)	1½ (3.81)	3	220
Over 9 (22.86)	16 (40.64)	2 (5.08)	4	390
Over 16 (40.64)	19 (48.26)	3 (7.62)	4½	500
Over 19 (48.26)	24 (60.96)	4 (10.16)	5	610
Over 24 (60.96)	30 (76.2)	5 (12.7)	6	880

Wheel dimension, inches (centimeters)		Exhaust outlet, inches (centimeters) E	Volume of air at 4,500 ft/min
Diameter	Width		
Min= d Max=4d			
36 Max	6 (15.24) Max	7	
Over 30 (76.2)			1,200

Entry loss = 0.45 velocity pressure for tapered takeoff 0.65 velocity pressure for straight takeoff.

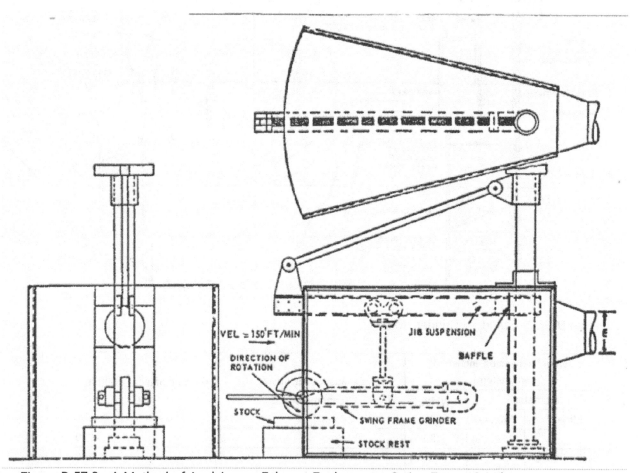

Figure D-57.3 - A Method of Applying an Exhaust Enclosure to Swing-Frame Grinders

Note: Baffle to reduce front opening as much as possible

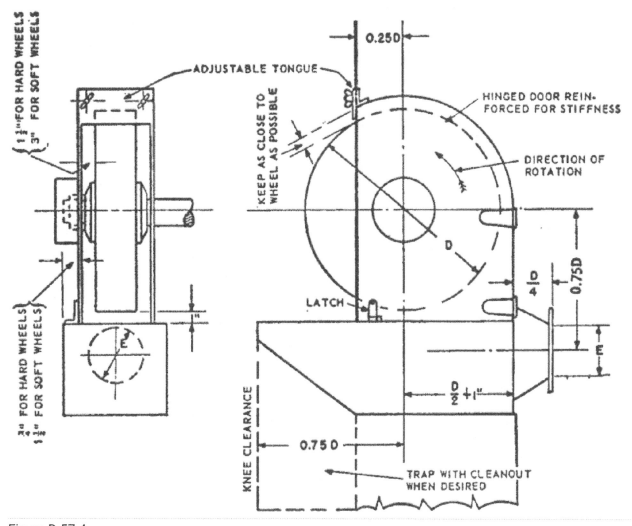

Figure D-57.4

Standard Buffing and Polishing Hood

Wheel dimension, inches (centimeters)		Width, Max	Exhaust outlet, inches E	Volume of air at 4,500 ft/min
Diameter				
Min= d	Max= D			
	9 (22.86)	2 (5.08)	3½ (3.81)	300
Over 9 (22.86)	16 (40.64)	3 (5.08)	4	500
Over 16 (40.64)	19 (48.26)	4 (11.43)	5	610
Over 19 (48.26)	24 (60.96)	5 (12.7)	5½	740
Over 24 (60.96)	30 (76.2)	6 (15.24)	6½	1.040
Over 30 (76.2)	36 (91.44)	6 (15.24)	7	1.200

Entry loss = 0.15 velocity pressure for tapered takeoff; 0.65 velocity pressure for straight takeoff.

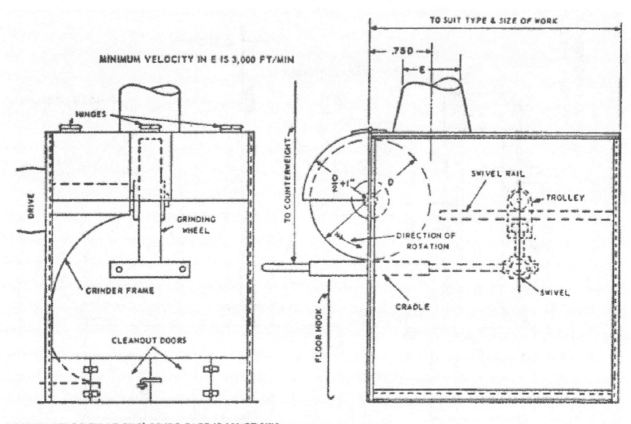

Figure D-57.5 - Cradle Polishing or Grinding Enclosure

Entry loss = 0.45 velocity pressure for tapered takeoff

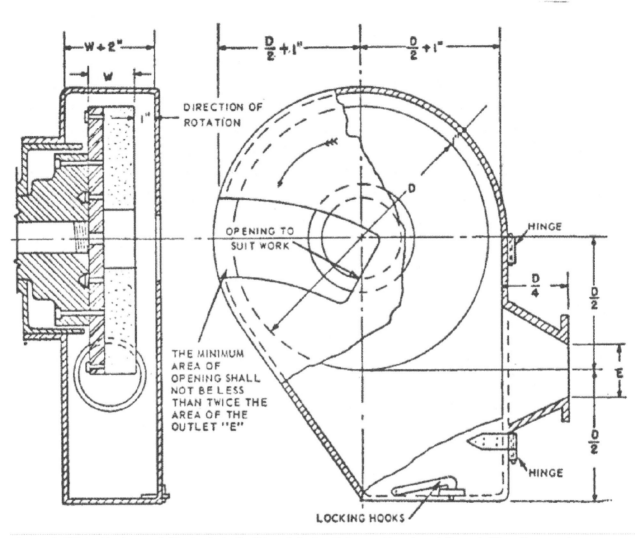

Figure D-57.6 - Horizontal Single-Spindle Disc Grinder Exhaust Hood and Branch Pipe Connections

Dia *D*, inches (centimeters)		Exhaust *E*, dia. inches (cm)	Volume exhausted at 4,500 ft/min ft³/min
Min.	Max.		
	12 (30.48)	3 (7.6)	220
Over 12 (30.48)	19 (48.26)	4 (10.16)	390
Over 19 (48.26)	30 (76.2)	5 (12.7)	610
Over 30 (76.2)	36 (91.44)	6 (15.24)	880

NOTE: If grinding wheels are used for disc grinding purposes, hoods must conform to structural strength and materials as described in 9.1.

Entry loss = 0.45 velocity pressure for tapered takeoff.

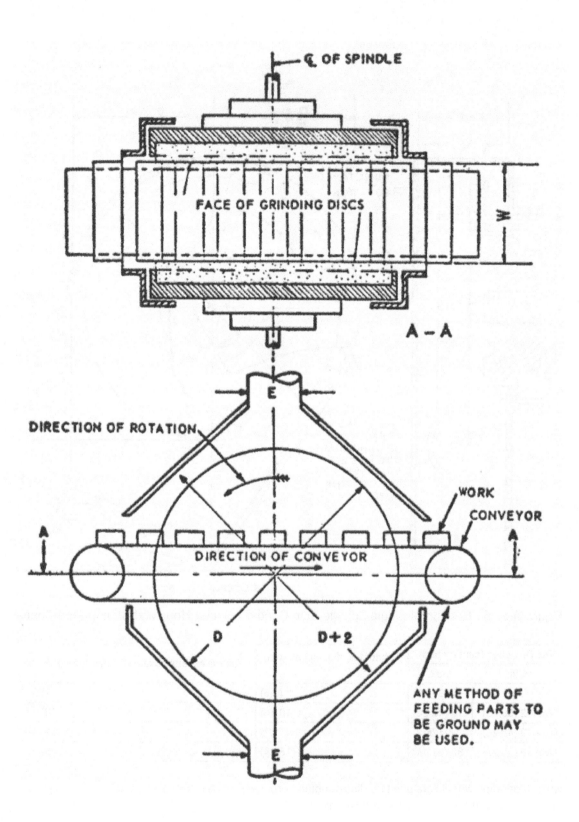

Figure D-57.7 - Horizontal Double-Spindle Disc Grinder Exhaust Hood and Branch Pipe Connections

Disc dia. inches (centimeters)		Exhaust E		Volume exhaust at 4,500 ft/min. ft³/min	Note
Min.	Max.	No Pipes	Dia.		
	19 (48.26)	1	5	610	
Over 19 (48.26)	25 (63.5)	1	6	880	When width "W" permits, exhaust ducts shouldbe as near heaviest grinding as possible.
Over 25 (63.5)	30 (76.2)	1	7	1,200	
Over 30 (76.2)	53 (134.62)	2	6	1,770	
Over 53 (134.62)	72 (182.88)	4	8	6,280	

Entry loss = 0.45 velocity pressure for tapered takeoff.

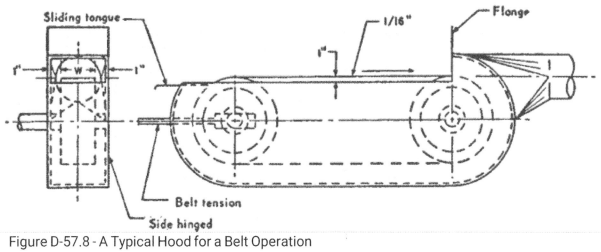

Figure D-57.8 - A Typical Hood for a Belt Operation

Entry loss = 0.45 velocity pressure for tapered takeoff

Belt width W. inches (centimeters)	Exhaust volume. ft.^{1}/min
Up to 3 (7.62)	220
3 to 5 (7.62 to 12.7)	300
5 to 7 (12.7 to 17.78)	390
7 to 9 (17.78 to 22.86)	500
9 to 11 (22.86 to 27.94)	610
11 to 13 (27.94 to 33.02)	740

Minimum duct velocity = 4,500 ft/min branch, 3,500 ft/min main.

Entry loss = 0.45 velocity pressure for tapered takeoff; 0.65 velocity pressure for straight takeoff.

    (6)   **Scope.** This paragraph (g), prescribes the use of exhaust hood enclosures and systems in removingdust, dirt, fumes, and gases generated through the grinding, polishing, or buffing of ferrous and nonferrous metals.

(h)   **Spray finishing operations -**

    (1)   **Definitions applicable to this paragraph -**

        (i)   **Spray-finishing operations.** Spray-finishing operations are employment of methods wherein organic or inorganic materials are utilized in dispersed form for deposit on surfaces to be coated, treated, or cleaned. Such methods of deposit may involve either automatic, manual, or electrostatic deposition but do not include metal spraying or metallizing, dipping, flow coating, roller coating, tumbling, centrifuging, or spray washing and degreasing as conducted in self- contained washing and degreasing machines or systems.

        (ii)   **Spray booth.** Spray booths are defined and described in § 1926.66(a). (See sections 103, 104, and 105 of the Standard for Spray Finishing Using Flammable and Combustible Materials, NFPANo. 33-1969).

        (iii)   **Spray room.** A spray room is a room in which spray-finishing operations not conducted in aspray booth are performed separately from other areas.

        (iv)   **Minimum maintained velocity.** Minimum maintained velocity is the velocity of air movement which must be maintained in order to meet minimum specified requirements for health andsafety.

    (2)   **Location and application.** Spray booths or spray rooms are to be used to enclose or confine all operations. Spray-finishing operations shall be located as provided in sections 201 through 206 of the Standard for Spray Finishing Using Flammable and Combustible Materials, NFPA No. 33-1969.

    (3)   **Design and construction of spray booths.**

(i) Spray booths shall be designed and constructed in accordance with § 1926.66(b) (1) through
(4) and (6) through (10) (see sections 301-304 and 306-310 of the Standard for Spray FinishingUsing Flammable and Combustible Materials, NFPA No. 33-1969), for general construction specifications. For a more detailed discussion of fundamentals relating to this subject, see ANSI Z9.2-1960

    (A) Lights, motors, electrical equipment, and other sources of ignition shall conform to the requirements of § 1926.66(b)(10) and (c). (See section 310 and chapter 4 of the Standardfor Spray Finishing Using Flammable and Combustible Materials NFPA No. 33-1969.)

    (B) In no case shall combustible material be used in the construction of a spray booth andsupply or exhaust duct connected to it.

(ii) Unobstructed walkways shall not be less than $6^1/_2$ feet (1.976 m) high and shall be maintained clear of obstruction from any work location in the booth to a booth exit or open booth front. In booths where the open front is the only exit, such exits shall be not less than 3 feet (0.912 m) wide. In booths having multiple exits, such exits shall not be less than 2 feet (0.608 m) wide, provided that the maximum distance from the work location to the exit is 25 feet (7.6 m) or less. Where booth exits are provided with doors, such doors shall open outward from the booth.

(iii) Baffles, distribution plates, and dry-type overspray collectors shall conform to the requirements of § 1926.66(b) (4) and (5). (See sections 304 and 305 of the Standard for Spray Finishing Using Flammable and Combustible Materials, NFPA No. 33-1969.)

    (A) Overspray filters shall be installed and maintained in accordance with the requirements of
§ 1926.66(b)(5), (see section 305 of the Standard for Spray Finishing Using Flammableand Combustible Materials, NFPA No. 33-1969), and shall only be in a location easily accessible for inspection, cleaning, or replacement.

    (B) Where effective means, independent of the overspray filters, are installed which will resultin design air distribution across the booth cross section, it is permissible to operate the booth without the filters in place.

(iv)

    (A) For wet or water-wash spray booths, the water-chamber enclosure, within which intimate contact of contaminated air and cleaning water or other cleaning medium is maintained, ifmade of steel, shall be 18 gage or heavier and adequately protected against corrosion.

    (B) Chambers may include scrubber spray nozzles, headers, troughs, or other devices. Chambers shall be provided with adequate means for creating and maintaining scrubbingaction for removal of particulate matter from the exhaust air stream.

(v) Collecting tanks shall be of welded steel construction or other suitable non-combustible material. If pits are used as collecting tanks, they shall be concrete, masonry, or other materialhaving similar properties.

    (A) Tanks shall be provided with weirs, skimmer plates, or screens to prevent sludge and floating paint from entering the pump suction box. Means for automatically maintaining the proper water level shall also be provided. Fresh water inlets shall not be submerged. They shall terminate at least one pipe diameter above the safety overflow level of the tank.

  (B) Tanks shall be so constructed as to discourage accumulation of hazardous deposits.

 (vi) Pump manifolds, risers, and headers shall be adequately sized to insure sufficient water flow to provide efficient operation of the water chamber.

(4) **Design and construction of spray rooms.**

 (i) Spray rooms, including floors, shall be constructed of masonry, concrete, or other noncombustible material.

 (ii) Spray rooms shall have noncombustible fire doors and shutters.

 (iii) Spray rooms shall be adequately ventilated so that the atmosphere in the breathing zone of the operator shall be maintained in accordance with the requirements of paragraph (h)(6)(ii) of this section.

 (iv) Spray rooms used for production spray-finishing operations shall conform to the requirements for spray booths.

(5) **Ventilation.**

 (i) Ventilation shall be provided in accordance with provisions of § 1926.66(d) (see chapter 5 of the Standard for Spray Finishing Using Flammable or Combustible Materials, NFPA No.
33-1969), and in accordance with the following:

  (A) Where a fan plenum is used to equalize or control the distribution of exhaust air movement through the booth, it shall be of sufficient strength or rigidity to withstand the differential air pressure or other superficially imposed loads for which the equipment is designed and also to facilitate cleaning. Construction specifications shall be at least equivalent to those of paragraph (h)(5)(iii) of this section.

  (B) [Reserved]

 (ii) Inlet or supply ductwork used to transport makeup air to spray booths or surrounding areas shall be constructed of noncombustible materials.

  (A) If negative pressure exists within inlet ductwork, all seams and joints shall be sealed if there is a possibility of infiltration of harmful quantities of noxious gases, fumes, or mists from areas through which ductwork passes.

  (B) Inlet ductwork shall be sized in accordance with volume flow requirements and provide design air requirements at the spray booth.

  (C) Inlet ductwork shall be adequately supported throughout its length to sustain at least its own weight plus any negative pressure which is exerted upon it under normal operating conditions.

 (iii) [Reserved]

  (A) Exhaust ductwork shall be adequately supported throughout its length to sustain its weight plus any normal accumulation in interior during normal operating conditions and any negative pressure exerted upon it.

- (B) Exhaust ductwork shall be sized in accordance with good design practice which shall include consideration of fan capacity, length of duct, number of turns and elbows, variation in size, volume, and character of materials being exhausted. See American National Standard Z9.2-1960 for further details and explanation concerning elements of design.

- (C) Longitudinal joints in sheet steel ductwork shall be either lock-seamed, riveted, or welded. For other than steel construction, equivalent securing of joints shall be provided.

- (D) Circumferential joints in ductwork shall be substantially fastened together and lapped in the direction of airflow. At least every fourth joint shall be provided with connecting flanges, bolted together, or of equivalent fastening security.

- (E) Inspection or clean-out doors shall be provided for every 9 to 12 feet (2.736 to 3.648 m) of running length for ducts up to 12 inches (0.304 m) in diameter, but the distance between cleanout doors may be greater for larger pipes. (See 8.3.21 of American National Standard Z9.1-1951.) A clean-out door or doors shall be provided for servicing the fan, and where necessary, a drain shall be provided.

- (F) Where ductwork passes through a combustible roof or wall, the roof or wall shall be protected at the point of penetration by open space or fire-resistive material between the duct and the roof or wall. When ducts pass through firewalls, they shall be provided with automatic fire dampers on both sides of the wall, except that three-eighth-inch steel plates may be used in lieu of automatic fire dampers for ducts not exceeding 18 inches (45.72 cm) in diameter.

- (G) Ductwork used for ventilating any process covered in this standard shall not be connected to ducts ventilating any other process or any chimney or flue used for conveying any products of combustion.

(6) *Velocity and air flow requirements.*

  (i) Except where a spray booth has an adequate air replacement system, the velocity of air into all openings of a spray booth shall be not less than that specified in Table D-57.7 for the operating conditions specified. An adequate air replacement system is one which introduces replacement air upstream or above the object being sprayed and is so designed that the velocity of air in the booth cross section is not less than that specified in Table D-57.7 when measured upstream or above the object being sprayed.

Operating conditions for objects completely inside booth	Crossdraft, f.p.m.	Airflow velocities, f.p.m.	
		Design	Range
Electrostatic and automatic airless operation contained in booth without operator	Negligible	50 large booth	50-75
		100 small booth	75-125
Air-operated guns, manual or automatic	Up to 50	100 large booth	75-125

Table D-57.7 - Minimum Maintained Velocities Into Spray Booths

Operating conditions for objects completely inside booth	Crossdraft, f.p.m.	Airflow velocities, f.p.m.	
		Design	Range
		150 small booth	125-175
Air-operated guns, manual or automatic	Up to 100	150 large booth	125-175
		200 small booth	150-250

NOTES:
(1) Attention is invited to the fact that the effectiveness of the spray booth is dependent upon the relationship of the depth of the booth to its height and width.
(2) Crossdrafts can be eliminated through proper design and such design should be sought. Crossdrafts in excess of 100fpm (feet per minute) should not be permitted.
(3) Excessive air pressures result in loss of both efficiency and material waste in addition to creating a backlash that may carry overspray and fumes into adjacent work areas.
(4) Booths should be designed with velocities shown in the column headed "Design." However, booths operating with velocities shown in the column headed "Range" are in compliance with this standard.

(ii) In addition to the requirements in paragraph (h)(6)(i) of this section the total air volume exhausted through a spray booth shall be such as to dilute solvent vapor to at least 25 percent of the lower explosive limit of the solvent being sprayed. An example of the method of calculating this volume is given below.

Example: To determine the lower explosive limits of the most common solvents used in spray finishing, see Table D-57.8. Column 1 gives the number of cubic feet of vapor per gallon of solvent and column 2 gives the lower explosive limit (LEL) in percentage by volume of air. Note that the quantity of solvent will be diminished by the quantity of solids and nonflammables contained in the finish.

To determine the volume of air in cubic feet necessary to dilute the vapor from 1 gallon of solvent to 25 percent of the lower explosive limit, apply the following formula:

Dilution volume required per gallon of solvent = 4 (100-LEL) (cubic feet of vapor per gallon) ÷ LEL

Using toluene as the solvent.

(1) LEL of toluene from Table D-57.8, column 2, is 1.4 percent.

(2) Cubic feet of vapor per gallon from Table D-57.8, column 1, is 30.4 cubic feet pergallon.

(3) Dilution volume required =

4 (100-1.4) 30.4 ÷ 1.4 = 8,564 cubic feet.

(4) To convert to cubic feet per minute of required ventilation, multiply the dilution volume required per gallon of solvent by the number of gallons of solvent evaporated perminute.

Table D-57.8 - Lower Explosive Limit of Some Commonly Used Solvents

Solvent	Cubic feet per gallon of vapor of liquid at 70 °F (21.11 °C).	Lower explosive limit in percent by volume of air at 70 °F (21.11 °C)
	Column 1	Column 2
Acetone	44.0	2.6
Amyl Acetate (iso)	21.6	[1] 1.0
Amyl Alcohol (n)	29.6	1.2
Amyl Alcohol (iso)	29.6	1.2
Benzene	36.8	[1] 1.4
Butyl Acetate (n)	24.8	1.7
Butyl Alcohol (n)	35.2	1.4
Butyl Cellosolve	24.8	1.1
Cellosolve	33.6	1.8
Cellosolve Acetate	23.2	1.7
Cyclohexanone	31.2	[1] 1.1
1,1 Dichloroethylene	42.4	5.9
1,2 Dichloroethylene	42.4	9.7
Ethyl Acetate	32.8	2.5
Ethyl Alcohol	55.2	4.3
Ethyl Lactate	28.0	[1] 1.5
Methyl Acetate	40.0	3.1
Methyl Alcohol	80.8	7.3
Methyl Cellosolve	40.8	2.5
Methyl Ethyl Ketone	36.0	1.8
Methyl n-Propyl Ketone	30.4	1.5
Naphtha (VM&P) (76°Naphtha)	22.4	0.9
Naphtha (100°Flash) Safety Solvent - Stoddard Solvent	23.2	1.0
Propyl Acetate (n)	27.2	2.8

Solvent	Cubic feet per gallon of vapor of liquid at 70 °F (21.11 °C).	Lower explosive limit in percent by volume of air at 70 °F (21.11 °C)
Propyl Acetate (iso)	28.0	1.1
Propyl Alcohol (n)	44.8	2.1
Propyl Alcohol (iso)	44.0	2.0
Toluene	30.4	1.4
Turpentine	20.8	0.8
Xylene (o)	26.4	1.0

[1] At 212 °F (100 °C).

(iii)

    (A) When an operator is in a booth downstream of the object being sprayed, an air-suppliedrespirator or other type of respirator approved by NIOSH under 42 CFR part 84 for the material being sprayed should be used by the operator.

    (B) Where downdraft booths are provided with doors, such doors shall be closed when spraypainting.

(7) **Make-up air.**

  (i) Clean fresh air, free of contamination from adjacent industrial exhaust systems, chimneys, stacks, or vents, shall be supplied to a spray booth or room in quantities equal to the volume ofair exhausted through the spray booth.

  (ii) Where a spray booth or room receives make-up air through self-closing doors, dampers, or louvers, they shall be fully open at all times when the booth or room is in use for spraying. The velocity of air through such doors, dampers, or louvers shall not exceed 200 feet per minute. Ifthe fan characteristics are such that the required air flow through the booth will be provided, higher velocities through the doors, dampers, or louvers may be used.

  (iii)

    (A) Where the air supply to a spray booth or room is filtered, the fan static pressure shall becalculated on the assumption that the filters are dirty to the extent that they require cleaning or replacement.

    (B) The rating of filters shall be governed by test data supplied by the manufacturer of the filter. A pressure gage shall be installed to show the pressure drop across the filters. This gage shall be marked to show the pressure drop at which the filters require cleaning or replacement. Filters shall be replaced or cleaned whenever the pressure drop across thembecomes excessive or whenever the air flow through the face of the booth falls below that specified in Table D-57.7.

  (iv)

    (A) Means for heating make-up air to any spray booth or room, before or at the time spraying is normally performed, shall be provided in all places where the outdoor temperature maybe expected to remain below 55 °F. (12.77 °C.) for appreciable periods of time during the operation of the booth except where adequate and safe means of radiant heating for all

operating personnel affected is provided. The replacement air during the heating seasonsshall be maintained at not less than 65 °F. (18.33 °C.) at the point of entry into the spray booth or spray room. When otherwise unheated make-up air would be at a temperature ofmore than 10 °F. below room temperature, its temperature shall be regulated as provided in section 3.6.3 of ANSI Z9.2-1960.

(B) As an alternative to an air replacement system complying with the preceding section, general heating of the building in which the spray room or booth is located may be employed provided that all occupied parts of the building are maintained at not less than65 °F. (18.33 °C.) when the exhaust system is in operation or the general heating system supplemented by other sources of heat may be employed to meet this requirement.

(C) No means of heating make-up air shall be located in a spray booth.

(D) Where make-up air is heated by coal or oil, the products of combustion shall not be allowed to mix with the make-up air, and the products of combustion shall be conductedoutside the building through a flue terminating at a point remote from all points where make-up air enters the building.

(E) Where make-up air is heated by gas, and the products of combustion are not mixed with the make-up air but are conducted through an independent flue to a point outside the building remote from all points where make-up air enters the building, it is not necessaryto comply with paragraph (h)(7)(iv)(F) of this section.

(F) Where make-up air to any manually operated spray booth or room is heated by gas and theproducts of combustion are allowed to mix with the supply air, the following precautions must be taken:

(*1*) The gas must have a distinctive and strong enough odor to warn workmen in a spraybooth or room of its presence if in an unburned state in the make-up air.

(*2*) The maximum rate of gas supply to the make-up air heater burners must not exceedthat which would yield in excess of 200 p.p.m. (parts per million) of carbon monoxideor 2,000 p.p.m. of total combustible gases in the mixture if the unburned gas upon the occurrence of flame failure were mixed with all of the make-up air supplied.

(*3*) A fan must be provided to deliver the mixture of heated air and products of combustion from the plenum chamber housing the gas burners to the spray booth orroom.

(8) **Scope.** Spray booths or spray rooms are to be used to enclose or confine all spray finishing operations covered by this paragraph (h). This paragraph does not apply to the spraying of the exteriors of buildings, fixed tanks, or similar structures, nor to small portable spraying apparatus notused repeatedly in the same location.

(i) *Open surface tanks* -

(1) *General.*

(i) This paragraph applies to all operations involving the immersion of materials in liquids, or in thevapors of such liquids, for the purpose of cleaning or altering the surface or adding to or imparting a finish thereto or changing the character of the materials, and their subsequent

removal from the liquid or vapor, draining, and drying. These operations include washing, electroplating, anodizing, pickling, quenching, dying, dipping, tanning, dressing, bleaching, degreasing, alkaline cleaning, stripping, rinsing, digesting, and other similar operations.

(ii) Except where specific construction specifications are prescribed in this section, hoods, ducts, elbows, fans, blowers, and all other exhaust system parts, components, and supports thereof shall be so constructed as to meet conditions of service and to facilitate maintenance and shall conform in construction to the specifications contained in American National Standard Fundamentals Governing the Design and Operation of Local Exhaust Systems, Z9.2-1960.

(2) **Classification of open-surface tank operations.**

(i) Open-surface tank operations shall be classified into 16 classes, numbered A-1 to D-4, inclusive.

(ii) **Determination of class.** Class is determined by two factors, hazard potential designated by a letter from A to D, inclusive, and rate of gas, vapor, or mist evolution designated by a number from 1 to 4, inclusive (for example, B.3).

(iii) Hazard potential is an index, on a scale of from A to D, inclusive, of the severity of the hazard associated with the substance contained in the tank because of the toxic, flammable, or explosive nature of the vapor, gas, or mist produced therefrom. The toxic hazard is determined from the concentration, measured in parts by volume of a gas or vapor, per million parts by volume of contaminated air (p.p.m.), or in milligrams of mist per cubic meter of air (mg./m.^{3}), below which ill effects are unlikely to occur to the exposed worker. The concentrations shall be those in § 1926.55 or other pertinent sections of this part.

(iv) The relative fire or explosion hazard is measured in degrees Fahrenheit in terms of the closed-cup flash point of the substance in the tank. Detailed information on the prevention of fire hazards in dip tanks may be found in Dip Tanks Containing Flammable or Combustible Liquids, NFPA No. 34-1966, National Fire Protection Association. Where the tank contains a mixture of liquids, other than organic solvents, whose effects are additive, the hygienic standard of the most toxic component (for example, the one having the lowest p.p.m. or mg./m.^{3}) shall be used, except where such substance constitutes an insignificantly small fraction of the mixture. For mixtures of organic solvents, their combined effect, rather than that of either individually, shall determine the hazard potential. In the absence of information to the contrary, the effects shall be considered as additive. If the sum of the ratios of the airborne concentration of each contaminant to the toxic concentration of that contaminant exceeds unity, the toxic concentration shall be considered to have been exceeded. (See Note A to paragraph (i)(2)(v) of this section.)

(v) Hazard potential shall be determined from Table D-57.9, with the value indicating greater hazard being used. When the hazardous material may be either a vapor with a threshold limit value (*TLV*) in p.p.m. or a mist with a *TLV* in mg./m.$_3$, the *TLV* indicating the greater hazard shall be used (for example, A takes precedence over B or C; B over C; C over D).

NOTE A:

$$(c_1 \div TLV_1) + (c_2 \div TLV_2) + (c_3 \div TLV_3) + ; \ldots (c_N \div TLV_N)1$$

Where:

c = Concentration measured at the operation in p.p.m.

Table D-57.9 - Determination of Hazard Potential

Hazard potential	Toxicity group		
	Gas or vapor (p.p.m.)	Mist (mg./m^3)	Flash point in degrees F. (C.)
A	0-10	0-0.1	
B	11-100	0.11-1.0	Under 100 (37.77)
C	101-500	1.1-10	100 200 (37.77-93.33)
D	Over 500	Over 10	Over 200 (93.33)

(vi) Rate of gas, vapor, or mist evolution is a numerical index, on a scale of from 1 to 4, inclusive, both of the relative capacity of the tank to produce gas, vapor, or mist and of the relative energywith which it is projected or carried upwards from the tank. Rate is evaluated in terms of

(A) The temperature of the liquid in the tank in degrees Fahrenheit;

(B) The number of degrees Fahrenheit that this temperature is below the boiling point of theliquid in degrees Fahrenheit;

(C) The relative evaporation of the liquid in still air at room temperature in an arbitrary scale -fast, medium, slow, or nil; and

(D) The extent that the tank gases or produces mist in an arbitrary scale - high, medium, low,and nil. (See Table D-57.10, Note 2.) Gassing depends upon electrochemical or mechanical processes, the effects of which have to be individually evaluated for each installation (see Table D-57.10, Note 3).

(vii) Rate of evolution shall be determined from Table D-57.10. When evaporation and gassing yielddifferent rates, the lowest numerical value shall be used.

Table D-57.10 - Determination of Rate of Gas, Vapor, or Mist Evolution[1]

Rate	Liquid temperature, °F. (C.)	Degrees below boiling point	Relative evaporation[2]	Gassing[3]
1	Over 200 (93.33)	0-20	Fast	High.
2	150-200 (65.55-93.33)	21-50	Medium	Medium.
3	94-149 (34.44-65)	51-100	Slow	Low.
4	Under 94 (34.44)	Over 100	Nil	Nil.

[1] In certain classes of equipment, specifically vapor degreasers, an internal condenser or vaporlevel thermostat is used to prevent the vapor from leaving the tank during normal operation. In such cases, rate of vapor evolution from the tank into the workroom is not dependent upon the factors listed in the table, but rather upon abnormalities of operating procedure, such as carryout of vapors from excessively fast action, dragout of

liquid by entrainment in parts, contamination of solvent by water and other materials, or improper heat balance. When operating procedure is excellent, effective rate of evolution may be taken as 4. When operating procedure is average, the effective rate of evolution may be taken as 3. When operation is poor, a rate of 2 or 1 is indicated, depending upon observed conditions.[2] Relative evaporation rate is determined according to the methods described by A. K. Doolittlein Industrial and Engineering Chemistry, vol. 27, p. 1169, (3) where time for 100-percent evaporation is as follows: Fast: 0-3 hours; Medium: 3-12 hours; Slow: 12-50 hours; Nil: more than 50 hours.

[3] Gassing means the formation by chemical or electrochemical action of minute bubbles of gas under the surface of the liquid in the tank and is generally limited to aqueous solutions.

(3) **Ventilation.** Where ventilation is used to control potential exposures to workers as defined in paragraph (i)(2)(iii) of this section, it shall be adequate to reduce the concentration of the air contaminant to the degree that a hazard to the worker does not exist. Methods of ventilation arediscussed in American National Standard Fundamentals Governing the Design and Operation of Local Exhaust Systems, Z9.2-1960.

(4) **Control requirements.**

   (i) Control velocities shall conform to Table D-57.11 in all cases where the flow of air past the breathing or working zone of the operator and into the hoods is undisturbed by local environmental conditions, such as open windows, wall fans, unit heaters, or moving machinery.

   (ii) All tanks exhausted by means of hoods which

      (A) Project over the entire tank;

      (B) Are fixed in position in such a location that the head of the workman, in all his normal operating positions while working at the tank, is in front of all hood openings; and

      (C) Are completely enclosed on at least two sides, shall be considered to be exhausted through an enclosing hood.

      (D) The quantity of air in cubic feet per minute necessary to be exhausted through an enclosing hood shall be not less than the product of the control velocity times the net areaof all openings in the enclosure through which air can flow into the hood.

Table D-57.11 - Control Velocities in Feet Per Minute (f.p.m.) forUndisturbed Locations

Class	Enclosing hood		Lateral exhaust[1]	Canopy hood[2]	
	One open side	Two open sides		Three open sides	Four open sides
B-1 and A-2	100	150	150	Do not use	Do not use
A-3[2], B-1, B-2, and C-1	75	100	100	125	175
A-3, C-2, and D-1[3]	65	90	75	100	150
B-4[2], C-3, and D-2[3]	50	75	50	75	125
A-4, C-4, D-3[3], and D-4[4]					

[1] See Table D-57.12 for computation of ventilation rate.

[2] Do not use canopy hood for Hazard Potential A processes.

[3] Where complete control of hot water is desired, design as next highest class.

[4] General room ventilation required.

(iii) All tanks exhausted by means of hoods which do not project over the entire tank, and in which the direction of air movement into the hood or hoods is substantially horizontal, shall be considered to be laterally exhausted. The quantity of air in cubic feet per minute necessary to be laterally exhausted per square foot of tank area in order to maintain the required control velocity shall be determined from Table D-57.12 for all variations in ratio of tank width (W) to tank length (L). The total quantity of air in cubic feet per minute required to be exhausted per tank shall be not less than the product of the area of tank surface times the cubic feet per minute per square foot of tank area, determined from Table D-57.12.

(A) For lateral exhaust hoods over 42 inches (1.06 m) wide, or where it is desirable to reduce the amount of air removed from the workroom, air supply slots or orifices shall be provided along the side or the center of the tank opposite from the exhaust slots. The design of such systems shall meet the following criteria:

(1) The supply air volume plus the entrained air shall not exceed 50 percent of the exhaust volume.

(2) The velocity of the supply airstream as it reaches the effective control area of the exhaust slot shall be less than the effective velocity over the exhaust slot area.

Table D-57.12 - Minimum Ventilation Rate in Cubic Feet of Air PerMinute Per Square Foot of Tank Area for Lateral Exhaust

Required minimum control velocity, f.p.m. (from Table D-57.11)	C.f.m. per sq. ft. to maintain required minimum velocities at following ratios (tank width (W)/tank length (L)).[1][2]				
	0.0-0.09	0.1-0.24	0.25-0.49	0.5-0.99	1.0-2.0
Hood along one side or two parallel sides of tank when one hood is against a wall or baffle.[2]					
50	50	60	75	90	100
75	75	90	110	130	150
100	100	125	150	175	200
150	150	190	225	260	300
50	75	90	100	110	125
75	110	130	150	170	190
100	150	175	200	225	250

| Also for a manifold along tank centerline.³ | 150 | | 225 | 260 | 300 | 340 | 375 |

³Hood along one side or two parallel sides of free standing tank not against wall or baffle.

[1] It is not practicable to ventilate across the long dimension of a tank whose ratio W/L exceeds 2.0.

It is undesirable to do so when W/L exceeds 1.0. For circular tanks with lateral exhaust along up to $1/2$ the circumference, use $W/L = 1.0$; for over one-half the circumference use $W/L = 0.5$.

[2] Baffle is a vertical plate the same length as the tank, and with the top of the plate as high as the tank is wide. If the exhaust hood is on the side of a tank against a building wall or close to it, it is perfectly baffled.

[3] Use $W/2$ as tank width in computing when manifold is along centerline, or when hoods are used on two parallel sides of a tank.

Tank Width (W) means the effective width over which the hood must pull air to operate (for example, where the hood face is set back from the edge of the tank, this set back must be added in measuring tank width). The surface area of tanks can frequently be reduced and better control obtained (particularly on conveyorized systems) by using covers extending from the upper edges of the slots toward the center of the tank.

(*3*) The vertical height of the receiving exhaust hood, including any baffle, shall not be less than one-quarter the width of the tank.

(*4*) The supply airstream shall not be allowed to impinge on obstructions between it and the exhaust slot in such a manner as to significantly interfere with the performance of the exhaust hood.

(*5*) Since most failure of push-pull systems result from excessive supply air volumes and pressures, methods of measuring and adjusting the supply air shall be provided.
When satisfactory control has been achieved, the adjustable features of the hood shall be fixed so that they will not be altered.

(iv) All tanks exhausted by means of hoods which project over the entire tank, and which do not conform to the definition of enclosing hoods, shall be considered to be overhead canopy hoods. The quantity of air in cubic feet per minute necessary to be exhausted through a canopy hood shall be not less than the product of the control velocity times the net area of all openings between the bottom edges of the hood and the top edges of the tank.

(v) The rate of vapor evolution (including steam or products of combustion) from the process shall be estimated. If the rate of vapor evolution is equal to or greater than 10 percent of the calculated exhaust volume required, the exhaust volume shall be increased in equal amount.

(5) **Spray cleaning and degreasing.** Wherever spraying or other mechanical means are used to

disperse aliquid above an open-surface tank, control must be provided for the airborne spray. Such operations shall be enclosed as completely as possible. The inward air velocity into the enclosure shall be sufficient to prevent the discharge of spray into the workroom. Mechanical baffles may be used to help prevent the discharge of spray. Spray painting operations are covered by paragraph (h) of this section.

(6) ***Control means other than ventilation.*** Tank covers, foams, beads, chips, or other materials floating on the tank surface so as to confine gases, mists, or vapors to the area under the cover or to the foam, bead, or chip layer; or surface tension depressive agents added to the liquid in the tank to minimize mist formation, or any combination thereof, may all be used as gas, mist, or vapor control means for open-surface tank operations, provided that they effectively reduce the concentrations of hazardous materials in the vicinity of the worker below the limits set in accordance with paragraph (i)(2) of this section.

(7) ***System design.***

    (i) The equipment for exhausting air shall have sufficient capacity to produce the flow of air required in each of the hoods and openings of the system.

    (ii) The capacity required in paragraph (i)(7)(i) of this section shall be obtained when the airflow producing equipment is operating against the following pressure losses, the sum of which is the static pressure:

        (A) Entrance losses into the hood.

        (B) Resistance to airflow in branch pipe including bends and transformations.

        (C) Entrance loss into the main pipe.

        (D) Resistance to airflow in main pipe including bends and transformations.

        (E) Resistance of mechanical equipment; that is, filters, washers, condensers, absorbers, etc., plus their entrance and exit losses.

        (F) Resistance in outlet duct and discharge stack.

    (iii) Two or more operations shall not be connected to the same exhaust system where either one or the combination of the substances removed may constitute a fire, explosion, or chemical reaction hazard in the duct system. Traps or other devices shall be provided to insure that condensate in ducts does not drain back into any tank.

    (iv) The exhaust system, consisting of hoods, ducts, air mover, and discharge outlet, shall be designed in accordance with American National Standard Fundamentals Governing the Design and Operation of Local Exhaust Systems, Z9.2-1960, or the manual, Industrial Ventilation, published by the American Conference of Governmental Industrial Hygienists 1970. Airflow and pressure loss data provided by the manufacturer of any air cleaning device shall be included in the design calculations.

(8) ***Operation.***

    (i) The required airflow shall be maintained at all times during which gas, mist, or vapor is emitted from the tank, and at all times the tank, the draining, or the drying area is in operation or use.
When the system is first installed, the airflow from each hood shall be measured by means of a pitot traverse in the exhaust duct and corrective action taken if the flow is

less than that required. When the proper flow is obtained, the hood static pressure shall be measured and recorded. At intervals of not more than 3 months operation, or after a prolonged shutdown period, the hoods and duct system shall be inspected for evidence of corrosion or damage. In any case where the airflow is found to be less than required, it shall be increased to the requiredvalue. (Information on airflow and static pressure measurement and calculations may be found in American National Standard Fundamental Governing the Design and Operation of Local Exhaust Systems, Z9.2-1960, or in the manual, Industrial Ventilation, published by the American Conference of Governmental Industrial Hygienists.)

(ii) The exhaust system shall discharge to the outer air in such a manner that the possibility of its effluent entering any building is at a minimum. Recirculation shall only be through a device forcontaminant removal which will prevent the creation of a health hazard in the room or area to which the air is recirculated.

(iii) A volume of outside air in the range of 90 percent to 110 percent of the exhaust volume shall beprovided to each room having exhaust hoods. The outside air supply shall enter the workroomin such a manner as not to be detrimental to any exhaust hood. The airflow of the makeup air system shall be measured on installation. Corrective action shall be taken when the airflow is below that required. The makeup air shall be uncontaminated.

(9) **Personal protection.**

    (i) All employees working in and around open-surface tank operations must be instructed as to thehazards of their respective jobs, and in the personal protection and first aid procedures applicable to these hazards.

    (ii) All persons required to work in such a manner that their feet may become wet shall be providedwith rubber or other impervious boots or shoes, rubbers, or wooden-soled shoes sufficient to keep feet dry.

    (iii) All persons required to handle work wet with a liquid other than water shall be provided with gloves impervious to such a liquid and of a length sufficient to prevent entrance of liquid intothe tops of the gloves. The interior of gloves shall be kept free from corrosive or irritating contaminants.

    (iv) All persons required to work in such a manner that their clothing may become wet shall be provided with such aprons, coats, jackets, sleeves, or other garments made of rubber, or of other materials impervious to liquids other than water, as are required to keep their clothing dry.Aprons shall extend well below the top of boots to prevent liquid splashing into the boots.
Provision of dry, clean, cotton clothing along with rubber shoes or short boots and an apron impervious to liquids other than water shall be considered a satisfactory substitute where smallparts are cleaned, plated, or acid dipped in open tanks and rapid work is required.

    (v) Whenever there is a danger of splashing, for example, when additions are made manually to thetanks, or when acids and chemicals are removed from the tanks, the employees so engaged shall be required to wear either tight-fitting chemical goggles or an effective face shield. See § 1926.102.

    (vi) When, during the emergencies specified in paragraph (i)(11)(v) of this section, employees must be in areas where concentrations of air contaminants are greater than the limits set by paragraph (i)(2)(iii) of this section or oxygen concentrations are less than 19.5 percent, they must use respirators that reduce their exposure to a level below these limits or that provide adequate oxygen. Such respirators must also be provided in marked, quickly-accessible storagecompartments built for this purpose when the possibility exists of accidental release of hazardous concentrations of air contaminants. Respirators must be approved by NIOSH under 42 CFR part 84, selected by a competent industrial hygienist or other technically-qualified source, and used in accordance with 29 CFR 1926.103.

(vii) Near each tank containing a liquid which may burn, irritate, or otherwise be harmful to the skin ifsplashed upon the worker's body, there shall be a supply of clean cold water. The water pipe (carrying a pressure not exceeding 25 pounds (11.325 kg)) shall be provided with a quick opening valve and at least 48 inches (1.216 m) of hose not smaller than three-fourths inch, so that no time may be lost in washing off liquids from the skin or clothing. Alternatively, deluge showers and eye flushes shall be provided in cases where harmful chemicals may be splashed on parts of the body.

(viii) Operators with sores, burns, or other skin lesions requiring medical treatment shall not be allowed to work at their regular operations until so authorized by a physician. Any small skin abrasions, cuts, rash, or open sores which are found or reported shall be treated by a properlydesignated person so that chances of exposures to the chemicals are removed. Workers exposed to chromic acids shall have a periodic examination made of the nostrils and other parts of the body, to detect incipient ulceration.

(ix) Sufficient washing facilities, including soap, individual towels, and hot water, shall be providedfor all persons required to use or handle any liquids which may burn, irritate, or otherwise be harmful to the skin, on the basis of at least one basin (or its equivalent) with a hot water faucetfor every 10 employees. See § 1926.51(f).

(x) Locker space or equivalent clothing storage facilities shall be provided to prevent contamination of street clothing.

(xi) First aid facilities specific to the hazards of the operations conducted shall be readily available.

(10) *Special precautions for cyanide.* Dikes or other arrangements shall be provided to prevent the possibility of intermixing of cyanide and acid in the event of tank rupture.

(11) *Inspection, maintenance, and installation.*

(i) Floors and platforms around tanks shall be prevented from becoming slippery both by original type of construction and by frequent flushing. They shall be firm, sound, and of the design andconstruction to minimize the possibility of tripping.

(ii) Before cleaning the interior of any tank, the contents shall be drained off, and the cleanoutdoors shall be opened where provided. All pockets in tanks or pits, where it is possible for hazardous vapors to collect, shall be ventilated and cleared of such vapors.

(iii) Tanks which have been drained to permit employees to enter for the purposes of cleaning, inspection, or maintenance may contain atmospheres which are hazardous to life or health, through the presence of flammable or toxic air contaminants, or through the absence of sufficient oxygen. Before employees shall be permitted to enter any such tank, appropriate tests of the atmosphere shall be made to determine if the limits set by paragraph (i)(2)(iii) of this section are exceeded, or if the oxygen concentration is less than 19.5 percent.

(iv) If the tests made in accordance with paragraph (i)(11)(iii) of this section indicate that the atmosphere in the tank is unsafe, before any employee is permitted to enter the tank, the tank shall be ventilated until the hazardous atmosphere is removed, and ventilation shall be continued so as to prevent the occurrence of a hazardous atmosphere as long as an employee is in the tank.

(v) If, in emergencies, such as rescue work, it is necessary to enter a tank which may contain a hazardous atmosphere, suitable respirators, such as self-contained breathing apparatus; hose mask with blower, if there is a possibility of oxygen deficiency; or a gas mask, selected and operated in accordance with paragraph (i)(9)(vi) of this section, shall be used. If a contaminant in the tank can cause dermatitis, or be absorbed through the skin, the employee entering the tank shall also wear protective clothing. At least one trained standby employee, with suitable respirator, shall be present in the nearest uncontaminated area. The standby employee must be able to communicate with the employee in the tank and be able to haul him out of the tank with a lifeline if necessary.

(vi) Maintenance work requiring welding or open flame, where toxic metal fumes such as cadmium, chromium, or lead may be evolved, shall be done only with sufficient local exhaust ventilation to prevent the creation of a health hazard, or be done with respirators selected and used in accordance with paragraph (i)(9)(vi) of this section. Welding, or the use of open flames near any solvent cleaning equipment shall be permitted only after such equipment has first been thoroughly cleared of solvents and vapors.

(12) *Vapor degreasing tanks.*

(i) In any vapor degreasing tank equipped with a condenser or vapor level thermostat, the condenser or thermostat shall keep the level of vapors below the top edge of the tank by a distance at least equal to one-half the tank width, or at least 36 inches (0.912 m), whichever is shorter.

(ii) Where gas is used as a fuel for heating vapor degreasing tanks, the combustion chamber shall be of tight construction, except for such openings as the exhaust flue, and those that are necessary for supplying air for combustion. Flues shall be of corrosion-resistant construction and shall extend to the outer air. If mechanical exhaust is used on this flue, a draft diverter shall be used. Special precautions must be taken to prevent solvent fumes from entering the combustion air of this or any other heater when chlorinated or fluorinated hydrocarbon solvents (for example, trichloroethylene, Freon) are used.

(iii) Heating elements shall be so designed and maintained that their surface temperature will not cause the solvent or mixture to decompose, break down, or be converted into an excessive quantity of vapor.

(iv) Tanks or machines of more than 4 square feet (0.368 m^2) of vapor area, used for solvent cleaning or vapor degreasing, shall be equipped with suitable cleanout or sludge doors located near the bottom of each tank or still. These doors shall be so designed and gasketed that there will be no leakage of solvent when they are closed.

(13) *Scope.*

- (i) This paragraph (i) applies to all operations involving the immersion of materials in liquids, or in the vapors of such liquids, for the purpose of cleaning or altering their surfaces, or adding or imparting a finish thereto, or changing the character of the materials, and their subsequent removal from the liquids or vapors, draining, and drying. Such operations include washing, electroplating, anodizing, pickling, quenching, dyeing, dipping, tanning, dressing, bleaching, degreasing, alkaline cleaning, stripping, rinsing, digesting, and other similar operations, but do not include molten materials handling operations, or surface coating operations.

- (ii) *Molten materials handling operations* means all operations, other than welding, burning, and soldering operations, involving the use, melting, smelting, or pouring of metals, alloys, salts, or other similar substances in the molten state. Such operations also include heat treating baths, descaling baths, die casting stereotyping, galvanizing, tinning, and similar operations.

- (iii) *Surface coating operations* means all operations involving the application of protective, decorative, adhesive, or strengthening coating or impregnation to one or more surfaces, or into the interstices of any object or material, by means of spraying, spreading, flowing, brushing, rollcoating, pouring, cementing, or similar means; and any subsequent draining or drying operations, excluding open-tank operations.

[44 FR 8577, Feb. 9, 1979; 44 FR 20940, Apr. 6, 1979, as amended at 58 FR 35099, June 30, 1993; 61 FR 9250, Mar. 3, 1996; 63 FR 1295, Jan. 8, 1998]

§ 1926.58 COVID-19.

The requirements applicable to construction work under this section are identical to those set forth at 29 CFR 1910.501 subpart U.

[86 FR 61555, Nov. 5, 2021]

§ 1926.59 Hazard communication.

Note: The requirements applicable to construction work under this section are identical to those set forth at § 1910.1200 of this chapter.

[61 FR 31431, June 20, 1996]

§ 1926.60 Methylenedianiline.

(a) *Scope and application.*

(1) This section applies to all construction work as defined in 29 CFR 1910.12(b), in which there is exposure to MDA, including but not limited to the following:

- (i) Construction, alteration, repair, maintenance, or renovation of structures, substrates, or portions thereof, that contain MDA;
- (ii) Installation or the finishing of surfaces with products containing
- (iii) MDA;MDA spill/emergency cleanup at construction sites; and
- (iv)

Transportation, disposal, storage, or containment of MDA or products containing MDA on thesite or location at which construction activities are performed.

(2) Except as provided in paragraphs (a)(7) and (f)(5) of this section, this section does not apply to the processing, use, and handling of products containing MDA where initial monitoring indicates that the product is not capable of releasing MDA in excess of the action level under the expected conditionsof processing, use, and handling which will cause the greatest possible release; and where no "dermal exposure to MDA" can occur.

(3) Except as provided in paragraph (a)(7) of this section, this section does not apply to the processing,use, and handling of products containing MDA where objective data are reasonably relied upon which demonstrate the product is not capable of releasing MDA under the expected conditions of processing, use, and handling which will cause the greatest possible release; and where no "dermal exposure to MDA" can occur.

(4) Except as provided in paragraph (a)(7) of this section, this section does not apply to the storage, transportation, distribution or sale of MDA in intact containers sealed in such a manner as to containthe MDA dusts, vapors, or liquids, except for the provisions of 29 CFR 1910.1200 and paragraph (e) of this section.

(5) Except as provided in paragraph (a)(7) of this section, this section does not apply to materials in anyform which contain less than 0.1% MDA by weight or volume.

(6) Except as provided in paragraph (a)(7) of this section, this section does not apply to "finished articlescontaining MDA."

(7) Where products containing MDA are exempted under paragraphs (a)(2) through (a)(6) of this section,the employer shall maintain records of the initial monitoring results or objective data supporting that exemption and the basis for the employer's reliance on the data, as provided in the recordkeeping provision of paragraph (o) of this section.

(b) **Definitions.** For the purpose of this section, the following definitions shall apply:

Action level means a concentration of airborne MDA of 5 ppb as an eight (8)-hour time-weighted average.

Assistant Secretary means the Assistant Secretary of Labor for Occupational Safety and Health, U.S. Department of Labor, or designee.

Authorized person means any person specifically authorized by the employer whose duties require the person to enter a regulated area, or any person entering such an area as a designated representativeof employees for the purpose of exercising the right to observe monitoring and measuring procedures under paragraph (p) of this section, or any other person authorized by the Act or regulations issued under the Act.

Container means any barrel, bottle, can, cylinder, drum, reaction vessel, storage tank, commercial packaging or the like, but does not include piping systems.

Decontamination area means an area outside of but as near as practical to the regulated area, consistingof an equipment storage area, wash area, and clean change area, which is used for the decontamination of workers, materials, and equipment contaminated with MDA.

Dermal exposure to MDA occurs where employees are engaged in the handling, application or use ofmixtures or materials containing MDA, with any of the following non-airborne forms of MDA:

(i) Liquid, powdered, granular, or flaked mixtures containing MDA in concentrations greater than 0.1% by weight or volume; and

(ii) Materials other than "finished articles" containing MDA in concentrations greater than 0.1% by weight or volume.

Director means the Director of the National Institute for Occupational Safety and Health, U.S. Department of Health and Human Services, or designee.

Emergency means any occurrence such as, but not limited to, equipment failure, rupture of containers, or failure of control equipment which results in an unexpected and potentially hazardous release of MDA.

Employee exposure means exposure to MDA which would occur if the employee were not using respirators or protective work clothing and equipment.

Finished article containing MDA is defined as a manufactured item:

(i) Which is formed to a specific shape or design during manufacture;

(ii) Which has end use function(s) dependent in whole or part upon its shape or design during end use; and

(iii) Where applicable, is an item which is fully cured by virtue of having been subjected to the conditions (temperature, time) necessary to complete the desired chemical reaction.

Historical monitoring data means monitoring data for construction jobs that meet the following conditions:

(i) The data upon which judgments are based are scientifically sound and were collected using methods that are sufficiently accurate and precise;

(ii) The processes and work practices that were in use when the historical monitoring data were obtained are essentially the same as those to be used during the job for which initial monitoring will not be performed;

(iii) The characteristics of the MDA-containing material being handled when the historical monitoring data were obtained are the same as those on the job for which initial monitoring will not be performed;

(iv) Environmental conditions prevailing when the historical monitoring data were obtained are the same as those on the job for which initial monitoring will not be performed; and

(v) Other data relevant to the operations, materials, processing, or employee exposures covered by the exception are substantially similar. The data must be scientifically sound, the characteristics of the MDA containing material must be similar and the environmental conditions comparable.

4,4 'Methylenedianiline or MDA means the chemical; 4,4'-diaminodiphenylmethane, Chemical Abstract Service Registry number 101-77-9, in the form of a vapor, liquid, or solid. The definition also includes the salts of MDA.

Regulated Areas means areas where airborne concentrations of MDA exceed or can reasonably be expected to exceed, the permissible exposure limits, or where "dermal exposure to MDA" can occur.

STEL means short term exposure limit as determined by any 15-minute sample period.

(c) **Permissible exposure limits.** The employer shall assure that no employee is exposed to an airborne concentration of MDA in excess of ten parts per billion (10 ppb) as an 8-hour time-weighted average and a STEL of one hundred parts per billion (100 ppb).

(d) **Communication among employers.** On multi-employer worksites, an employer performing work involving the application of MDA or materials containing MDA for which establishment of one or more regulated areas is required shall inform other employers on the site of the nature of the employer's work with MDA and of the existence of, and requirements pertaining to, regulated areas.

(e) *Emergency situations* -

    (1) **Written plan.**

        (i) A written plan for emergency situations shall be developed for each construction operation where there is a possibility of an emergency. The plan shall include procedures where the employer identifies emergency escape routes for his employees at each construction site before the construction operation begins. Appropriate portions of the plan shall be implemented in the event of an emergency.

        (ii) The plan shall specifically provide that employees engaged in correcting emergency conditions shall be equipped with the appropriate personal protective equipment and clothing as required in paragraphs (i) and (j) of this section until the emergency is abated.

        (iii) The plan shall specifically include provisions for alerting and evacuating affected employees as well as the applicable elements prescribed in 29 CFR 1910.38 and 29 CFR 1910.39, "Emergency action plans" and "Fire prevention plans," respectively.

    (2) **Alerting employees.** Where there is the possibility of employee exposure to MDA due to an emergency, means shall be developed to promptly alert employees who have the potential to be directly exposed. Affected employees not engaged in correcting emergency conditions shall be evacuated immediately in the event that an emergency occurs. Means shall also be developed for alerting other employees who may be exposed as a result of the emergency.

(f) *Exposure monitoring* -

    (1) **General.**

        (i) Determinations of employee exposure shall be made from breathing zone air samples that are representative of each employee's exposure to airborne MDA over an eight (8) hour period.
Determination of employee exposure to the STEL shall be made from breathing zone air samples collected over a 15 minute sampling period.

        (ii) Representative employee exposure shall be determined on the basis of one or more samples representing full shift exposure for each shift for each job classification in each work area where exposure to MDA may occur.

        (iii) Where the employer can document that exposure levels are equivalent for similar operations in different work shifts, the employer shall only be required to determine representative employee exposure for that operation during one shift.

    (2) **Initial monitoring.** Each employer who has a workplace or work operation covered by this standard shall perform initial monitoring to determine accurately the airborne concentrations of MDA to which employees may be exposed unless:

- (i) The employer can demonstrate, on the basis of objective data, that the MDA-containing producter material being handled cannot cause exposures above the standard's action level, even under worst-case release conditions; or
- (ii) The employer has historical monitoring or other data demonstrating that exposures on a particular job will be below the action level.

(3) *Periodic monitoring and monitoring frequency.*

- (i) If the monitoring required by paragraph (f)(2) of this section reveals employee exposure at or above the action level, but at or below the PELs, the employer shall repeat such monitoring for each such employee at least every six (6) months.
- (ii) If the monitoring required by paragraph (f)(2) of this section reveals employee exposure above the PELs, the employer shall repeat such monitoring for each such employee at least every three (3) months.
- (iii) Employers who are conducting MDA operations within a regulated area can forego periodic monitoring if the employees are all wearing supplied-air respirators while working in the regulated area.
- (iv) The employer may alter the monitoring schedule from every three months to every six months for any employee for whom two consecutive measurements taken at least 7 days apart indicate that the employee exposure has decreased to below the PELs but above the action level.

(4) *Termination of monitoring.*

- (i) If the initial monitoring required by paragraph (f)(2) of this section reveals employee exposure to be below the action level, the employer may discontinue the monitoring for that employee, except as otherwise required by paragraph (f)(5) of this section.
- (ii) If the periodic monitoring required by paragraph (f)(3) of this section reveals that employee exposures, as indicated by at least two consecutive measurements taken at least 7 days apart, are below the action level the employer may discontinue the monitoring for that employee, except as otherwise required by paragraph (f)(5) of this section.

(5) *Additional monitoring.* The employer shall institute the exposure monitoring required under paragraphs (f)(2) and (f)(3) of this section when there has been a change in production process, chemicals present, control equipment, personnel, or work practices which may result in new or additional exposures to MDA, or when the employer has any reason to suspect a change which may result in new or additional exposures.

(6) *Accuracy of monitoring.* Monitoring shall be accurate, to a confidence level of 95 percent, to within plus or minus 25 percent for airborne concentrations of MDA.

(7) *Employee notification of monitoring results.*

- (i) The employer must, as soon as possible but no later than 5 working days after the receipt of the results of any monitoring performed under this section, notify each affected employee of these results either individually in writing or by posting the results in an appropriate location that is accessible to employees.
- (ii) The written notification required by paragraph (f)(7)(i) of this section shall contain the

correctiveaction being taken by the employer or any other protective measures which have been implemented to reduce the employee exposure to or below the PELs, wherever the PELs are exceeded.

(8) **Visual monitoring.** The employer shall make routine inspections of employee hands, face and forearms potentially exposed to MDA. Other potential dermal exposures reported by the employeemust be referred to the appropriate medical personnel for observation. If the employer determinesthat the employee has been exposed to MDA the employer shall:

   (i) Determine the source of exposure;

   (ii) Implement protective measures to correct the hazard; and

   (iii) Maintain records of the corrective actions in accordance with paragraph (o) of this section.

(g) **Regulated areas** -

  (1) **Establishment** -

   (i) **Airborne exposures.** The employer shall establish regulated areas where airborne concentrations of MDA exceed or can reasonably be expected to exceed, the permissibleexposure limits.

   (ii) **Dermal exposures.** Where employees are subject to "dermal exposure to MDA" the employershall establish those work areas as regulated areas.

  (2) **Demarcation.** Regulated areas shall be demarcated from the rest of the workplace in a manner thatminimizes the number of persons potentially exposed.

  (3) **Access.** Access to regulated areas shall be limited to authorized persons.

  (4) **Personal protective equipment and clothing.** Each person entering a regulated area shall be supplied with, and required to use, the appropriate personal protective clothing and equipment in accordancewith paragraphs (i) and (j) of this section.

  (5) **Prohibited activities.** The employer shall ensure that employees do not eat, drink, smoke, chewtobacco or gum, or apply cosmetics in regulated areas.

(h) **Methods of compliance** -

  (1) **Engineering controls and work practices and respirators.**

   (i) The employer shall use one or any combination of the following control methods to achieve compliance with the permissible exposure limits prescribed by paragraph (c) of this section:

      (A) Local exhaust ventilation equipped with HEPA filter dust collection

      (B) systems;General ventilation systems;

      (C) Use of workpractices; or

      (D) Other engineering controls such as isolation and enclosure that the Assistant Secretarycan show to be feasible.

   (ii) Wherever the feasible engineering controls and work practices "which can be instituted are notsufficient to reduce employee exposure to or below the PELs, the employer shall use them to reduce employee exposure to the lowest levels achievable by these controls and shall supplement them by the use of respiratory protective devices which comply with the requirements of paragraph (i) of this section.

(2) **Special Provisions.** For workers engaged in spray application methods, respiratory protection must be used in addition to feasible engineering controls and work practices to reduce employee exposureto or below the PELs.

(3) **Prohibitions.** Compressed air shall not be used to remove MDA, unless the compressed air is used inconjunction with an enclosed ventilation system designed to capture the dust cloud created by thecompressed air.

(4) **Employee rotation.** The employer shall not use employee rotation as a means of compliance with theexposure limits prescribed in paragraph (c) of this section.

(5) **Compliance program.**

    (i) The employer shall establish and implement a written program to reduce employee exposure to or below the PELs by means of engineering and work practice controls, as required by paragraph (h)(1) of this section, and by use of respiratory protection where permitted under thissection.

    (ii) Upon request this written program shall be furnished for examination and copying to the Assistant Secretary, the Director, affected employees and designated employee representatives. The employer shall review and, as necessary, update such plans at least onceevery 12 months to make certain they reflect the current status of the program.

(i) **Respiratory protection -**

(1) **General.** For employees who use respirators required by this section, the employer must provide eachemployee an appropriate respirator that complies with the requirements of this paragraph. Respirators must be used during:

    (i) Periods necessary to install or implement feasible engineering and work-practice controls.

    (ii) Work operations, such as maintenance and repair activities and spray-application processes, forwhich engineering and work-practice controls are not feasible.

    (iii) Work operations for which feasible engineering and work-practice controls are not yet sufficientto reduce employee exposure to or below the PELs.

    (iv) Emergencies.

(2) **Respirator program.** The employer must implement a respiratory protection program in accordancewith § 1910.134 (b) through (d) (except (d)(1)(iii)), and (f) through (m), which covers each employee required by this section to use a respirator.

(3) **Respirator selection.**

    (i) Employers must:

        (A) Select, and provide to employees, the appropriate respirators specified in paragraph(d)(3)(i)(A) of 29 CFR 1910.134.

        (B) Provide HEPA filters for powered and non-powered air-purifying respirators.

        (C) For escape, provide employees with one of the following respirator options: Any self- contained breathing apparatus with a full facepiece or hood operated in the positive-pressure or continuous-flow mode; or a full facepiece air-purifying respirator.

        (D) Provide a combination HEPA filter and organic vapor canister or cartridge with air-purifying respirators when MDA is in liquid form or used as part of a process requiring heat.

(ii) An employee who cannot use a negative-pressure respirator must be given the option of using apositive-pressure respirator, or a supplied-air respirator operated in the continuous-flow or pressure-demand mode.

(j) **Protective work clothing and equipment -**

(1) **Provision and use.** Where employees are subject to dermal exposure to MDA, where liquids containing MDA can be splashed into the eyes, or where airborne concentrations of MDA are in excess of the PEL, the employer shall provide, at no cost to the employee, and ensure that the employee uses, appropriate protective work clothing and equipment which prevent contact with MDAsuch as, but not limited to:

(i) Aprons, coveralls or other full-body work

(ii) clothing;Gloves, head coverings, and foot

(iii) coverings; and Face shields, chemical

(iv) goggles; or

Other appropriate protective equipment which comply with 29 CFR 1910.133.

(2) **Removal and storage.**

(i) The employer shall ensure that, at the end of their work shift, employees remove MDA-contaminated protective work clothing and equipment that is not routinely removed throughoutthe day in change areas provided in accordance with the provisions in paragraph (k) of this section.

(ii) The employer shall ensure that, during their work shift, employees remove all other MDA-contaminated protective work clothing or equipment before leaving a regulated area.

(iii) The employer shall ensure that no employee takes MDA-contaminated work clothing or equipment out of the decontamination areas, except those employees authorized to do so forthe purpose of laundering, maintenance, or disposal.

(iv) MDA-contaminated work clothing or equipment shall be placed and stored and transported insealed, impermeable bags, or other closed impermeable containers.

(v) Containers of MDA-contaminated protective work clothing or equipment which are to be takenout of decontamination areas or the workplace for cleaning, maintenance, or disposal, shall bear labels warning of the hazards of MDA.

(3) **Cleaning and replacement.**

(i) The employer shall provide the employee with clean protective clothing and equipment. The employer shall ensure that protective work clothing or equipment required by this paragraph iscleaned, laundered, repaired, or replaced at intervals appropriate to maintain its effectiveness.

(ii) The employer shall prohibit the removal of MDA from protective work clothing or equipment byblowing, shaking, or any methods which allow MDA to re-enter the workplace.

(iii) The employer shall ensure that laundering of MDA-contaminated clothing shall be done so as toprevent the release of MDA in the workplace.

- (iv) Any employer who gives MDA-contaminated clothing to another person for laundering shall inform such person of the requirement to prevent the release of MDA.
- (v) The employer shall inform any person who launders or cleans protective clothing or equipment contaminated with MDA of the potentially harmful effects of exposure.

(4) **Visual Examination.**
- (i) The employer shall ensure that employees' work clothing is examined periodically for rips or tears that may occur during performance of work.
- (ii) When rips or tears are detected, the protective equipment or clothing shall be repaired and replaced immediately.

(k) **Hygiene facilities and practices -**

(1) **General.**
- (i) The employer shall provide decontamination areas for employees required to work in regulated areas or required by paragraph (j)(1) of this section to wear protective clothing. *Exception:* In lieu of the decontamination area requirement specified in paragraph (k)(1)(i) of this section, the employer may permit employees engaged in small scale, short duration operations, to clean their protective clothing or dispose of the protective clothing before such employees leave the area where the work was performed.
- (ii) **Change areas.** The employer shall ensure that change areas are equipped with separate storage facilities for protective clothing and street clothing, in accordance with 29 CFR 1910.141(e).
- (iii) **Equipment area.** The equipment area shall be supplied with impermeable, labeled bags and containers for the containment and disposal of contaminated protective clothing and equipment.

(2) **Shower area.**
- (i) Where feasible, shower facilities shall be provided which comply with 29 CFR 1910.141(d)(3) wherever the possibility of employee exposure to airborne levels of MDA in excess of the permissible exposure limit exists.
- (ii) Where dermal exposure to MDA occurs, the employer shall ensure that materials spilled or deposited on the skin are removed as soon as possible by methods which do not facilitate the dermal absorption of MDA.

(3) **Lunch Areas.**
- (i) Whenever food or beverages are consumed at the worksite and employees are exposed to MDA the employer shall provide clean lunch areas were MDA levels are below the action level and where no dermal exposure to MDA can occur.
- (ii) The employer shall ensure that employees wash their hands and faces with soap and water prior to eating, drinking, smoking, or applying cosmetics.
- (iii) The employer shall ensure that employees do not enter lunch facilities with contaminated protective work clothing or equipment.

(l) **Communication of hazards to employees -**

(1) **Hazard communication.** The employer shall include Methylenedianiline (MDA) in the program established to comply with the Hazard Communication Standard (HCS) (§ 1910.1200). The employer shall ensure that each employee has access to labels on containers of MDA and safety data sheets, and is trained in accordance with the provisions of HCS and paragraph (l)(3) of this section. The employer shall ensure that at least the following hazards are addressed: Cancer; liver effects; and skin sensitization.

(2) **Signs and labels -**

   (i) **Signs.**

      (A) The employer shall post and maintain legible signs demarcating regulated areas and entrances or access-ways to regulated areas that bear the following legend:

MAY CAUSE CANCER

DANGER
Pure MDA:

CAUSES DAMAGE TO THE LIVER

RESPIRATORY PROTECTION AND PROTECTIVE CLOTHING MAY BE REQUIRED IN THIS AREA

A

      (B) Prior to June 1, 2016, employers may use the following legend in lieu of that specified in paragraph (l)(2)(i)(A) of this section:

:

MAY CAUSE

Pure MDA:

DANGER

CANCERLIVER

TOXIN

RESPIRATORS AND PROTECTIVE CLOTHING MAY BE REQUIRED TO BE WORN IN THIS AREA

   (ii) **Labels.**

      (A) The employer shall ensure that labels or other appropriate forms of warning are provided for containers of MDA within the workplace. The labels shall comply with the requirements of § 1910.1200(f) and shall include at least the following information for pure MDA and mixtures containing MDA:

>   DANGER
>
>   ONTAINS
>
>   DA
>
>   MAY CAUSE CANCER
>
>   CAUSES DAMAGE TO THE LIVER

- (B) Prior to June 1, 2015, employers may include the following information workplace labels inlieu of the labeling requirements in paragraph (l)(2)(ii)(A) of this section:

    - (1) For

        Pure MDA:

        DANGER

        CONTAINS

        MDA MAY

        CAUSE

        CANCERLIVER

        TOXIN

    - (2) For mixtures

        containing MDA:DANGER

        CONTAINS MDA

        CONTAINS MATERIALS WHICH MAY CAUSE

        CANCERLIVER TOXIN

- (3) *Information and training.*
    - (i) The employer shall provide employees with information and training on MDA, in accordancewith 29 CFR 1910.1200(h), at the time of initial assignment and at least annually thereafter.
    - (ii) ı addition to the information required under 29 CFR 1910.1200, the employer shall:
        - (A) Provide an explanation of the contents of this section, including appendices A and

(B) Describe the medical surveillance program required under paragraph (n) of this section, and explain the information contained in appendix C of this section; and

(C) Describe the medical removal provision required under paragraph (n) of this section.

(4) **Access to training materials.**

(i) The employer shall make readily available to all affected employees, without cost, all written materials relating to the employee training program, including a copy of this regulation.

(ii) The employer shall provide to the Assistant Secretary and the Director, upon request, all information and training materials relating to the employee information and training program.

(m) **Housekeeping.**

(1) ll surfaces shall be maintained as free as practicable of visible accumulations of MDA.

(2) The employer shall institute a program for detecting MDA leaks, spills, and discharges, including regular visual inspections of operations involving liquid or solid MDA.

(3) All leaks shall be repaired and liquid or dust spills cleaned up promptly.

(4) Surfaces contaminated with MDA may not be cleaned by the use of compressed air.

(5) Shoveling, dry sweeping, and other methods of dry clean-up of MDA may be used where HEPA filtered vacuuming and/or wet cleaning are not feasible or practical.

(6) Waste, scrap, debris, bags, containers, equipment, and clothing contaminated with MDA shall be collected and disposed of in a manner to prevent the re-entry of MDA into the workplace.

(n) **Medical surveillance** -

(1) **General.**

(i) The employer shall make available a medical surveillance program for employees exposed to MDA under the following circumstances:

(A) Employees exposed at or above the action level for 30 or more days per year;

(B) Employees who are subject to dermal exposure to MDA for 15 or more days per

(C) year; Employees who have been exposed in an emergency situation;

(D) Employees whom the employer, based on results from compliance with paragraph (f)(8) of this section, has reason to believe are being dermally exposed; and

(E) Employees who show signs or symptoms of MDA exposure.

(ii) The employer shall ensure that all medical examinations and procedures are performed by or under the supervision of a licensed physician at a reasonable time and place, and provided without cost to the employee.

(2) **Initial examinations.**

(i) Within 150 days of the effective date of this standard, or before the time of initial assignment, the employer shall provide each employee covered by paragraph (n)(1)(i) of this section with amedical examination including the following elements:

- (A) A detailed history which includes:
    - (1) Past work exposure to MDA or any other toxic substances;
    - (2) A history of drugs, alcohol, tobacco, and medication routinely taken (duration and quantity); and
    - (3) A history of dermatitis, chemical skin sensitization, or previous hepatic disease.
- (B) A physical examination which includes all routine physical examination parameters, skinexamination, and examination for signs of liver disease.
- (C) Laboratory tests
    - (1) including: Liver
    - (2) function tests
- (D) andUrinalysis.

Additional tests as necessary in the opinion of the physician.

(ii) No initial medical examination is required if adequate records show that the employee has beenexamined in accordance with the requirements of this section within the previous six months prior to the effective date of this standard or prior to the date of initial assignment.

(3) **Periodic examinations.**

(i) The employer shall provide each employee covered by this section with a medical examination at least annually following the initial examination. These periodic examinations shall include atleast the following elements:

- (A) A brief history regarding any new exposure to potential liver toxins, changes in drug, tobacco, and alcohol intake, and the appearance of physical signs relating to the liver, andthe skin;
- (B) The appropriate tests and examinations including liver function tests and skin examinations; and
- (C) Appropriate additional tests or examinations as deemed necessary by the physician.

(ii) If in the physician's opinion the results of liver function tests indicate an abnormality, the employee shall be removed from further MDA exposure in accordance with paragraph (n)(9) ofthis section. Repeat liver function tests shall be conducted on advice of the physician.

(4) **Emergency examinations.** If the employer determines that the employee has been exposed to a potentially hazardous amount of MDA in an emergency situation under paragraph (e) of this section,the employer shall provide medical examinations in accordance with paragraphs (n)(3) (i) and (ii) of this section. If the results of liver function testing indicate an abnormality, the employee shall be removed in accordance with paragraph (n)(9) of this section. Repeat liver function tests shall be conducted on the advice of the physician. If the results of the tests are normal, tests must be repeated two to three weeks from the initial testing. If the results of the second set of tests are normal and on the advice of the physician, no additional

testing is required.

(5) **Additional examinations.** Where the employee develops signs and symptoms associated with exposure to MDA, the employer shall provide the employee with an additional medical examination including liver function tests. Repeat liver function tests shall be conducted on the advice of the physician. If the results of the tests are normal, tests must be repeated two to three weeks from the initial testing. If the results of the second set of tests are normal and on the advice of the physician, no additional testing is required.

(6) **Multiple physician review mechanism.**

  (i) If the employer selects the initial physician who conducts any medical examination or consultation provided to an employee under this section, and the employee has signs or symptoms of occupational exposure to MDA (which could include an abnormal liver function test), and the employee disagrees with the opinion of the examining physician, and this opinion could affect the employee's job status, the employee may designate an appropriate and mutually acceptable second physician:

   (A) To review any findings, determinations or recommendations of the initial physician; and

   (B) To conduct such examinations, consultations, and laboratory tests as the second physician deems necessary to facilitate this review.

  (ii) The employer shall promptly notify an employee of the right to seek a second medical opinion after each occasion that an initial physician conducts a medical examination or consultation pursuant to this section. The employer may condition its participation in, and payment for, the multiple physician review mechanism upon the employee doing the following within fifteen (15) days after receipt of the foregoing notification, or receipt of the initial physician's written opinion, whichever is later:

   (A) The employee informing the employer that he or she intends to seek a second medical opinion, and

   (B) The employee initiating steps to make an appointment with a second physician.

  (iii) If the findings, determinations, or recommendations of the second physician differ from those of the initial physician, then the employer and the employee shall assure that efforts are made for the two physicians to resolve any disagreement.

  (iv) If the two physicians have been unable to quickly resolve their disagreement, then the employer and the employee through their respective physicians shall designate a third physician:

   (A) To review any findings, determinations, or recommendations of the prior physicians; and

   (B) To conduct such examinations, consultations, laboratory tests, and discussions with the prior physicians as the third physician deems necessary to resolve the disagreement of the prior physicians.

  (v) The employer shall act consistent with the findings, determinations, and recommendations of the second physician, unless the employer and the employee reach a mutually acceptable agreement.

(7) **Information provided to the examining physician.**

- (i) The employer shall provide the following information to the examining physician:
  - (A) A copy of this regulation and its appendices;
  - (B) A description of the affected employee's duties as they relate to the employee's potential exposure to MDA;
  - (C) The employee's current actual or representative MDA exposure level;
  - (D) A description of any personal protective equipment used or to be used; and
  - (E) Information from previous employment related medical examinations of the affected employee.
- (ii) The employer shall provide the foregoing information to a second physician under this section upon request either by the second physician, or by the employee.

(8) **Physician's written opinion.**

- (i) For each examination under this section, the employer shall obtain, and provide the employee with a copy of, the examining physician's written opinion within 15 days of its receipt. The written opinion shall include the following:
  - (A) The occupationally pertinent results of the medical examination and tests;
  - (B) The physician's opinion concerning whether the employee has any detected medical conditions which would place the employee at increased risk of material impairment of health from exposure to MDA;
  - (C) The physician's recommended limitations upon the employee's exposure to MDA or upon the employee's use of protective clothing or equipment and respirators; and
  - (D) A statement that the employee has been informed by the physician of the results of the medical examination and any medical conditions resulting from MDA exposure which require further explanation or treatment.
- (ii) The written opinion obtained by the employer shall not reveal specific findings or diagnoses unrelated to occupational exposures.

(9) **Medical removal -**

- (i) **Temporary medical removal of an employee -**
  - (A) **Temporary removal resulting from occupational exposure.** The employee shall be removed from work environments in which exposure to MDA is at or above the action level or where dermal exposure to MDA may occur, following an initial examination (paragraph (n)(2) of

    this section), periodic examinations (paragraph (n)(3) of this section), an emergency situation (paragraph (n)(4) of this section), or an additional examination (paragraph (n)(5) of this section) in the following circumstances:
    - (1) When the employee exhibits signs and/or symptoms indicative of acute exposure to MDA; or
    - (2) When the examining physician determines that an employee's abnormal liver function tests are not associated with MDA exposure but that the abnormalities may be exacerbated as a result of occupational exposure to MDA.

(B) **Temporary removal due to a final medical determination.**

    (1) The employer shall remove an employee from work having an exposure to MDA at or above the action level or where the potential for dermal exposure exists on each occasion that a final medical determination results in a medical finding, determination, or opinion that the employee has a detected medical condition which places the employee at increased risk of material impairment to health from exposure to MDA.

    (2) For the purposes of this section, the phrase "final medical determination" shall mean the outcome of the physician review mechanism used pursuant to the medical surveillance provisions of this section.

    (3) Where a final medical determination results in any recommended special protective measures for an employee, or limitations on an employee's exposure to MDA, the employer shall implement and act consistent with the recommendation.

(ii) **Return of the employee to former job status.**

    (A) The employer shall return an employee to his or her former job status:

        (1) When the employee no longer shows signs or symptoms of exposure to MDA, or upon the advice of the physician.

        (2) When a subsequent final medical determination results in a medical finding, determination, or opinion that the employee no longer has a detected medical condition which places the employee at increased risk of material impairment to health from exposure to MDA.

    (B) For the purposes of this section, the requirement that an employer return an employee to his or her former job status is not intended to expand upon or restrict any rights an employee has or would have had, absent temporary medical removal, to a specific job classification or position under the terms of a collective bargaining agreement.

(iii) **Removal of other employee special protective measure or limitations.** The employer shall remove any limitations placed on an employee or end any special protective measures provided to an employee pursuant to a final medical determination when a subsequent final medical determination indicates that the limitations or special protective measures are no longer necessary.

(iv) **Employer options pending a final medical determination.** Where the physician review mechanism used pursuant to the medical surveillance provisions of this section, has not yet resulted in a final medical determination with respect to an employee, the employer shall act as follows:

    (A) **Removal.** The employer may remove the employee from exposure to MDA, provide special protective measures to the employee, or place limitations upon the employee, consistent with the medical findings, determinations, or recommendations of the physician who has reviewed the employee's health status.

    (B) **Return.** The employer may return the employee to his or her former job status, and end any special protective measures provided to the employee, consistent with the medical findings, determinations, or recommendations of any of the physicians who have reviewed the employee's health status, with two exceptions:

(1) If the initial removal, special protection, or limitation of the employee resulted from a final medical determination which differed from the findings, determinations, or recommendations of the initial physician; or

(2) The employee has been on removal status for the preceding six months as a result of exposure to MDA, then the employer shall await a final medical determination.

(v) *Medical removal protection benefits* -

(A) *Provisions of medical removal protection benefits.* The employer shall provide to an employee up to six (6) months of medical removal protection benefits on each occasion that an employee is removed from exposure to MDA or otherwise limited pursuant to this section.

(B) *Definition of medical removal protection benefits.* For the purposes of this section, the requirement that an employer provide medical removal protection benefits means that the employer shall maintain the earnings, seniority, and other employment rights and benefits of an employee as though the employee had not been removed from normal exposure to MDA or otherwise limited.

(C) *Follow-up medical surveillance during the period of employee removal or limitations.* During the period of time that an employee is removed from normal exposure to MDA or otherwise limited, the employer may condition the provision of medical removal protection benefits upon the employee's participation in follow-up medical surveillance made available pursuant to this section.

(D) *Workers' compensation claims.* If a removed employee files a claim for workers' compensation payments for a MDA-related disability, then the employer shall continue to provide medical removal protection benefits pending disposition of the claim. To the extent that an award is made to the employee for earnings lost during the period of removal, the employer's medical removal protection obligation shall be reduced by such amount. The employer shall receive no credit for workers' compensation payments received by the employee for treatment-related expenses.

(E) *Other credits.* The employer's obligation to provide medical removal protection benefits to a removed employee shall be reduced to the extent that the employee receives compensation for earnings lost during the period of removal either from a publicly or employer-funded compensation program, or receives income from employment with any employer made possible by virtue of the employee's removal.

(F) *Employees who do not recover within the 6 months of removal.* The employer shall take the following measures with respect to any employee removed from exposure to MDA:

(1) The employer shall make available to the employee a medical examination pursuant to this section to obtain a final medical determination with respect to the employee;

(2) The employer shall assure that the final medical determination obtained

indicates whether or not the employee may be returned to his or her former job status, and, if not, what steps should be taken to protect the employee's health;

(3) Where the final medical determination has not yet been obtained, or once obtained indicates that the employee may not yet be returned to his or her former job status, the employer shall continue to provide medical removal protection benefits to the employee until either the employee is returned to former job status, or a final medical determination is made that the employee is incapable of ever safely returning to his or her former job status; and

(4) Where the employer acts pursuant to a final medical determination which permits the return of the employee to his or her former job status despite what would otherwise be an unacceptable liver function test, later questions concerning removing the employee again shall be decided by a final medical determination. The employer need not automatically remove such an employee pursuant to the MDA removal criteria provided by this section.

(vi) **Voluntary removal or restriction of an employee.** Where an employer, although not required by this section to do so, removes an employee from exposure to MDA or otherwise places limitations on an employee due to the effects of MDA exposure on the employee's medical condition, the employer shall provide medical removal protection benefits to the employee equal to that required by paragraph (n)(9)(v) of this section.

(o) *Recordkeeping -*

(1) *Objective data for exempted operations.*

(i) Where the employer has relied on objective data that demonstrate that products made from or containing MDA are not capable of releasing MDA or do not present a dermal exposure problem under the expected conditions of processing, use, or handling to exempt such operations from the initial monitoring requirements under paragraph (f)(2) of this section, the employer shall establish and maintain an accurate record of objective data reasonably relied upon in support of the exemption.

(ii) The record shall include at least the following

(A) information: The product qualifying for

(B) exemption;

The source of the objective data;

(C) The testing protocol, results of testing, and/or analysis of the material for the release of MDA;

(D) A description of the operation exempted and how the data support the exemption; and

(E) Other data relevant to the operations, materials, processing, or employee exposures covered by the exemption.

(iii) The employer shall maintain this record for the duration of the employer's reliance upon such objective data.

(2) *Historical monitoring data.*

(i) Where the employer has relied on historical monitoring data that demonstrate that exposures on a particular job will be below the action level to exempt such operations from the initial monitoring requirements under paragraph (f)(2) of this section, the employer shall establish and maintain an accurate record of historical monitoring data reasonably relied upon in supportof the exception.

(ii) The record shall include information that reflect the following conditions:

(A) The data upon which judgments are based are scientifically sound and were collectedusing methods that are sufficiently accurate and precise;

(B) The processes and work practices that were in use when the historical monitoring data were obtained are essentially the same as those to be used during the job for which initialmonitoring will not be performed;

(C) The characteristics of the MDA-containing material being handled when the historicalmonitoring data were obtained are the same as those on the job for which initial monitoring will not be performed;

(D) Environmental conditions prevailing when the historical monitoring data were obtained arethe same as those on the job for which initial monitoring will not be performed; and

(E) Other data relevant to the operations, materials, processing, or employee exposurescovered by the exception.

(iii) The employer shall maintain this record for the duration of the employer's reliance upon suchhistorical monitoring data.

(3) The employer may utilize the services of competent organizations such as industry trade associations and employee associations to maintain the records required by this section.

(4) **Exposure measurements.**

(i) The employer shall keep an accurate record of all measurements taken to monitor employeeexposure to MDA.

(ii) This record shall include at least the following

(A) information:The date of measurement;

(B) The operation involving exposure to MDA;

(C) Sampling and analytical methods used and evidence of their accuracy;

(D) Number, duration, and results of samples

(E) taken;Type of protective devices worn, if

(F) any; and

Name and exposure of the employees whose exposures are represented.

(iii) The employer shall maintain this record for at least thirty (30) years, in accordance with 29 CFR1910.33.

(5) **Medical surveillance.**

(i) The employer shall establish and maintain an accurate record for each employee subject to medical surveillance by paragraph (n) of this section, in accordance with 29 CFR 1910.33.

(ii) The record shall include at least the following information:

(A) The name of the employee;

(B) A copy of the employee's medical examination results, including the medical history, questionnaire responses, results of any tests, and physician's recommendations.

(C) Physician's written opinions;

(D) Any employee medical complaints related to exposure to MDA; and

(E) A copy of the information provided to the physician as required by paragraph (n) of this section.

(iii) The employer shall ensure that this record is maintained for the duration of employment plus thirty (30) years, in accordance with 29 CFR 1910.33.

(iv) A copy of the employee's medical removal and return to work status.

(6) **Training records.** The employer shall maintain all employee training records for one (1) year beyond the last date of employment.

(7) **Availability.**

(i) The employer, upon written request, shall make all records required to be maintained by this section available to the Assistant Secretary and the Director for examination and copying.

(ii) The employer, upon request, shall make any exposure records required by paragraphs (f) and (n) of this section available for examination and copying to affected employees, former employees, designated representatives, and the Assistant Secretary, in accordance with 29 CFR 1910.33(a)-(e) and (g)-(i).

(iii) The employer, upon request, shall make employee medical records required by paragraphs (n) and (o) of this section available for examination and copying to the subject employee, anyone having the specific written consent of the subject employee, and the Assistant Secretary, in accordance with 29 CFR 1910.33.

(8) **Transfer of records.** The employer shall comply with the requirements concerning transfer of records set forth in 29 CFR 1910.1020(h).

(p) **Observation of monitoring -**

(1) **Employee observation.** The employer shall provide affected employees, or their designated representatives, an opportunity to observe the measuring or monitoring of employee exposure to MDA conducted pursuant to paragraph (f) of this section.

(2) **Observation procedures.** When observation of the measuring or monitoring of employee exposure to MDA requires entry into areas where the use of protective clothing and equipment or respirators is required, the employer shall provide the observer with personal protective clothing and equipment or respirators required to be worn by employees working in the area, assure the use of such clothing and equipment or respirators, and require the observer to comply with all other applicable safety and health procedures.

(q) **Appendices.** The information contained in appendices A, B, C, and D of this section is not intended, by itself, to create any additional obligations not otherwise imposed by this standard nor detract from any existing obligation.

Appendix A to § 1926.60 - Substance Data Sheet, for 4-4'Methylenedianiline

Note: The requirements applicable to construction work under this appendix A are identical to those set forth in appendix A to § 1910.1050 of this chapter.

Appendix B to § 1926.60 - Substance Technical Guidelines, MDA

Note: The requirements applicable to construction work under this appendix B are identical to those set forth in appendix B to § 1910.1050 of this chapter.

Appendix C to § 1926.60 - Medical Surveillance Guidelines for MDA

Note: The requirements applicable to construction work under this appendix C are identical to those set forth in appendix C to § 1910.1050 of this chapter.

Appendix D to § 1926.60 - Sampling and Analytical Methods for MDAMonitoring and Measurement Procedures

Note: The requirements applicable to construction work under this appendix D are identical to those set forth in appendix D to § 1910.1050 of this chapter.

[57 FR 35681, Aug. 10, 1992, as amended at 57 FR 49649, Nov. 3, 1992; 61 FR 5510, Feb. 13, 1996; 61 FR 31431, June 20, 1996;
63 FR 1296, Jan. 8, 1998; 69 FR 70373, Dec. 6, 2004; 70 FR 1143, Jan. 5, 2005; 71 FR 16674, Apr. 3, 2006; 71 FR 50191, Aug. 24,
2006; 73 FR 75588, Dec. 12, 2008; 76 FR 33611, June 8, 2011; 77 FR 17889, Mar. 26, 2012]

§ 1926.61 Retention of DOT markings, placards and labels.

Note: The requirements applicable to construction work under this section are identical to thoseset forth at § 1910.1201 of this chapter.

[61 FR 31432, June 20, 1996]

§ 1926.62 Lead.

(a) **Scope.** This section applies to all construction work where an employee may be occupationally exposed to lead. All construction work excluded from coverage in the general industry standard for lead by 29 CFR 1910.1025(a)(2) is covered by this standard. Construction work is defined as work for construction, alteration and/or repair, including painting and decorating. It includes but is not limited to the following:

(1) Demolition or salvage of structures where lead or materials containing lead are

(2) present;Removal or encapsulation of materials containing lead;

(3) New construction, alteration, repair, or renovation of structures, substrates, or portions thereof, that contain lead, or materials containing lead;

(5)
(4) Installation of products containing

lead; Lead contamination/emergency

cleanup;

(6) Transportation, disposal, storage, or containment of lead or materials containing lead on the site orlocation at which construction activities are performed, and

(7) Maintenance operations associated with the construction activities described in this paragraph.

(b) **Definitions.**

Action level means employee exposure, without regard to the use of respirators, to an airborne concentration of lead of 30 micrograms per cubic meter of air (30 µg/m^3) calculated as an 8-hourtime-weighted average (TWA).

Assistant Secretary means the Assistant Secretary of Labor for Occupational Safety and Health, U.S. Department of Labor, or designee.

Competent person means one who is capable of identifying existing and predictable lead hazards in the surroundings or working conditions and who has authorization to take prompt corrective measuresto eliminate them.

Director means the Director, National Institute for Occupational Safety and Health (NIOSH), U.S. Department of Health and Human Services, or designee.

Lead means metallic lead, all inorganic lead compounds, and organic lead soaps. Excluded from this definition are all other organic lead compounds.

This section means this standard.

(c) **Permissible exposure limit.**

(1) The employer shall assure that no employee is exposed to lead at concentrations greater than fiftymicrograms per cubic meter of air (50 µg/m^3) averaged over an 8-hour period.

(2) If an employee is exposed to lead for more than 8 hours in any work day the employees' allowable exposure, as a time weighted average (TWA) for that day, shall be reduced according to the followingformula:

Allowable employee exposure (in µg/m^3) = 400 divided by hours worked in the day.

(3) When respirators are used to limit employee exposure as required under paragraph (c) of this sectionand all the requirements of paragraphs (e)(1) and (f) of this section have been met, employee exposure may be considered to be at the level provided by the protection factor of

the respirator forthose periods the respirator is worn. Those periods may be averaged with exposure levels during periods when respirators are not worn to determine the employee's daily TWA exposure.

(d) **Exposure assessment** -

(1) **General.**

(i) Each employer who has a workplace or operation covered by this standard shall initially determine if any employee may be exposed to lead at or above the action level.

(ii) For the purposes of paragraph (d) of this section, employee exposure is that exposure whichwould occur if the employee were not using a respirator.

(iii) With the exception of monitoring under paragraph (d)(3), where monitoring is required under this section, the employer shall collect personal samples representative of a full shift including at least one sample for each job classification in each work area either for each shift or for the shift with the highest exposure level.

(iv) Full shift personal samples shall be representative of the monitored employee's regular, daily exposure to lead.

(2) **Protection of employees during assessment of exposure.**

(i) With respect to the lead related tasks listed in paragraph (d)(2)(i) of this section, where lead is present, until the employer performs an employee exposure assessment as required in paragraph (d) of this section and documents that the employee performing any of the listed tasks is not exposed above the PEL, the employer shall treat the employee as if the employeewere exposed above the PEL, and not in excess of ten (10) times the PEL, and shall implementemployee protective measures prescribed in paragraph (d)(2)(v) of this section. The tasks covered by this requirement are:

(A) Where lead containing coatings or paint are present: Manual demolition of structures (e.g,dry wall), manual scraping, manual sanding, heat gun applications, and power tool cleaning with dust collection systems;

(B) Spray painting with lead paint.

(ii) In addition, with regard to tasks not listed in paragraph (d)(2)(i), where the employee has any reason to believe that an employee performing the task may be exposed to lead in excess ofthe PEL, until the employer performs an employee exposure assessment as required by paragraph (d) of this section and documents that the employee's lead exposure is not above

the PEL the employer shall treat the employee as if the employee were exposed above the PELand shall implimemt employee protective measures as prescribed in paragraph (d)(2)(v) of thissection.

(iii) With respect to the tasks listed in this paragraph (d)(2)(iii) of this section, where lead is present, until the employer performs an employee exposure assessment as required in this paragraph (d), and documents that the employee performing any of the listed tasks is not exposed in excess of 500 µg/m^3, the employer shall treat the employee as if the employee were exposed tolead in excess of 500 µg/m^3 and shall implement employee protective measures as prescribedin paragraph (d)(2)(v) of this section. Where the employer does establish that the employee is exposed to levels of lead below 500 µg/m³, the employer may provide the exposed employee with the appropriate respirator prescribed for such use at such lower exposures, in accordance with paragraph (f) of this section. The tasks covered by this requirement are:

- (A) Using lead containing mortar; lead burning
- (B) Where lead containing coatings or paint are present: rivet busting; power tool cleaning without dust collection systems; cleanup activities where dry expendable abrasives areused; and abrasive blasting enclosure movement and removal.

(iv) With respect to the tasks listed in this paragraph (d)(2)(iv), where lead is present, until the employer performs an employee exposure assessment as required in this paragraph (d) and documents that the employee performing any of the listed tasks is not exposed to lead in excess of 2,500 µg/m³ (50×PEL), the employer shall treat the employee as if the employee wereexposed to lead in excess of 2,500 µg/m³ and shall implement employee protective measures as prescribed in paragraph (d)(2)(v) of this section. Where the employer does establish that the employee is exposed to levels of lead below 2,500 µg/m³, the employer may provide the exposed employee with the appropriate respirator prescribed for use at such lower exposures,in accordance with paragraph (f) of this section. Interim protection as described in this paragraph is required where lead containing coatings or paint are present on structures when performing:

- (A) Abrasive
- (B) blasting,
- (C) Welding,

    Cutting,
- (D) and Torch

(v) burning.

Until the employer performs an employee exposure assessment as required under paragraph
(d) of this section and determines actual employee exposure, the employer shall provide to employees performing the tasks described in paragraphs (d)(2)(i), (d)(2)(ii), (d)(2)(iii), and (d)(2)(iv) of this section with interim protection as follows:

- (A) Appropriate respiratory protection in accordance with paragraph (f) of this section.
- (B) Appropriate personal protective clothing and equipment in accordance with paragraph (g)of this section.
- (C) Change areas in accordance with paragraph (i)(2) of this section.
- (D) Hand washing facilities in accordance with paragraph (i)(5) of this section.
- (E) Biological monitoring in accordance with paragraph (j)(1)(i) of this section, to consist ofblood sampling and analysis for lead and zinc protoporphyrin levels, and
- (F) Training as required under paragraph (l)(1)(i) of this section regarding 29 CFR 1926.59, Hazard Communication; training as required under paragraph (1)(2)(iii) of this section, regarding use of respirators; and training in accordance with 29 CFR 1926.21, Safety

training and education.

(3) **Basis of initial determination.**

 (i) Except as provided under paragraphs (d)(3)(iii) and (d)(3)(iv) of this section the employer shall monitor employee exposures and shall base initial determinations on the employee exposure monitoring results and any of the following, relevant considerations:

  (A) Any information, observations, or calculations which would indicate employee exposure to lead;

  (B) Any previous measurements of airborne lead; and

  (C) Any employee complaints of symptoms which may be attributable to exposure to lead.

 (ii) Monitoring for the initial determination where performed may be limited to a representative sample of the exposed employees who the employer reasonably believes are exposed to the greatest airborne concentrations of lead in the workplace.

 (iii) Where the employer has previously monitored for lead exposures, and the data were obtained within the past 12 months during work operations conducted under workplace conditions closely resembling the processes, type of material, control methods, work practices, and environmental conditions used and prevailing in the employer's current operations, the employer may rely on such earlier monitoring results to satisfy the requirements of paragraphs (d)(3)(i) and (d)(6) of this section if the sampling and analytical methods meet the accuracy and confidence levels of paragraph (d)(9) of this section.

 (iv) Where the employer has objective data, demonstrating that a particular product or material containing lead or a specific process, operation or activity involving lead cannot result in employee exposure to lead at or above the action level during processing, use, or handling, the employer may rely upon such data instead of implementing initial monitoring.

  (A) The employer shall establish and maintain an accurate record documenting the nature and relevancy of objective data as specified in paragraph (n)(4) of this section, where used in assessing employee exposure in lieu of exposure monitoring.

  (B) Objective data, as described in paragraph (d)(3)(iv) of this section, is not permitted to be used for exposure assessment in connection with paragraph (d)(2) of this section.

(4) **Positive initial determination and initial monitoring.**

 (i) Where a determination conducted under paragraphs (d) (1), (2) and (3) of this section shows the possibility of any employee exposure at or above the action level the employer shall conduct monitoring which is representative of the exposure for each employee in the workplace who is exposed to lead.

 (ii) Where the employer has previously monitored for lead exposure, and the data were

(5) **Negative initial determination.** Where a determination, conducted under paragraphs (d) (1), (2), and
(3) of this section is made that no employee is exposed to airborne concentrations of lead at or above the action level the employer shall make a written record of such determination. The record shall include at least the information specified in paragraph (d)(3)(i) of this section and shall also include the date of determination, location within the worksite, and the name of each employee monitored.

(6) **Frequency.**

    (i) If the initial determination reveals employee exposure to be below the action level further exposure determination need not be repeated except as otherwise provided in paragraph (d)(7) of this section.

    (ii) If the initial determination or subsequent determination reveals employee exposure to be at or above the action level but at or below the PEL the employer shall perform monitoring in accordance with this paragraph at least every 6 months. The employer shall continue monitoring at the required frequency until at least two consecutive measurements, taken at least 7 days apart, are below the action level at which time the employer may discontinue monitoring for that employee except as otherwise provided in paragraph (d)(7) of this section.

    (iii) If the initial determination reveals that employee exposure is above the PEL the employer shall perform monitoring quarterly. The employer shall continue monitoring at the required frequency until at least two consecutive measurements, taken at least 7 days apart, are at or below the PEL but at or above the action level at which time the employer shall repeat monitoring for that employee at the frequency specified in paragraph (d)(6)(ii) of this section, except as otherwise provided in paragraph (d)(7) of this section. The employer shall continue monitoring at the required frequency until at least two consecutive measurements, taken at least 7 days apart, are below the action level at which time the employer may discontinue monitoring for that employee except as otherwise provided in paragraph (d)(7) of this section.

(7) **Additional exposure assessments.** Whenever there has been a change of equipment, process, control, personnel or a new task has been initiated that may result in additional employees being exposed to lead at or above the action level or may result in employees already exposed at or above the action level being exposed above the PEL, the employer shall conduct additional monitoring in accordance with this paragraph.

(8) **Employee notification.**

    (i) The employer must, as soon as possible but no later than 5 working days after the receipt of

[Note: The beginning of page continues from previous page with text about monitoring results obtained within the past 12 months during work operations conducted under workplace conditions closely resembling the processes, type of material, control methods, work practices, and environmental conditions used and prevailing in the employer's current operations, the employer may rely on such earlier monitoring results to satisfy the requirements of paragraph (d)(4)(i) of this section if the sampling and analytical methods meet the accuracy and confidence levels of paragraph (d)(9) of this section.]

theresults of any monitoring performed under this section, notify each affected employee of these results either individually in writing or by posting the results in an appropriate location that is accessible to employees.

    (ii) Whenever the results indicate that the representative employee exposure, without regard to respirators, is at or above the PEL the employer shall include in the written notice a statementthat the employees exposure was at or above that level and a description of the corrective action taken or to be taken to reduce exposure to below that level.

(9) **Accuracy of measurement.** The employer shall use a method of monitoring and analysis which hasan accuracy (to a confidence level of 95%) of not less than plus or minus 25 percent for airborne concentrations of lead equal to or greater than 30 µg/m^3.

(e) **Methods of compliance** -

  (1) **Engineering and work practice controls.** The employer shall implement engineering and work practicecontrols, including administrative controls, to reduce and maintain employee exposure to lead to orbelow the permissible exposure limit to the extent that such controls are feasible. Wherever all feasible engineering and work practices controls that can be instituted are not sufficient to reduce employee exposure to or below the permissible exposure limit prescribed in paragraph (c) of this section, the employer shall nonetheless use them to reduce employee exposure to the lowest feasible level and shall supplement them by the use of respiratory protection that complies with the requirements of paragraph (f) of this section.

  (2) **Compliance program.**

    (i) Prior to commencement of the job each employer shall establish and implement a written compliance program to achieve compliance with paragraph (c) of this section.

    (ii) Written plans for these compliance programs shall include at least the following:

        (A) A description of each activity in which lead is emitted; e.g. equipment used, material involved, controls in place, crew size, employee job responsibilities, operating proceduresand maintenance practices;

        (B) A description of the specific means that will be employed to achieve compliance and, where engineering controls are required engineering plans and studies used to determinemethods selected for controlling exposure to lead;

        (C) A report of the technology considered in meeting the PEL;

        (D) Air monitoring data which documents the source of lead emissions;

        (E) A detailed schedule for implementation of the program, including documentation such ascopies of purchase orders for equipment, construction contracts, etc.;

        (F) A work practice program which includes items required under paragraphs (g), (h) and (i) ofthis section and incorporates other relevant work practices such as those specified in paragraph (e)(5) of this section;

- (G) An administrative control schedule required by paragraph (e)(4) of this section, if applicable;
- (H) A description of arrangements made among contractors on multi-contractor sites with respect to informing affected employees of potential exposure to lead and with respect to responsibility for compliance with this section as set-forth in § 1926.16.
- (I) Other relevant information.

(iii) The compliance program shall provide for frequent and regular inspections of job sites, materials, and equipment to be made by a competent person.

(iv) Written programs shall be submitted upon request to any affected employee or authorized employee representatives, to the Assistant Secretary and the Director, and shall be available at the worksite for examination and copying by the Assistant Secretary and the Director.

(v) Written programs must be revised and updated at least annually to reflect the current status of the program.

(3) **Mechanical ventilation.** When ventilation is used to control lead exposure, the employer shall evaluate the mechanical performance of the system in controlling exposure as necessary to maintain its effectiveness.

(4) **Administrative controls.** If administrative controls are used as a means of reducing employees TWA exposure to lead, the employer shall establish and implement a job rotation schedule which includes:

- (i) Name or identification number of each affected employee;
- (ii) Duration and exposure levels at each job or work station where each affected employee is located; and
- (iii) Any other information which may be useful in assessing the reliability of administrative controls to reduce exposure to lead.

(5) The employer shall ensure that, to the extent relevant, employees follow good work practices such as described in appendix B of this section.

(f) **Respiratory protection** -

(1) **General.** For employees who use respirators required by this section, the employer must provide each employee an appropriate respirator that complies with the requirements of this paragraph. Respirators must be used during:

- (i) Periods when an employee's exposure to lead exceeds the PEL.
- (ii) Work operations for which engineering and work-practice controls are not sufficient to reduce employee exposures to or below the PEL.
- (iii) Periods when an employee requests a respirator.
- (iv) Periods when respirators are required to provide interim protection of employees while they perform the operations specified in paragraph (d)(2) of this section.

(2) **Respirator program.**

- (i) The employer must implement a respiratory protection program in accordance with § 1910.134(b) through (d) (except (d)(1)(iii)), and (f) through (m), which covers each employee required by this section to use a respirator.
- (ii) If an employee has breathing difficulty during fit testing or respirator use, the employer must provide the employee with a medical examination in accordance with paragraph (j)(3)(i)(B) of this section to determine whether or not the employee can use a respirator while performing the required duty.

(3) **Respirator selection.**

- (i) Employers must:
    - (A) Select, and provide to employees, the appropriate respirators specified in paragraph (d)(3)(i)(A) of 29 CFR 1910.134.
    - (B) Provide employees with a full facepiece respirator instead of a half mask respirator for protection against lead aerosols that may cause eye or skin irritation at the use concentrations.
    - (C) Provide HEPA filters for powered and non-powered air-purifying respirators.
- (ii) The employer must provide a powered air-purifying respirator when an employee chooses to use such a respirator and it will provide adequate protection to the employee.

(g) **Protective work clothing and equipment** -

(1) **Provision and use.** Where an employee is exposed to lead above the PEL without regard to the use of respirators, where employees are exposed to lead compounds which may cause skin or eye irritation (e.g. lead arsenate, lead azide), and as interim protection for employees performing tasks as specified in paragraph (d)(2) of this section, the employer shall provide at no cost to the employee and assure that the employee uses appropriate protective work clothing and equipment that prevents contamination of the employee and the employee's garments such as, but not limited to:

- (i) Coveralls or similar full-body work clothing;
- (ii) Gloves, hats, and shoes or disposable shoe coverlets; and
- (iii) Face shields, vented goggles, or other appropriate protective equipment which complies with §1910.133 of this chapter.

(2) **Cleaning and replacement.**

- (i) The employer shall provide the protective clothing required in paragraph (g)(1) of this section in a clean and dry condition at least weekly, and daily to employees whose exposure levels without regard to a respirator are over 200 µg/m^3 of lead as an 8-hour TWA.
- (ii) The employer shall provide for the cleaning, laundering, and disposal of protective clothing and equipment required by paragraph (g)(1) of this section.
- (iii) The employer shall repair or replace required protective clothing and equipment as needed to maintain their effectiveness.

(iv) The employer shall assure that all protective clothing is removed at the completion of a workshift only in change areas provided for that purpose as prescribed in paragraph (i)(2) of this section.

(v) The employer shall assure that contaminated protective clothing which is to be cleaned, laundered, or disposed of, is placed in a closed container in the change area which preventsdispersion of lead outside the container.

(vi) The employer shall inform in writing any person who cleans or launders protective clothing orequipment of the potentially harmful effects of exposure to lead.

(A) The employer shall ensure that the containers of contaminated protective clothing and equipment required by paragraph (g)(2)(v) of this section are labeled as follows:

DANGER: CLOTHING AND EQUIPMENT CONTAMINATED WITH LEAD. MAY DAMAGE FERTILITYOR THE UNBORN CHILD. CAUSES DAMAGE TO THE CENTRAL NERVOUS SYSTEM. DO NOT EAT, DRINK OR SMOKE WHEN HANDLING. DO NOT REMOVE DUST BY BLOWING OR SHAKING. DISPOSE OF LEAD CONTAMINATED WASH WATER IN ACCORDANCE WITH APPLICABLE LOCAL, STATE, OR FEDERAL REGULATIONS.

(B) Prior to June 1, 2015, employers may include the following information on bags or containers of contaminated protective clothing and equipment required by paragraph (g)(2)(v) in lieu of the labeling requirements in paragraph (g)(2)(vii)(A) of this section:

Caution: Clothing contaminated with lead. Do not remove dust by blowing or shaking. Disposeof lead contaminated wash water in accordance with applicable local, state, or federal regulations.

(viii) The employer shall prohibit the removal of lead from protective clothing or equipment by blowing, shaking, or any other means which disperses lead into the air.

(h) *Housekeeping* -

(1) All surfaces shall be maintained as free as practicable of accumulations of lead.

(2) Clean-up of floors and other surfaces where lead accumulates shall wherever possible, be cleaned byvacuuming or other methods that minimize the likelihood of lead becoming airborne.

(3) Shoveling, dry or wet sweeping, and brushing may be used only where vacuuming or other equallyeffective methods have been tried and found not to be effective.

(4) Where vacuuming methods are selected, the vacuums shall be equipped with HEPA filters and usedand emptied in a manner which minimizes the reentry of lead into the workplace.

(5) Compressed air shall not be used to remove lead from any surface unless the compressed air is used in conjunction with a ventilation system designed to capture the airborne dust created by thecompressed air.

(i) *Hygiene facilities and practices.*

(1) The employer shall assure that in areas where employees are exposed to lead above the PEL withoutregard to the use of respirators, food or beverage is not present or consumed, tobacco products are not present or used, and cosmetics are not applied.

(2) *Change areas.*

    (i) The employer shall provide clean change areas for employees whose airborne exposure to lead is above the PEL, and as interim protection for employees performing tasks as specified in paragraph (d)(2) of this section, without regard to the use of respirators.

    (ii) The employer shall assure that change areas are equipped with separate storage facilities for protective work clothing and equipment and for street clothes which prevent cross-contamination.

    (iii) The employer shall assure that employees do not leave the workplace wearing any protective clothing or equipment that is required to be worn during the work shift.

(3) *Showers.*

    (i) The employer shall provide shower facilities, where feasible, for use by employees whose airborne exposure to lead is above the PEL.

    (ii) The employer shall assure, where shower facilities are available, that employees shower at the end of the work shift and shall provide an adequate supply of cleansing agents and towels for use by affected employees.

(4) *Eating facilities.*

    (i) The employer shall provide lunchroom facilities or eating areas for employees whose airborne exposure to lead is above the PEL, without regard to the use of respirators.

    (ii) The employer shall assure that lunchroom facilities or eating areas are as free as practicable from lead contamination and are readily accessible to employees.

    (iii) The employer shall assure that employees whose airborne exposure to lead is above the PEL, without regard to the use of a respirator, wash their hands and face prior to eating, drinking, smoking or applying cosmetics.

    (iv) The employer shall assure that employees do not enter lunchroom facilities or eating areas with protective work clothing or equipment unless surface lead dust has been removed by vacuuming, downdraft booth, or other cleaning method that limits dispersion of lead dust.

(5) *Hand washing facilities.*

    (i) The employer shall provide adequate handwashing facilities for use by employees exposed to lead in accordance with 29 CFR 1926.51(f).

    (ii) Where showers are not provided the employer shall assure that employees wash their hands and face at the end of the work-shift.

(j) *Medical surveillance -*

(1) *General.*

    (i) The employer shall make available initial medical surveillance to employees occupationally exposed on any day to lead at or above the action level. Initial medical surveillance consists of biological monitoring in the form of blood sampling and analysis for lead and zinc protoporphyrin levels.

    (ii) The employer shall institute a medical surveillance program in accordance with paragraphs (j)(2) and (j)(3) of this section for all employees who are or may be exposed

(ii) by the employer ator above the action level for more than 30 days in any consecutive 12 months;

(iii) The employer shall assure that all medical examinations and procedures are performed by orunder the supervision of a licensed physician.

(iv) The employer shall make available the required medical surveillance including multiple physician review under paragraph (j)(3)(iii) without cost to employees and at a reasonable timeand place.

(2) **Biological monitoring** -

(i) **Blood lead and ZPP level sampling and analysis.** The employer shall make available biological monitoring in the form of blood sampling and analysis for lead and zinc protoporphyrin levels toeach employee covered under paragraphs (j)(1)(i) and (ii) of this section on the following schedule:

(A) For each employee covered under paragraph (j)(1)(ii) of this section, at least every 2months for the first 6 months and every 6 months thereafter;

(B) For each employee covered under paragraphs (j)(1) (i) or (ii) of this section whose last blood sampling and analysis indicated a blood lead level at or above 40 µg/dl, at least every two months. This frequency shall continue until two consecutive blood samples andanalyses indicate a blood lead level below 40 µg/dl; and

(C) For each employee who is removed from exposure to lead due to an elevated blood leadlevel at least monthly during the removal period.

(ii) **Follow-up blood sampling tests.** Whenever the results of a blood lead level test indicate that anemployee's blood lead level is at or above the numerical criterion for medical removal under paragraph (k)(1)(i) of this section, the employer shall provide a second (follow-up) blood sampling test within two weeks after the employer receives the results of the first blood sampling test.

(iii) **Accuracy of blood lead level sampling and analysis.** Blood lead level sampling and analysis provided pursuant to this section shall have an accuracy (to a confidence level of 95 percent)within plus or minus 15 percent or 6 µg/dl, whichever is greater, and shall be conducted by a laboratory approved by OSHA.

(iv) **Employee notification.**

(A) Within five working days after the receipt of biological monitoring results, the employershall notify each employee in writing of his or her blood lead level; and

(B) The employer shall notify each employee whose blood lead level is at or above 40 µg/dl that the standard requires temporary medical removal with Medical Removal Protectionbenefits when an employee's blood lead level is at or above the numerical criterion for medical removal under paragraph (k)(1)(i) of this section.

(3) **Medical examinations and consultations** -

(i) **Frequency.** The employer shall make available medical examinations and consultations to eachemployee covered under paragraph (j)(1)(ii) of this section on the following schedule:

(A) At least annually for each employee for whom a blood sampling test conducted at any time during the preceding 12 months indicated a blood lead level at or above 40 µg/dl;

(B) As soon as possible, upon notification by an employee either that the employee has developed signs or symptoms commonly associated with lead intoxication, that the employee desires medical advice concerning the effects of current or past exposure to lead on the employee's ability to procreate a healthy child, that the employee is pregnant,or that the employee has demonstrated difficulty in breathing during a respirator fittingtest or during use; and

(C) As medically appropriate for each employee either removed from exposure to lead due to a risk of sustaining material impairment to health, or otherwise limited pursuant to a finalmedical determination.

(ii) **Content.** The content of medical examinations made available pursuant to paragraph (j)(3)(i)(B)-(C) of this section shall be determined by an examining physician and, if requestedby an employee, shall include pregnancy testing or laboratory evaluation of male fertility.
Medical examinations made available pursuant to paragraph (j)(3)(i)(A) of this section shall include the following elements:

(A) A detailed work history and a medical history, with particular attention to past lead exposure (occupational and non-occupational), personal habits (smoking, hygiene), and past gastrointestinal, hematologic, renal, cardiovascular, reproductive and neurological problems;

(B) A thorough physical examination, with particular attention to teeth, gums, hematologic, gastrointestinal, renal, cardiovascular, and neurological systems. Pulmonary status should be evaluated if respiratory protection will be used;

(C) A blood pressure measurement;

(D) A blood sample and analysis which

(1) determines:Blood lead level;

(2) Hemoglobin and hematocrit determinations, red cell indices, and examination ofperipheral smear morphology;

(3) Zinc

(4) protoporphyrin;

(5) Blood urea

(E) nitrogen; and,

Serum creatinine;

A routine urinalysis with microscopic examination; and

(F) Any laboratory or other test relevant to lead exposure which the examining physician deems necessary by sound medical practice.

(iii) **Multiple physician review mechanism.**

- (A) If the employer selects the initial physician who conducts any medical examination or consultation provided to an employee under this section, the employee may designate a second physician:
  - (1) To review any findings, determinations or recommendations of the initial physician; and
  - (2) To conduct such examinations, consultations, and laboratory tests as the second physician deems necessary to facilitate this review.
- (B) The employer shall promptly notify an employee of the right to seek a second medical opinion after each occasion that an initial physician conducts a medical examination or consultation pursuant to this section. The employer may condition its participation in, and payment for, the multiple physician review mechanism upon the employee doing the following within fifteen (15) days after receipt of the foregoing notification, or receipt of the initial physician's written opinion, whichever is later:
  - (1) The employee informing the employer that he or she intends to seek a second medical opinion, and
  - (2) The employee initiating steps to make an appointment with a second physician.
- (C) If the findings, determinations or recommendations of the second physician differ from those of the initial physician, then the employer and the employee shall assure that efforts are made for the two physicians to resolve any disagreement.
- (D) If the two physicians have been unable to quickly resolve their disagreement, then the employer and the employee through their respective physicians shall designate a third physician:
  - (1) To review any findings, determinations or recommendations of the prior physicians; and
  - (2) To conduct such examinations, consultations, laboratory tests and discussions with the prior physicians as the third physician deems necessary to resolve the disagreement of the prior physicians.
- (E) The employer shall act consistent with the findings, determinations and recommendations of the third physician, unless the employer and the employee reach an agreement which is otherwise consistent with the recommendations of at least one of the three physicians.

(iv) *Information provided to examining and consulting physicians.*

- (A) The employer shall provide an initial physician conducting a medical examination or consultation under this section with the following information:
  - (1) A copy of this regulation for lead including all Appendices;
  - (2) A description of the affected employee's duties as they relate to the employee's exposure;
  - (3) The employee's exposure level or anticipated exposure level to lead and to any other toxic substance (if applicable);
  - (4) A description of any personal protective equipment used or to be
  - (5)
  - (6)

used; Prior blood lead determinations; and

All prior written medical opinions concerning the employee in the employer's possession or control.

- (B) The employer shall provide the foregoing information to a second or third physician conducting a medical examination or consultation under this section upon request either by the second or third physician, or by the employee.

(v) *Written medical opinions.*

- (A) The employer shall obtain and furnish the employee with a copy of a written medical opinion from each examining or consulting physician which contains only the following information:
  - (*1*) The physician's opinion as to whether the employee has any detected medical condition which would place the employee at increased risk of material impairment of the employee's health from exposure to lead;
  - (*2*) Any recommended special protective measures to be provided to the employee, or limitations to be placed upon the employee's exposure to lead;
  - (*3*) Any recommended limitation upon the employee's use of respirators, including a determination of whether the employee can wear a powered air purifying respirator if a physician determines that the employee cannot wear a negative pressure respirator; and
  - (*4*) The results of the blood lead determinations.
- (B) The employer shall instruct each examining and consulting physician to:
  - (*1*) Not reveal either in the written opinion or orally, or in any other means of communication with the employer, findings, including laboratory results, or diagnoses unrelated to an employee's occupational exposure to lead; and
  - (*2*) Advise the employee of any medical condition, occupational or nonoccupational, which dictates further medical examination or treatment.

(vi) *Alternate physician determination mechanisms.* The employer and an employee or authorized employee representative may agree upon the use of any alternate physician determination mechanism in lieu of the multiple physician review mechanism provided by paragraph (j)(3)(iii) of this section so long as the alternate mechanism is as expeditious and protective as the requirements contained in this paragraph.

(4) *Chelation.*

- (i) The employer shall assure that any person whom he retains, employs, supervises or controls does not engage in prophylactic chelation of any employee at any time.
- (ii) If therapeutic or diagnostic chelation is to be performed by any person in paragraph (j)(4)(i) of this section, the employer shall assure that it be done under the supervision of a licensed physician in a clinical setting with thorough and appropriate medical monitoring and that the employee is notified in writing prior to its occurrence.

(k) *Medical removal protection* -

(1) *Temporary medical removal and return of an employee* -

(i) ***Temporary removal due to elevated blood lead level.*** The employer shall remove an employee from work having an exposure to lead at or above the action level on each occasion that a periodic and a follow-up blood sampling test conducted pursuant to this section indicate that the employee's blood lead level is at or above 50 μg/dl; and,

(ii) ***Temporary removal due to a final medical determination.***

    (A) The employer shall remove an employee from work having an exposure to lead at or above the action level on each occasion that a final medical determination results in a medical finding, determination, or opinion that the employee has a detected medical condition which places the employee at increased risk of material impairment to health from exposure to lead.

    (B) For the purposes of this section, the phrase *final medical determination* means the written medical opinion on the employees' health status by the examining physician or, where relevant, the outcome of the multiple physician review mechanism or alternate medical determination mechanism used pursuant to the medical surveillance provisions of this section.

    (C) Where a final medical determination results in any recommended special protective measures for an employee, or limitations on an employee's exposure to lead, the employer shall implement and act consistent with the recommendation.

(iii) ***Return of the employee to former job status.***

    (A) The employer shall return an employee to his or her former job status:

        (*1*) For an employee removed due to a blood lead level at or above 50 μg/dl when two consecutive blood sampling tests indicate that the employee's blood lead level is below 40 μg/dl;

        (*2*) For an employee removed due to a final medical determination, when a subsequent final medical determination results in a medical finding, determination, or opinion that the employee no longer has a detected medical condition which places the employee at increased risk of material impairment to health from exposure to lead.

    (B) For the purposes of this section, the requirement that an employer return an employee to his or her former job status is not intended to expand upon or restrict any rights an employee has or would have had, absent temporary medical removal, to a specific job classification or position under the terms of a collective bargaining agreement.

(iv) ***Removal of other employee special protective measure or limitations.*** The employer shall remove any limitations placed on an employee or end any special protective measures provided to an employee pursuant to a final medical determination when a subsequent final medical determination indicates that the limitations or special protective measures are no longer necessary.

(v) ***Employer options pending a final medical determination.*** Where the multiple physician review mechanism, or alternate medical determination mechanism used pursuant to the medical surveillance provisions of this section, has not yet resulted in a final medical determination with respect to an employee, the employer shall act as follows:

(A) ***Removal.*** The employer may remove the employee from exposure to lead, provide special protective measures to the employee, or place limitations upon the employee, consistent with the medical findings, determinations, or recommendations of any of the physicians who have reviewed the employee's health status.

(B) ***Return.*** The employer may return the employee to his or her former job status, end any special protective measures provided to the employee, and remove any limitations placed upon the employee, consistent with the medical findings, determinations, or recommendations of any of the physicians who have reviewed the employee's health status, with two exceptions.

(*1*) If the initial removal, special protection, or limitation of the employee resulted from a final medical determination which differed from the findings, determinations, or recommendations of the initial physician or;

(*2*) If the employee has been on removal status for the preceding eighteen months due to an elevated blood lead level, then the employer shall await a final medical determination.

(2) ***Medical removal protection benefits -***

(i) ***Provision of medical removal protection benefits.*** The employer shall provide an employee up to eighteen (18) months of medical removal protection benefits on each occasion that an employee is removed from exposure to lead or otherwise limited pursuant to this section.

(ii) ***Definition of medical removal protection benefits.*** For the purposes of this section, the requirement that an employer provide medical removal protection benefits means that, as long as the job the employee was removed from continues, the employer shall maintain the total normal earnings, seniority and other employment rights and benefits of an employee, including the employee's right to his or her former job status as though the employee had not been medically removed from the employee's job or otherwise medically limited.

(iii) ***Follow-up medical surveillance during the period of employee removal or limitation.*** During the period of time that an employee is medically removed from his or her job or otherwise medically limited, the employer may condition the provision of medical removal protection benefits upon the employee's participation in follow-up medical surveillance made available pursuant to this section.

(iv) ***Workers' compensation claims.*** If a removed employee files a claim for workers' compensation payments for a lead-related disability, then the employer shall continue to provide medical removal protection benefits pending disposition of the claim. To the extent that an award is made to the employee for earnings lost during the period of removal, the employer's medical removal protection obligation shall be reduced by such amount. The employer shall receive no credit for workers' compensation payments received by the employee for treatment-related expenses.

(v) ***Other credits.*** The employer's obligation to provide medical removal protection benefits to a removed employee shall be reduced to the extent that the employee receives compensation for earnings lost during the period of removal either from a publicly or employer-funded compensation program, or receives income from employment with another employer made possible by virtue of the employee's removal.

(vi) **Voluntary removal or restriction of an employee.** Where an employer, although not required by this section to do so, removes an employee from exposure to lead or otherwise places limitations on an employee due to the effects of lead exposure on the employee's medical condition, the employer shall provide medical removal protection benefits to the employee equal to that required by paragraph (k)(2) (i) and (ii) of this section.

(l) *Communication of hazards -*

(1) *General -*

(i) *Hazard communication.* The employer shall include lead in the program established to comply with the Hazard Communication Standard (HCS) (§ 1910.1200). The employer shall ensure that each employee has access to labels on containers of lead and safety data sheets, and is trained in accordance with the provisions of HCS and paragraph (l) of this section. The employer shall ensure that at least the following hazards are addressed:

(A) Reproductive/developmental

(B) toxicity; Central nervous system

(C) effects; Kidney effects;

(D) Blood effects;

(E) and Acute

(ii) toxicity

effects.

The employer shall train each employee who is subject to exposure to lead at or above the action level on any day, or who is subject to exposure to lead compounds which may cause skin or eye irritation (*e.g.*, lead arsenate, lead azide), in accordance with the requirements of this section. The employer shall institute a training program and ensure employee participation in the program.

(iii) The employer shall provide the training program as initial training prior to the time of job assignment or prior to the start up date for this requirement, whichever comes last.

(iv) The employer shall also provide the training program at least annually for each employee who is subject to lead exposure at or above the action level on any day.

(2) *Training program.* The employer shall assure that each employee is trained in the

(i) following: The content of this standard and its appendices;

(ii) The specific nature of the operations which could result in exposure to lead above the action level;

(iii) The purpose, proper selection, fitting, use, and limitations of respirators;

(iv) The purpose and a description of the medical surveillance program, and the medical removal protection program including information concerning the adverse health effects associated with excessive exposure to lead (with particular attention to the adverse reproductive effects on both males and females and hazards to the fetus and additional precautions for employees who are pregnant);

(v) The engineering controls and work practices associated with the employee's job assignment including training of employees to follow relevant good work practices described in appendix B of this section;

(vi) The contents of any compliance plan in effect;

(vii) Instructions to employees that chelating agents should not routinely be used to remove lead from their bodies and should not be used at all except under the direction of a licensed physician; and

(viii) The employee's right of access to records under 29 CFR 1910.20.

(3) *Access to information and training materials.*

(i) The employer shall make readily available to all affected employees a copy of this standard and its appendices.

(ii) The employer shall provide, upon request, all materials relating to the employee information and training program to affected employees and their designated representatives, and to the Assistant Secretary and the Director.

(m) *Signs -*

(1) *General.*

(i) The employer shall post the following warning signs in each work area where an employee's exposure to lead is above the PEL.

DANGER

LEAD WORK AREA

MAY DAMAGE FERTILITY OR THE UNBORN CHILD

CAUSES DAMAGE TO THE CENTRAL NERVOUS

SYSTEM DO NOT EAT, DRINK OR SMOKE IN THIS

AREA

(ii) The employer shall ensure that no statement appears on or near any sign required by this paragraph (m) that contradicts or detracts from the meaning of the required sign.

(iii) The employer shall ensure that signs required by this paragraph (m) are illuminated and cleaned as necessary so that the legend is readily visible.

(iv) The employer may use signs required by other statutes, regulations or ordinances in addition to, or in combination with, signs required by this paragraph (m).

(v) Prior to June 1, 2016, employers may use the following legend in lieu of that specified in paragraph (m)(1)(i) of this section:

> WARNING
>
> LEAD
>
> WORK
>
> AREA
>
> POISON
>
> NO SMOKING OR EATING

(n) *Recordkeeping* -

  (1) *Exposure assessment.*

    (i) The employer shall establish and maintain an accurate record of all monitoring and other dataused in conducting employee exposure assessments as required in paragraph (d) of this section.

    (ii) Exposure monitoring records shall include:

      (A) The date(s), number, duration, location and results of each of the samples taken if any,including a description of the sampling procedure used to determine representative employee exposure where applicable;

      (B) A description of the sampling and analytical methods used and evidence of their

      (C) accuracy;The type of respiratory protective devices worn, if any;

      (D) Name and job classification of the employee monitored and of all other employees whoseexposure the measurement is intended to represent; and

      (E) The environmental variables that could affect the measurement of employee exposure.

    (iii) The employer shall maintain monitoring and other exposure assessment records in accordancewith the provisions of 29 CFR 1910.33.

  (2) *Medical surveillance.*

    (i) The employer shall establish and maintain an accurate record for each employee subject to medical surveillance as required by paragraph (j) of this section.

    (ii) This record shall include:

      (A) The name and description of the duties of the

      (B) employee;A copy of the physician's written

      (C) opinions;

        Results of any airborne exposure monitoring done on or for that employee and provided tothe physician; and

      (D) Any employee medical complaints related to exposure to lead.

(iii) The employer shall keep, or assure that the examining physician keeps, the following medical records:

    (A) A copy of the medical examination results including medical and work history required under paragraph (j) of this section;

    (B) A description of the laboratory procedures and a copy of any standards or guidelines used to interpret the test results or references to that information;

    (C) A copy of the results of biological monitoring.

(iv) The employer shall maintain or assure that the physician maintains medical records in accordance with the provisions of 29 CFR 1910.33.

(3) **Medical removals.**

  (i) The employer shall establish and maintain an accurate record for each employee removed from current exposure to lead pursuant to paragraph (k) of this section.

  (ii) Each record shall include:

    (A) The name of the employee;

    (B) The date of each occasion that the employee was removed from current exposure to lead as well as the corresponding date on which the employee was returned to his or her former job status;

    (C) A brief explanation of how each removal was or is being accomplished; and

    (D) A statement with respect to each removal indicating whether or not the reason for the removal was an elevated blood lead level.

  (iii) The employer shall maintain each medical removal record for at least the duration of an employee's employment.

(4) **Objective data for exemption from requirement for initial monitoring.**

  (i) For purposes of this section, objective data are information demonstrating that a particular product or material containing lead or a specific process, operation, or activity involving lead cannot release dust or fumes in concentrations at or above the action level under any expected conditions of use. Objective data can be obtained from an industry-wide study or from laboratory product test results from manufacturers of lead containing products or materials.
The data the employer uses from an industry-wide survey must be obtained under workplace conditions closely resembling the processes, types of material, control methods, work practices and environmental conditions in the employer's current operations.

  (ii) The employer shall maintain the record of the objective data relied upon for at least 30 years.

(5) **Availability.** The employer shall make available upon request all records required to be maintained by paragraph (n) of this section to affected employees, former employees, and their designated representatives, and to the Assistant Secretary and the Director for examination and copying.

(6) **Transfer of records.**

(i) Whenever the employer ceases to do business, the successor employer shall receive and retain all records required to be maintained by paragraph (n) of this section.

(ii) The employer shall also comply with any additional requirements involving the transfer of records set forth in 29 CFR 1910.1020(h).

(o) **Observation of monitoring** -

(1) **Employee observation.** The employer shall provide affected employees or their designated representatives an opportunity to observe any monitoring of employee exposure to lead conducted pursuant to paragraph (d) of this section.

(2) **Observation procedures.**

(i) Whenever observation of the monitoring of employee exposure to lead requires entry into an area where the use of respirators, protective clothing or equipment is required, the employer shall provide the observer with and assure the use of such respirators, clothing and equipment, and shall require the observer to comply with all other applicable safety and health procedures.

(ii) Without interfering with the monitoring, observers shall be entitled

(A) to: Receive an explanation of the measurement procedures;

(B) Observe all steps related to the monitoring of lead performed at the place of exposure; and

(C) Record the results obtained or receive copies of the results when returned by the laboratory.

(p) **Appendices.** The information contained in the appendices to this section is not intended by itself, to create any additional obligations not otherwise imposed by this standard nor detract from any existing obligation.

Appendix A to § 1926.62 - Substance Data Sheet for Occupational Exposure to Lead

I. Substance Identification

A. *Substance:* Pure lead (Pb) is a heavy metal at room temperature and pressure and is a basic chemical element. It can combine with various other substances to form numerous lead compounds.

B. *Compounds covered by the standard:* The word *lead* when used in this interim final standard means elemental lead, all inorganic lead compounds and a class of organic lead compounds called lead soaps. This standard does not apply to other organic lead compounds.

C. *Uses:* Exposure to lead occurs in several different occupations in the construction industry, including demolition or salvage of structures where lead or lead-containing materials are present; removal or encapsulation of lead-containing materials, new construction, alteration, repair, or renovation of structures that contain lead or materials containing lead; installation of products containing lead. In addition, there are construction related activities where exposure to lead may occur, including transportation, disposal, storage, or containment of lead or materials containing lead on construction sites, and maintenance operations associated with construction activities.

D. *Permissible exposure:* The permissible exposure limit (PEL) set by the standard is 50 micrograms of lead percubic meter of air (50 µg/m^3), averaged over an 8-hour workday.

E. *Action level:* The interim final standard establishes an action level of 30 micrograms of lead per cubic meter of air (30 µg/m^3), averaged over an 8-hour workday. The action level triggers several ancillary provisions of thestandard such as exposure monitoring, medical surveillance, and training.

II. Health Hazard Data

A. *Ways in which lead enters your body.* When absorbed into your body in certain doses, lead is a toxic substance. The object of the lead standard is to prevent absorption of harmful quantities of lead. The standard is intended to protect you not only from the immediate toxic effects of lead, but also from the serious toxic effects that may not become apparent until years of exposure have passed. Lead can be absorbed into your body by inhalation (breathing) and ingestion (eating). Lead (except for certain organic lead compounds not covered by the standard, such as tetraethyl lead) is not absorbed through your skin. When lead is scattered in the air as a dust, fume respiratory tract. Inhalation of airborne lead is generally the most important source of occupational lead absorption. You can also absorb lead through your digestive system if lead gets into your mouth and is swallowed. If you handle food, cigarettes, chewing tobacco, or make-up which have lead on them or handle them with hands contaminated with lead, this will contribute to ingestion. A significant portion of the

lead that you inhale or ingest gets into your blood stream. Once in your blood stream, lead is circulated throughout your body and stored in various organs and body tissues. Some of this lead is quickly filtered out of your body and excreted, but some remains in the blood and other tissues. As exposure to lead continues, the amount stored in your body will increase if you are absorbing more lead than your body is excreting. Even though you may not be aware of any immediate symptoms of disease, this lead stored in your tissues can be slowly causing irreversible damage, first to individual cells, then to your organs and whole body systems.

B. *Effects of overexposure to lead* - (1) *Short term (acute) overexposure.* Lead is a potent, systemic poison thatserves no known useful function once absorbed by your body. Taken in large enough doses, lead can kill you in a matter of days. A condition affecting the brain called acute encephalopathy may arise which develops quickly to seizures, coma, and death from cardiorespiratory arrest. A short term dose of lead can lead to acute encephalopathy. Short term occupational exposures of this magnitude are highly unusual, but not impossible.Similar forms of encephalopathy may, however, arise from extended, chronic exposure to lower doses of lead.There is no sharp dividing line between rapidly developing acute effects of lead, and chronic effects which take longer to acquire. Lead adversely affects numerous body systems, and causes forms of health impairment and disease which arise after periods of exposure as short as days or as long as several years.

(2) *Long-term (chronic) overexposure.* Chronic overexposure to lead may result in severe damage to your blood- forming, nervous, urinary and reproductive systems. Some common symptoms of chronic overexposure include loss of appetite, metallic taste in the mouth, anxiety, constipation, nausea, pallor, excessive tiredness, weakness, insomnia, headache, nervous irritability, muscle and joint pain or soreness, fine tremors, numbness,dizziness, hyperactivity and colic. In lead colic there may be severe abdominal pain. Damage to the central nervous system in general and the brain (encephalopathy) in

particular is one of the most severe forms of lead poisoning. The most severe, often fatal, form of encephalopathy may be preceded by vomiting, a feeling of dullness progressing to drowsiness and stupor, poor memory, restlessness, irritability, tremor, and convulsions. It may arise suddenly with the onset of seizures, followed by coma, and death. There is a tendency for muscular weakness to develop at the same time. This weakness may progress to paralysis often observed as acharacteristic "wrist drop" or "foot drop" and is a manifestation of a disease to the nervous system called peripheral neuropathy. Chronic overexposure to lead also results in kidney disease with few, if any, symptomsappearing until extensive and most likely permanent kidney damage has occurred. Routine laboratory tests reveal the presence of this kidney disease only after about two-thirds of kidney function is lost. When overt symptoms of urinary dysfunction arise, it is often too late to correct or prevent worsening conditions, and progression to kidney dialysis or death is possible. Chronic overexposure to lead impairs the reproductive systems of both men and women. Overexposure to lead may result in decreased sex drive, impotence and sterility in men. Lead can alter the structure of sperm cells raising the risk of birth defects. There is evidence of miscarriage and stillbirth in women whose husbands were exposed to lead or who were exposed to lead themselves. Lead exposure also may result in decreased fertility, and abnormal menstrual cycles in women. The course of pregnancy may be adversely affected by exposure to lead since lead crosses the placental barrier and poses risks to developing fetuses. Children born of parents either one of whom were exposed to excess lead levels are more likely to have birth defects, mental retardation, behavioral disorders or die during the first year of childhood. Overexposure to lead also disrupts the blood-forming system resulting in decreased hemoglobin (the substance in the blood that carries oxygen to the cells) and ultimately anemia. Anemia is characterized by weakness, pallor and fatigability as a result of decreased oxygen carrying capacity in the blood.

(3) *Health protection goals of the standard.* Prevention of adverse health effects for most workers from exposure to lead throughout a working lifetime requires that a worker's blood lead level (BLL, also expressed as PbB) be maintained at or below forty micrograms per deciliter of whole blood (40 μg/dl). The blood lead levels

of workers (both male and female workers) who intend to have children should be maintained below 30 μg/dl to minimize adverse reproductive health effects to the parents and to the developing fetus. The measurementof your blood lead level (BLL) is the most useful indicator of the amount of lead being absorbed by your body.Blood lead levels are most often reported in units of milligrams (mg) or micrograms (μg) of lead (1 mg = 1000μg) per 100 grams (100g), 100 milliliters (100 ml) or deciliter (dl) of blood. These three units are essentially the same. Sometime BLLs are expressed in the form of mg% or μg%. This is a shorthand notation for 100g, 100 ml, or dl. (References to BLL measurements in this standard are expressed in the form of μg/dl.)

BLL measurements show the amount of lead circulating in your blood stream, but do not give any information about the amount of lead stored in your various tissues. BLL measurements merely show current absorption oflead, not the effect that lead is having on your body or the effects that past lead exposure may have already caused. Past research into lead-related diseases, however, has focused heavily on associations between BLLs and various diseases. As a result, your BLL is an important indicator of the likelihood that you will gradually acquire a lead-related health impairment or disease.

Once your blood lead level climbs above 40 μg/dl, your risk of disease increases. There is a wide variability of individual response to lead, thus it is difficult to say that a particular BLL in a given person will cause a particular effect. Studies have associated fatal encephalopathy with BLLs as low as 150 μg/dl. Other studies have shown other forms of diseases in some workers with BLLs well below 80 μg/dl. Your BLL is a crucial indicator of the risks to your health, but one other factor is also extremely important. This factor is the length of time you have had elevated BLLs. The longer you have an elevated BLL, the greater the risk that large quantities of lead are being gradually stored in your organs and tissues (body burden). The greater your overall body burden, the greater the chances of substantial permanent damage. The best way to prevent all forms of lead- related impairments and diseases - both short term and long term - is to maintain your BLL below 40 µg/dl. Theprovisions of the standard are designed with this end in mind.

Your employer has prime responsibility to assure that the provisions of the standard are complied with both by the company and by individual workers. You, as a worker, however, also have a responsibility to assist your employer in complying with the standard. You can play a key role in protecting your own health by learning about the lead hazards and their control, learning what the standard requires, following the standard where it governs your own actions, and seeing that your employer complies with provisions governing his or her actions.

(4) *Reporting signs and symptoms of health problems.* You should immediately notify your employer if you develop signs or symptoms associated with lead poisoning or if you desire medical advice concerning the effects of current or past exposure to lead or your ability to have a healthy child. You should also notify your employer if you have difficulty breathing during a respirator fit test or while wearing a respirator. In each of these cases, your employer must make available to you appropriate medical examinations or consultations. These must be provided at no cost to you and at a reasonable time and place. The standard contains a procedure whereby you can obtain a second opinion by a physician of your choice if your employer selected the initial physician.

Appendix B to § 1926.62 - Employee Standard Summary

This appendix summarizes key provisions of the interim final standard for lead in construction that you as a worker should become familiar with.

I. Permissible Exposure Limit (PEL) - Paragraph (C)

The standard sets a permissible exposure limit (PEL) of 50 micrograms of lead per cubic meter of air (50 µg/m^3), averaged over an 8-hour workday which is referred to as a time-weighted average (TWA). This is thehighest level of lead in air to which you may be permissibly exposed over an 8-hour workday. However, since this is an 8-hour average, short exposures above the PEL are permitted so long as for each 8-hour work day your average exposure does not exceed this level. This interim final standard, however, takes into account the fact that your daily exposure to lead can extend beyond a typical 8-hour workday as the result of overtime or other alterations in your work schedule. To deal with this situation, the standard contains a formula which reduces your permissible exposure when you are exposed more than 8 hours. For example, if you are exposed to lead for 10 hours a day, the maximum permitted average exposure would be 40 µg/m^3.

II. Exposure Assessment - Paragraph (D)

If lead is present in your workplace in any quantity, your employer is required to make an initial determination ofwhether any employee's exposure to lead exceeds the action level (30 µg/m^3 averaged over an 8-hour day).
Employee exposure is that exposure which would occur if the employee were not using a respirator. This initial determination requires your employer to monitor workers' exposures unless he or she has objective data which can demonstrate conclusively that no employee will be exposed to lead in excess of the

action level. Where objective data is used in lieu of actual monitoring the employer must establish and maintain an accurate record, documenting its relevancy in assessing exposure levels for current job conditions. If such objective data is available, the employer need proceed no further on employee exposure assessment until such time thatconditions have changed and the determination is no longer valid.

Objective data may be compiled from various sources, e.g., insurance companies and trade associations and information from suppliers or exposure data collected from similar operations. Objective data may also comprise previously-collected sampling data including area monitoring. If it cannot be determined through using objective data that worker exposure is less than the action level, your employer must conduct monitoring or must rely on relevant previous personal sampling, if available. Where monitoring is required for the initial determination, it may be limited to a representative number of employees who are reasonably expected to havethe highest exposure levels. If your employer has conducted appropriate air sampling for lead in the past 12 months, he or she may use these results, provided they are applicable to the same employee tasks and exposure conditions and meet the requirements for accuracy as specified in the standard. As with objective data, if such results are relied upon for the initial determination, your employer must establish and maintain arecord as to the relevancy of such data to current job conditions.

If there have been any employee complaints of symptoms which may be attributable to exposure to lead or ifthere is any other information or observations which would indicate employee exposure to lead, this must also be considered as part of the initial determination.

If this initial determination shows that a reasonable possibility exists that any employee may be exposed, without regard to respirators, over the action level, your employer must set up an air monitoring program to determine the exposure level representative of each employee exposed to lead at your workplace. In carrying out this air monitoring program, your employer is not required to monitor the exposure of every employee, but he or she must monitor a representative number of employees and job types. Enough sampling must be doneto enable each employee's exposure level to be reasonably represent full shift exposure. In addition, these air

samples must be taken under conditions which represent each employee's regular, daily exposure to lead. Sampling performed in the past 12 months may be used to determine exposures above the action level if suchsampling was conducted during work activities essentially similar to present work conditions.

The standard lists certain tasks which may likely result in exposures to lead in excess of the PEL and, in some cases, exposures in excess of 50 times the PEL. If you are performing any of these tasks, your employer must provide you with appropriate respiratory protection, protective clothing and equipment, change areas, hand washing facilities, biological monitoring, and training until such time that an exposure assessment is conducted which demonstrates that your exposure level is below the PEL.

If you are exposed to lead and air sampling is performed, your employer is required to notify you in writing within 5 working days of the air monitoring results which represent your exposure. If the results indicate that your exposure exceeds the PEL (without regard to your use of a respirator), then your employer must also notify you of this in writing, and provide you with a description of the corrective action that has been taken or will betaken to reduce your exposure.

Your exposure must be rechecked by monitoring, at least every six months if your exposure is at or over the action level but below the PEL. Your employer may discontinue monitoring for you if 2 consecutive measurements, taken at least 7 days apart, are at or below the action level. Air monitoring must be repeated every 3 months if you are exposed over the PEL. Your employer must continue monitoring for you at this frequency until 2 consecutive measurements, taken at least 7 days apart, are below the PEL but above the action level, at which time your employer must repeat monitoring of your exposure every six months and may discontinue monitoring only after your exposure drops to or below the action level. However, whenever there is a change of equipment, process, control, or personnel or a new type of job is added at your workplace which may result in new or additional exposure to lead, your employer must perform additional monitoring.

III. Methods of Compliance - Paragraph (E)

Your employer is required to assure that no employee is exposed to lead in excess of the PEL as an 8-hour TWA. The interim final standard for lead in construction requires employers to institute engineering and work practice controls including administrative controls to the extent feasible to reduce employee exposure to lead. Where such controls are feasible but not adequate to reduce exposures below the PEL they must be used nonetheless to reduce exposures to the lowest level that can be accomplished by these means and then supplemented with appropriate respiratory protection.

Your employer is required to develop and implement a written compliance program prior to the commencement of any job where employee exposures may reach the PEL as an 8-hour TWA. The interim final standard identifies the various elements that must be included in the plan. For example, employers are required to include a description of operations in which lead is emitted, detailing other relevant information about the operation such as the type of equipment used, the type of material involved, employee job responsibilities, operating procedures and maintenance practices. In addition, your employer's compliance plan must specify the means that will be used to achieve compliance and, where engineering controls are required, include any engineering plans or studies that have been used to select the control methods. If administrative controls involving job rotation are used to reduce employee exposure to lead, the job rotation schedule must be included in the compliance plan. The plan must also detail the type of protective clothing and equipment, including respirators, housekeeping and hygiene practices that will be used to protect you from the adverse effects of exposure to lead.

The written compliance program must be made available, upon request, to affected employees and their designated representatives, the Assistant Secretary and the Director.

Finally, the plan must be reviewed and updated at least every 6 months to assure it reflects the current status in exposure control.

IV. Respiratory Protection - Paragraph (F)

Your employer is required to provide and assure your use of respirators when your exposure to lead is not controlled below the PEL by other means. The employer must pay the cost of the respirator. Whenever you request one, your employer is also required to provide you a respirator even if your air exposure level is not above the PEL. You might desire a respirator when, for example, you have received medical advice that your lead absorption should be decreased. Or, you may intend to have children in the near future, and want to reduce the level of lead in your body to minimize adverse reproductive effects. While respirators are the least satisfactory means of controlling your exposure, they are capable of providing significant protection if properly chosen, fitted, worn, cleaned, maintained, and replaced when they stop providing adequate protection.

Your employer is required to select your respirator according to the requirements of 29 CFR 1926.62(f)(3), including the requirements referenced in 29 CFR 1910.134(d)(3)(i)(A) of this chapter. Any respirator chosen must be approved by NIOSH under the provisions of 42 CFR part 84. These respirator selection references will enable your employer to choose a type of respirator that will give you a proper amount of protection based on your airborne lead exposure. Your employer may select a type of respirator that provides greater protection than that required by the standard; that is, one recommended for a higher concentration of lead than is presentin your workplace. For example, a powered air-purifying respirator (PAPR) is much more protective than a typical negative pressure respirator, and may also be more comfortable to wear. A PAPR has a filter, cartridge, or canister to clean the air, and a power source that continuously blows filtered air into your breathing zone.
Your employer might make a PAPR available to you to ease the burden of having to wear a respirator for longperiods of time. The standard provides that you can obtain a PAPR upon request.

Your employer must also start a Respiratory Protection Program. This program must include written procedures for the proper selection, use, cleaning, storage, and maintenance of respirators.

Your employer must ensure that your respirator facepiece fits properly. Proper fit of a respirator facepiece is critical to your protection from airborne lead. Obtaining a proper fit on each employee may require your employer to make available several different types of respirator masks. To ensure that your respirator fits properly and that facepiece leakage is minimal, your employer must give you either a qualitative or quantitative fit test as specified in appendix A of the Respiratory Protection standard located at 29 CFR 1910.134.

You must also receive from your employer proper training in the use of respirators. Your employer is required toteach you how to wear a respirator, to know why it is needed, and to understand its limitations.

The standard provides that if your respirator uses filter elements, you must be given an opportunity to change the filter elements whenever an increase in breathing resistance is detected. You also must be permitted to periodically leave your work area to wash your face and respirator facepiece whenever necessary to prevent skin irritation. If you ever have difficulty in breathing during a fit test or while using a respirator, your employer must make a medical examination available to you to determine whether you can safely wear a respirator. Theresult of this examination may be to give you a positive pressure respirator (which reduces breathing resistance) or to provide alternative means of protection.

V. Protective Work Clothing and Equipment - Paragraph (G)

If you are exposed to lead above the PEL as an 8-hour TWA, without regard to your use of a respirator, or if youare exposed to lead compounds such as lead arsenate or lead azide which can cause skin and eye irritation, your employer must provide you with protective work clothing and equipment appropriate for the hazard. If work clothing is provided, it must be provided in a clean and dry condition at least weekly, and daily if your airborne exposure to lead is greater than 200 µg/m^3. Appropriate protective work clothing and equipment can include coveralls or similar full-body work clothing, gloves, hats, shoes or disposable shoe coverlets, and face shields or vented goggles. Your employer is required to provide all such equipment at no cost to you. In addition, your employer is responsible for providing repairs and replacement as necessary, and also is responsible for the cleaning, laundering or disposal of protective clothing and equipment.

The interim final standard requires that your employer assure that you follow good work practices when you are working in areas where your exposure to lead may exceed the PEL. With respect to protective clothing and equipment, where appropriate, the following procedures should be observed prior to beginning work:

1. Change into work clothing and shoe covers in the clean section of the designated changing areas;

2. Use work garments of appropriate protective gear, including respirators before entering the work area; and

3. Store any clothing not worn under protective clothing in the designated changing area.

Workers should follow these procedures upon leaving the work area:

1. HEPA vacuum heavily contaminated protective work clothing while it is still being worn. At no time may lead be removed from protective clothing by any means which result in uncontrolled dispersal of lead into the air;

2. Remove shoe covers and leave them in the work area;

3. Remove protective clothing and gear in the dirty area of the designated changing area. Remove protective coveralls by carefully rolling down the garment to reduce exposure to dust.

4. Remove respirators last; and

5. Wash hands and face.

Workers should follow these procedures upon finishing work for the day (in addition to procedures described above):

1. Where applicable, place disposal coveralls and shoe covers with the abatement waste;

2. Contaminated clothing which is to be cleaned, laundered or disposed of must be placed in closed containers in the change room.

3. Clean protective gear, including respirators, according to standard procedures;

4. Wash hands and face again. If showers are available, take a shower and wash hair. If shower facilities are not available at the work site, shower immediately at home and wash hair.

VI. Housekeeping - Paragraph (H)

Your employer must establish a housekeeping program sufficient to maintain all surfaces as free as practicable of accumulations of lead dust. Vacuuming is the preferred method of meeting this requirement, and the use of compressed air to clean floors and other surfaces is generally prohibited unless removal with compressed air is done in conjunction with ventilation systems designed to contain dispersal of the lead dust. Dry or wet sweeping, shoveling, or brushing may not be used except where vacuuming or other equally effective methods have been tried and do not work. Vacuums must be used equipped with a special filter called a high-efficiency particulate air (HEPA) filter and emptied in a manner which minimizes the reentry of lead into the workplace.

VII. Hygiene Facilities and Practices - Paragraph (I)

The standard requires that hand washing facilities be provided where occupational exposure to lead occurs. In addition, change areas, showers (where feasible), and lunchrooms or eating areas are to be made available to workers exposed to lead above the PEL. Your employer must assure that except in these facilities, food and beverage is not present or consumed, tobacco products are not present or used, and cosmetics are not applied, where airborne exposures are above the PEL. Change rooms provided by your employer must be equipped with separate storage facilities for your protective clothing and equipment and street clothes to avoid cross- contamination. After showering, no required protective clothing or equipment worn during the shift may be worn home. It is important that contaminated clothing or equipment be removed in change areas and not be worn home or you will extend your exposure and expose your family since lead from your clothing can accumulate in your house, car, etc.

Lunchrooms or eating areas may not be entered with protective clothing or equipment unless surface dust has been removed by vacuuming, downdraft booth, or other cleaning method. Finally, workers exposed above thePEL must wash both their hands and faces prior to eating, drinking, smoking or applying cosmetics.

All of the facilities and hygiene practices just discussed are essential to minimize additional sources of lead absorption from inhalation or ingestion of lead that may accumulate on you, your clothes, or your possessions.Strict compliance with these provisions can virtually eliminate several sources of lead exposure which significantly contribute to excessive lead absorption.

VIII. Medical Surveillance - Paragraph (J)

The medical surveillance program is part of the standard's comprehensive approach to the prevention of lead- related disease. Its purpose is to supplement the main thrust of the standard which is aimed at minimizing airborne concentrations of lead and sources of ingestion. Only medical surveillance can determine if the other provisions of the standard have affectively protected you as an individual. Compliance with the standard's provision will protect most workers from the adverse effects of lead exposure, but may not be satisfactory to protect individual workers (1) who have high body burdens of lead acquired over past years, (2) who have additional uncontrolled sources of non-occupational lead exposure, (3) who exhibit unusual variations in lead absorption rates, or (4) who have specific non-work related medical conditions which could be aggravated by lead exposure (e.g., renal disease, anemia). In addition, control systems may fail, or hygiene and respirator

programs may be inadequate. Periodic medical surveillance of individual workers will help detect those failures. Medical surveillance will also be important to protect your reproductive ability-regardless of whether you are a man or woman.

All medical surveillance required by the interim final standard must be performed by or under the supervision of a licensed physician. The employer must provide required medical surveillance without cost to employees and at a reasonable time and place. The standard's medical surveillance program has two parts - periodic biologicalmonitoring and medical examinations. Your employer's obligation to offer you medical surveillance is triggered by the results of the air monitoring program. Full medical surveillance must be made available to all employees who are or may be exposed to lead in excess of the action level for more than 30 days a year and whose bloodlead level exceeds 40 μg/dl. Initial medical surveillance consisting of blood sampling and analysis for lead and zinc protoporphyrin must be provided to all employees exposed at any time (1 day) above the action level.

Biological monitoring under the standard must be provided at least every 2 months for the first 6 months and every 6 months thereafter until your blood lead level is below 40 µg/dl. A zinc protoporphyrin (ZPP) test is avery useful blood test which measures an adverse metabolic effect of lead on your body and is therefore anindicator of lead toxicity.

If your BLL exceeds 40 µg/dl the monitoring frequency must be increased from every 6 months to at least every 2 months and not reduced until two consecutive BLLs indicate a blood lead level below 40 µg/dl. Each time your BLL is determined to be over 40 µg/dl, your employer must notify you of this in writing within five working days of his or her receipt of the test results. The employer must also inform you that the standard requires temporary medical removal with economic protection when your BLL exceeds 50 µg/dl. (See Discussion of Medical Removal Protection-Paragraph (k).) Anytime your BLL exceeds 50 µg/dl your employer must make available to you within two weeks of receipt of these test results a second follow-up BLL test to confirm your BLL. If the two tests both exceed 50 µg/dl, and you are temporarily removed, then your employer must make successive BLL tests available to you on a monthly basis during the period of your removal.

Medical examinations beyond the initial one must be made available on an annual basis if your blood lead level exceeds 40 µg/dl at any time during the preceding year and you are being exposed above the airborne action level of 30 µg/m^3 for 30 or more days per year. The initial examination will provide information to establish a baseline to which subsequent data can be compared.

An initial medical examination to consist of blood sampling and analysis for lead and zinc protoporphyrin must also be made available (prior to assignment) for each employee being assigned for the first time to an area where the airborne concentration of lead equals or exceeds the action level at any time. In addition, a medicalexamination or consultation must be made available as soon as possible if you notify your employer that youare experiencing signs or symptoms commonly associated with lead poisoning or that you have difficulty breathing while wearing a respirator or during a respirator fit test. You must also be provided a medical examination or consultation if you notify your employer that you desire medical advice concerning the effectsof current or past exposure to lead on your ability to procreate a healthy child.

Finally, appropriate follow-up medical examinations or consultations may also be provided for employees who have been temporarily removed from exposure under the medical removal protection provisions of the standard. (See Part IX, below.)

The standard specifies the minimum content of pre-assignment and annual medical examinations. The content of other types of medical examinations and consultations is left up to the sound discretion of the examining physician. Pre-assignment and annual medical examinations must include (1) a detailed work history and medical history; (2) a thorough physical examination, including an evaluation of your pulmonary status if you will be required to use a respirator; (3) a blood pressure measurement; and (4) a series of laboratory tests designed to check your blood chemistry and your kidney function. In addition, at any time upon your request, a laboratory evaluation of male fertility will be made (microscopic examination of a sperm sample), or a pregnancy test will be given.

The standard does not require that you participate in any of the medical procedures, tests, etc. which your employer is required to make available to you. Medical surveillance can, however, play a very important role in protecting your health. You are strongly encouraged, therefore, to participate in a meaningful fashion. The standard contains a multiple physician review mechanism which will give you a chance to have a physician of your choice directly participate in the medical surveillance program. If you are dissatisfied with an examination by a physician chosen by your employer, you can select a second physician to conduct an independent analysis. The two doctors would attempt to resolve any differences of opinion, and select a third physician toresolve any firm dispute. Generally your employer will choose the physician who conducts medical surveillance under the lead standard-unless you and your employer can agree on the choice of a physician or physicians.

Some companies and unions have agreed in advance, for example, to use certain independent medical laboratories or panels of physicians. Any of these arrangements are acceptable so long as required medical surveillance is made available to workers.

The standard requires your employer to provide certain information to a physician to aid in his or her examination of you. This information includes (1) the standard and its appendices, (2) a description of your duties as they relate to occupational lead exposure, (3) your exposure level or anticipated exposure level, (4) a description of any personal protective equipment you wear, (5) prior blood lead level results, and (6) prior written medical opinions concerning you that the employer has. After a medical examination or consultation the physician must prepare a written report which must contain (1) the physician's opinion as to whether youhave any medical condition which places you at increased risk of material impairment to health from exposure to lead, (2) any recommended special protective measures to be provided to you, (3) any blood lead level determinations, and (4) any recommended limitation on your use of respirators. This last element must includea determination of whether you can wear a powered air purifying respirator (PAPR) if you are found unable to wear a negative pressure respirator.

The medical surveillance program of the interim lead standard may at some point in time serve to notify certain workers that they have acquired a disease or other adverse medical condition as a result of occupational leadexposure. If this is true, these workers might have legal rights to compensation from public agencies, their employers, firms that supply hazardous products to their employers, or other persons. Some states have laws, including worker compensation laws, that disallow a worker who learns of a job-related health impairment to sue, unless the worker sues within a short period of time after learning of the impairment. (This period of timemay be a matter of months or years.) An attorney can be consulted about these possibilities. It should be stressed that OSHA is in no way trying to either encourage or discourage claims or lawsuits. However, since results of the standard's medical surveillance program can significantly affect the legal remedies of a worker who has acquired a job-related disease or impairment, it is proper for OSHA to make you aware of this.

The medical surveillance section of the standard also contains provisions dealing with chelation. Chelation is the use of certain drugs (administered in pill form or injected into the body) to reduce the amount of lead absorbed in body tissues. Experience accumulated by the medical and scientific communities has largely

confirmed the effectiveness of this type of therapy for the treatment of very severe lead poisoning. On the other hand, it has also been established that there can be a long list of extremely harmful side effects associated with the use of chelating agents. The medical community has balanced the advantages and disadvantages resulting from the use of chelating agents in various circumstances and has established when the use of these agents is acceptable. The standard includes these accepted limitations due to a history of abuse of chelation therapy by some lead companies. The most widely used chelating agents are calcium disodium EDTA, (Ca Na2 EDTA), Calcium Disodium Versenate (Versenate), and d-penicillamine (pencillamine or Cupramine).

The standard prohibits "prophylactic chelation" of any employee by any person the employer retains, supervises or controls. *Prophylactic chelation* is the routine use of chelating or similarly acting drugs to prevent elevated blood levels in workers who are occupationally exposed to lead, or the use of these drugs to routinely lower blood lead levels to predesignated concentrations believed to be "safe". It should be emphasized that where an employer takes a worker who has no symptoms of lead poisoning and has chelation carried out by a physician (either inside or outside of a hospital) solely to reduce the worker's blood lead level, that will generally be considered prophylactic chelation. The use of a hospital and a physician does not mean that prophylactic chelation is not being performed. Routine chelation to prevent increased or reduce current blood lead levels isunacceptable whatever the setting.

The standard allows the use of "therapeutic" or "diagnostic" chelation if administered under the supervision of a licensed physician in a clinical setting with thorough and appropriate medical monitoring. Therapeutic chelation responds to severe lead poisoning where there are marked symptoms. Diagnostic chelation involved giving a patient a dose of the drug then collecting all urine excreted for some period of time as an aid to the diagnosis of lead poisoning.

In cases where the examining physician determines that chelation is appropriate, you must be notified in writing of this fact before such treatment. This will inform you of a potentially harmful treatment, and allow you to obtain a second opinion.

IX. Medical Removal Protection - Paragraph (K)

Excessive lead absorption subjects you to increased risk of disease. Medical removal protection (MRP) is a means of protecting you when, for whatever reasons, other methods, such as engineering controls, work practices, and respirators, have failed to provide the protection you need. MRP involves the temporary removalof a worker from his or her regular job to a place of significantly lower exposure without any loss of earnings,seniority, or other employment rights or benefits. The purpose of this program is to cease further lead absorption and allow your body to naturally excrete lead which has previously been absorbed. Temporary medical removal can result from an elevated blood lead level, or a medical opinion. For up to 18 months, or for as long as the job the employee was removed from lasts, protection is provided as a result of either form of removal. The vast majority of removed workers, however, will return to their former jobs long before this eighteen month period expires.

You may also be removed from exposure even if your blood lead level is below 50 μg/dl if a final medical determination indicates that you temporarily need reduced lead exposure for medical reasons. If the physician who is implementing your employers medical program makes a final written opinion recommending your removal or other special protective measures, your employer must implement the physician's recommendation.If you are removed in this manner, you may only be returned when the doctor indicates that it is safe for you to do so.

The standard does not give specific instructions dealing with what an employer must do with a removed worker. Your job assignment upon removal is a matter for you, your employer and your union (if any) to work out consistent with existing procedures for job assignments. Each removal must be accomplished in a mannerconsistent with existing collective bargaining relationships. Your employer is given broad discretion to implement temporary removals so long as no attempt is made to override existing agreements. Similarly, a removed worker is provided no right to veto an employer's choice which satisfies the standard.

In most cases, employers will likely transfer removed employees to other jobs with sufficiently low lead exposure. Alternatively, a worker's hours may be reduced so that the time weighted average exposure isreduced, or he or she may be temporarily laid off if no other alternative is feasible.

In all of these situation, MRP benefits must be provided during the period of removal - i.e., you continue to receive the same earnings, seniority, and other rights and benefits you would have had if you had not been removed. Earnings includes more than just your base wage; it includes overtime, shift differentials, incentives,and other compensation you would have earned if you had not been removed. During the period of removal you must also be provided with appropriate follow-up medical surveillance. If you were removed because your blood lead level was too high, you must be provided with a monthly blood test. If a medical opinion caused your removal, you must be provided medical tests or examinations that the doctor believes to be appropriate. If you do not participate in this follow up medical surveillance, you may lose your eligibility for MRP benefits.

When you are medically eligible to return to your former job, your employer must return you to your "former job status." This means that you are entitled to the position, wages, benefits, etc., you would have had if you hadnot been removed. If you would still be in your old job if no removal had occurred that is where you go back. If not, you are returned consistent with whatever job assignment discretion your employer would have had if no removal had occurred. MRP only seeks to maintain your rights, not expand them or diminish them.

If you are removed under MRP and you are also eligible for worker compensation or other compensation for lost wages, your employer's MRP benefits obligation is reduced by the amount that you actually receive fromthese other sources. This is also true if you obtain other employment during the time you are laid off with MRP benefits.

The standard also covers situations where an employer voluntarily removes a worker from exposure to lead due to the effects of lead on the employee's medical condition, even though the standard does not require removal. In these situations MRP benefits must still be provided as though the standard required removal. Finally, it is important to note that in all cases where removal is required, respirators cannot be used as a substitute. Respirators may be used before removal becomes necessary, but not as an alternative to a transferto a low exposure job, or to a lay-off with MRP benefits.

X. Employee Information and Training - Paragraph (L)

Your employer is required to provide an information and training program for all employees exposed to lead above the action level or who may suffer skin or eye irritation from lead compounds such as lead arsenate or lead azide. The program must train these employees regarding the specific hazards associated with their work environment, protective measures which can be taken, including the contents of any compliance plan in effect, the danger of lead to their bodies (including their reproductive systems), and their rights under the standard. All employees must be trained prior to initial assignment to areas where there is a possibility of exposure over the action level.

This training program must also be provided at least annually thereafter unless further exposure above theaction level will not occur.

XI. Signs - Paragraph (M)

The standard requires that the following warning sign be posted in work areas when the exposure to lead is above the PEL:

DANGER

LEAD WORK AREA

MAY DAMAGE FERTILITY OR THE UNBORN CHILD

CAUSES DAMAGE TO THE CENTRAL NERVOUS

SYSTEMDO NOT EAT, DRINK OR SMOKE IN THIS

AREA

Prior to June 1, 2016, employers may use the following legend in lieu of that specified

above:WARNING

LEAD

WORK

AREA

POISON

NO SMOKING OR EATING

XII. Recordkeeping - Paragraph (N)

Your employer is required to keep all records of exposure monitoring for airborne lead. These records mustinclude the name and job classification of employees measured, details of the sampling and analytical techniques, the results of this sampling, and the type of respiratory protection being worn by the person sampled. Such records are to be retained for at least 30 years. Your employer is also required to keep all records of biological monitoring and medical examination results. These records must include the names of the employees, the physician's written opinion, and a copy of the results of the examination. Medical recordsmust be preserved and maintained for the duration of employment plus 30 years. However, if the employee'sduration of employment is less than one year, the employer need not retain that employee's medical records beyond the period of employment if they are provided to the employee upon termination of employment.

Recordkeeping is also required if you are temporarily removed from your job under the medical removal protection program. This record must include your name, the date of your removal and return, how the removalwas or is being accomplished, and whether or not the reason for the removal was an elevated blood lead level. Your employer is required to keep each medical removal record only for as long as the duration of an employee's employment.

The standard requires that if you request to see or copy environmental monitoring, blood lead level monitoring, or medical removal records, they must be made available to you or to a representative that you authorize. Your union also has access to these records. Medical records other than BLL's must also be provided upon request to you, to your physician or to any other person whom you may specifically designate. Your union does not have access to your personal medical records unless you authorize their access.

XIII. Observation of Monitoring - Paragraph (O)

When air monitoring for lead is performed at your workplace as required by this standard, your employer must allow you or someone you designate to act as an observer of the monitoring. Observers are entitled to an explanation of the measurement procedure, and to record the results obtained. Since results will not normally be available at the time of the monitoring, observers are entitled to record or receive the results of the monitoring when returned by the laboratory. Your employer is required to provide the observer with any personal protective devices required to be worn by employees working in the area that is being monitored. The employer must require the observer to wear all such equipment and to comply with all other applicable safety and health procedures.

XIV. For Additional Information

A. A copy of the interim standard for lead in construction can be obtained free of charge by calling or writing the OSHA Office of Publications, room N-3101, United States Department of Labor, Washington, DC 20210:Telephone (202) 219-4667.

B. Additional information about the standard, its enforcement, and your employer's compliance can be obtained from the nearest OSHA Area Office listed in your telephone directory under United States Government/ Department of Labor.

Appendix C to § 1926.62 - Medical Surveillance Guidelines

Introduction

The primary purpose of the Occupational Safety and Health Act of 1970 is to assure, so far as possible, safe and healthful working conditions for every working man and woman. The interim final occupational health standard for lead in construction is designed to protect workers exposed to inorganic lead including metallic lead, all inorganic lead compounds and organic lead soaps.

Under this interim final standard occupational exposure to inorganic lead is to be limited to 50 $\mu g/m^3$ (micrograms per cubic meter) based on an 8 hour time-weighted average (TWA). This permissible exposure limit (PEL) must be achieved through a combination of engineering, work practice and administrative controls to the extent feasible. Where these controls are in place but are found not to reduce employee exposures to or below the PEL, they must be used nonetheless, and supplemented with respirators to meet the 50 $\mu g/m^3$ exposure limit.

The standard also provides for a program of biological monitoring for employees exposed to lead above the action level at any time, and additional medical surveillance for all employees exposed to levels of inorganiclead above 30 $\mu g/m^3$ (TWA) for more than 30 days per year and whose BLL exceeds 40 $\mu g/dl$.

The purpose of this document is to outline the medical surveillance provisions of the interim standard for inorganic lead in construction, and to provide further information to the physician regarding the examination and evaluation of workers exposed to inorganic lead.

Section 1 provides a detailed description of the monitoring procedure including the required frequency of bloodtesting for exposed workers, provisions for medical removal protection (MRP), the recommended right of the employee to a second medical opinion, and notification and recordkeeping requirements of the employer. A discussion of the requirements for respirator use and respirator monitoring and OSHA's position on prophylactic chelation therapy are also included in this section.

Section 2 discusses the toxic effects and clinical manifestations of lead poisoning and effects of lead intoxication on enzymatic pathways in heme synthesis. The adverse effects on both male and female reproductive capacity and on the fetus are also discussed.

Section 3 outlines the recommended medical evaluation of the worker exposed to inorganic lead, includingdetails of the medical history, physical examination, and recommended laboratory tests, which are based on the toxic effects of lead as discussed in Section 2.

Section 4 provides detailed information concerning the laboratory tests available for the monitoring of exposed workers. Included also is a discussion of the relative value of each test and the limitations and precautions which are necessary in the interpretation of the laboratory results.

I. Medical Surveillance and Monitoring Requirements for Workers Exposed toInorganic Lead

Under the interim final standard for inorganic lead in the construction industry, initial medical surveillance consisting of biological monitoring to include blood lead and ZPP level determination shall be provided to employees exposed to lead at or above the action level on any one day. In addition, a program of biological monitoring is to be made available to all employees exposed above the action level at any time and additionalmedical surveillance is to be made available to all employees exposed to lead above 30 µg/m^3 TWA for more than 30 days each year and whose BLL exceeds 40 µg/dl. This program consists of periodic blood samplingand medical evaluation to be performed on a schedule which is defined by previous laboratory results, worker complaints or concerns, and the clinical assessment of the examining physician.

Under this program, the blood lead level (BLL) of all employees who are exposed to lead above 30 µg/m^3 formore than 30 days per year or whose blood lead is above 40 µg/dl but exposed for no more than 30 days per year is to be determined at least every two months for the first six months of exposure and every six months thereafter. The frequency is increased to every two months for employees whose last blood lead level was 40 µg/dl or above. For employees who are removed from exposure to lead due to an elevated blood lead, a newblood lead level must be measured monthly. A zinc protoporphyrin (ZPP) measurement is strongly recommended on each occasion that a blood lead level measurement is made.

An annual medical examination and consultation performed under the guidelines discussed in Section 3 is tobe made available to each employee exposed above 30 µg/m^3 for more than 30 days per year for whom a blood test conducted at any time during the preceding 12 months indicated a blood lead level at or above 40 µg/dl. Also, an examination is to be given to all employees prior to their assignment to an area in which airborne lead concentrations reach or exceed the 30 µg/m^3 for more than 30 days per year. In addition, a medical examination must be provided as soon as possible after notification by an employee that the employeehas developed signs or symptoms commonly associated with lead intoxication, that the employee desires medical advice regarding lead exposure and the ability to procreate a healthy child, or that the employee has demonstrated difficulty in breathing during a respirator fitting test or during respirator use. An examination is also to be made available to each employee removed from exposure to lead due to a risk of sustaining material impairment to health, or otherwise limited or specially protected pursuant to medical recommendations.

Results of biological monitoring or the recommendations of an examining physician may necessitate removal of an employee from further lead exposure pursuant to the standard's medical removal protection (MRP) program. The object of the MRP program is to provide temporary medical removal to workers either with substantially elevated blood lead levels or otherwise at risk of sustaining material health impairment from continued substantial exposure to lead.

Under the standard's ultimate worker removal criteria, a worker is to be removed from any work having an eight hour TWA exposure to lead of 30 µg/m^3 when his or her blood lead level reaches 50 µg/dl and is confirmed bya second follow-up blood lead level performed within two weeks after the employer receives the results of thefirst blood sampling test. Return of the employee to his or her job status depends on a worker's blood lead level declining to 40 µg/dl.

As part of the interim standard, the employer is required to notify in writing each employee whose blood lead level exceeds 40 µg/dl. In addition each such employee is to be informed that the standard requires medicalremoval with MRP benefits, discussed below, when an employee's blood lead level exceeds the above definedlimit.

In addition to the above blood lead level criterion, temporary worker removal may also take place as a result of medical determinations and recommendations. Written medical opinions must be prepared after each examination pursuant to the standard. If the examining physician includes a medical finding, determination or opinion that the employee has a medical condition which places the employee at increased risk of material health impairment from exposure to lead, then the employee must be removed from exposure to lead at or above 30 µg/m^3. Alternatively, if the examining physician recommends special protective measures for an employee (e.g., use of a powered air purifying respirator) or recommends limitations on an employee's exposure to lead, then the employer must implement these recommendations.

Recommendations may be more stringent than the specific provisions of the standard. The examining physician, therefore, is given broad flexibility to tailor special protective procedures to the needs of individual employees. This flexibility extends to the evaluation and management of pregnant workers and male and female workers who are planning to raise children. Based on the history, physical examination, and laboratory studies, the physician might recommend special protective measures or medical removal for an employee who is pregnant or who is planning to conceive a child when, in the physician's judgment, continued exposure to lead at the current job would pose a significant risk. The return of the employee to his or her former job status, or the removal of special protections or limitations, depends upon the examining physician determining that the employee is no longer at increased risk of material impairment or that special measures are no longer needed.

During the period of any form of special protection or removal, the employer must maintain the worker's earnings, seniority, and other employment rights and benefits (as though the worker had not been removed) for a period of up to 18 months or for as long as the job the employee was removed from lasts if less than 18

months. This economic protection will maximize meaningful worker participation in the medical surveillance program, and is appropriate as part of the employer's overall obligation to provide a safe and healthful workplace. The provisions of MRP benefits during the employee's removal period may, however, be conditioned upon participation in medical surveillance.

The lead standard provides for a multiple physician review in cases where the employee wishes a second opinion concerning potential lead poisoning or toxicity. If an employee wishes a second opinion, he or she can make an appointment with a physician of his or her choice. This second physician will review the findings, recommendations or determinations of the first physician and conduct any examinations, consultations or tests deemed necessary in an attempt to make a final medical determination. If the first and second physiciansdo not agree in their assessment they must try to resolve their differences. If they cannot reach an agreementthen they must designate a third physician to resolve the dispute.

The employer must provide examining and consulting physicians with the following specific information: A copy of the lead regulations and all appendices, a description of the employee's duties as related to exposure,the exposure level or anticipated level to lead and any other toxic substances (if applicable), a description of personal protective equipment used, blood lead levels, and all prior written medical opinions regarding the employee in the employer's possession or control. The employer must also obtain from the physician and provide the employee with a written medical opinion containing blood lead levels, the physicians's opinion as towhether the employee is at risk of material impairment to health, any recommended protective measures for the employee if further exposure is permitted, as well as any recommended limitations upon an employee's use of respirators.

Employers must instruct each physician not to reveal to the employer in writing or in any other way his or her findings, laboratory results, or diagnoses which are felt to be unrelated to occupational lead exposure. Theymust also instruct each physician to advise the employee of any occupationally or non-occupationally related medical condition requiring further treatment or evaluation.

The standard provides for the use of respirators where engineering and other primary controls are not effective. However, the use of respirator protection shall not be used in lieu of temporary medical removal due to elevated blood lead levels or findings that an employee is at risk of material health impairment. This is based on the numerous inadequacies of respirators including skin rash where the facepiece makes contact with the skin, unacceptable stress to breathing in some workers with underlying cardiopulmonary impairment, difficulty in providing adequate fit, the tendency for respirators to create additional hazards by interfering with vision, hearing, and mobility, and the difficulties of assuring the maximum effectiveness of a complicated work practice program involving respirators. Respirators do, however, serve a useful function where engineering and work practice controls are inadequate by providing supplementary, interim, or short-term protection, provided they are properly selected for the environment in which the employee will be working, properly fitted to the employee, maintained and cleaned periodically, and worn by the employee when required.

In its interim final standard on occupational exposure to inorganic lead in the construction industry, OSHA has prohibited prophylactic chelation. Diagnostic and therapeutic chelation are permitted only under the supervision of a licensed physician with appropriate medical monitoring in an acceptable clinical setting. The decision to initiate chelation therapy must be made on an individual basis and take into account the severity of symptoms felt to be a result of lead toxicity along with blood lead levels, ZPP levels, and other laboratory tests as appropriate. EDTA and penicillamine which are the primary chelating agents used in the therapy of occupational lead poisoning have significant potential side effects and their use must be justified on the basis

of expected benefits to the worker. Unless frank and severe symptoms are present, therapeutic chelation is not recommended, given the opportunity to remove a worker from exposure and allow the body to naturally excreteaccumulated lead. As a diagnostic aid, the chelation mobilization test using CA-EDTA has limited applicability. According to some investigators, the test can differentiate between lead-induced and other nephropathies. The test may also provide an estimation of the mobile fraction of the total body lead burden.

Employers are required to assure that accurate records are maintained on exposure assessment, including environmental monitoring, medical surveillance, and medical removal for each employee. Exposure assessment records must be kept for at least 30 years. Medical surveillance records must be kept for the duration of employment plus 30 years except in cases where the employment was less than one year. If duration of employment is less than one year, the employer need not retain this record beyond the term of employment if the record is provided to the employee upon termination of employment. Medical removal records also must be maintained for the duration of employment. All records required under the standard must be made available upon request to the Assistant Secretary of Labor for Occupational Safety and Health and the Director of the National Institute for Occupational Safety and Health. Employers must also make environmental and biological monitoring and medical removal records available to affected employees and to former employees or their authorized employee representatives. Employees or their specifically designated representatives have access to their entire medical surveillance records.

In addition, the standard requires that the employer inform all workers exposed to lead at or above 30 µg/m^3 of the provisions of the standard and all its appendices, the purpose and description of medical surveillance andprovisions for medical removal protection if temporary removal is required. An understanding of the potential health effects of lead exposure by all exposed employees along with full understanding of their rights under the lead standard is essential for an effective monitoring program.

II. Adverse Health Effects of Inorganic Lead

Although the toxicity of lead has been known for 2,000 years, the knowledge of the complex relationship between lead exposure and human response is still being refined. Significant research into the toxic properties of lead continues throughout the world, and it should be anticipated that our understanding of thresholds of effects and margins of safety will be improved in future years. The provisions of the lead standard are founded on two prime medical judgments: First, the prevention of adverse health effects from exposure to lead throughout a working lifetime requires that worker blood lead levels be maintained at or below 40 µg/dl and second, the blood lead levels of workers, male or female, who intend to parent in the near future should be maintained below 30 µg/dl to minimize adverse reproductive health effects to the parents and developing fetus. The adverse effects of lead on reproduction are being actively researched and OSHA encourages the physician to remain abreast of recent developments in the area to best advise pregnant workers or workers planning to conceive children.

The spectrum of health effects caused by lead exposure can be subdivided into five developmental stages: Normal, physiological changes of uncertain significance, pathophysiological changes, overt symptoms (morbidity), and mortality. Within this process there are no sharp distinctions, but rather a continuum of effects. Boundaries between categories overlap due to the wide variation of individual responses and exposures in the working population. OSHA's development of the lead standard focused on pathophysiological changes as well as later stages of disease.

1. Heme Synthesis Inhibition. The earliest demonstrated effect of lead involves its ability to inhibit at least two enzymes of the heme synthesis pathway at very low blood levels. Inhibition of delta aminolevulinic acid dehydrase (ALA-D) which catalyzes the conversion of delta-aminolevulinic acid (ALA) to protoporphyrin is observed at a blood lead level below 20 µg/dl. At a blood lead level of 40 µg/dl, more than 20% of the population would have 70% inhibition of ALA-D. There is an exponential increase in ALA excretion at blood lead levels greater than 40 µg/dl.

Another enzyme, ferrochelatase, is also inhibited at low blood lead levels. Inhibition of ferrochelatase leads to increased free erythrocyte protoporphyrin (FEP) in the blood which can then bind to zinc to yield zinc protoporphyrin. At a blood lead level of 50 µg/dl or greater, nearly 100% of the population will have an increase in FEP. There is also an exponential relationship between blood lead levels greater than 40 µg/dl and the associated ZPP level, which has led to the development of the ZPP screening test for lead exposure.

While the significance of these effects is subject to debate, it is OSHA's position that these enzyme disturbances are early stages of a disease process which may eventually result in the clinical symptoms of lead poisoning. Whether or not the effects do progress to the later stages of clinical disease, disruption of these enzyme processes over a working lifetime is considered to be a material impairment of health.

One of the eventual results of lead-induced inhibition of enzymes in the heme synthesis pathway is anemia which can be asymptomatic if mild but associated with a wide array of symptoms including dizziness, fatigue, and tachycardia when more severe. Studies have indicated that lead levels as low as 50 µg/dl can be associated with a definite decreased hemoglobin, although most cases of lead-induced anemia, as well as shortened red-cell survival times, occur at lead levels exceeding 80 µg/dl. Inhibited hemoglobin synthesis is more common in chronic cases whereas shortened erythrocyte life span is more common in acute cases.

In lead-induced anemias, there is usually a reticulocytosis along with the presence of basophilic stippling, and ringed sideroblasts, although none of the above are pathognomonic for lead-induced anemia.

2. Neurological Effects. Inorganic lead has been found to have toxic effects on both the central and peripheral nervous systems. The earliest stages of lead-induced central nervous system effects first manifest themselves in the form of behavioral disturbances and central nervous system symptoms including irritability, restlessness, insomnia and other sleep disturbances, fatigue, vertigo, headache, poor memory, tremor, depression, and apathy. With more severe exposure, symptoms can progress to drowsiness, stupor, hallucinations, delirium, convulsions and coma.

The most severe and acute form of lead poisoning which usually follows ingestion or inhalation of large amounts of lead is acute encephalopathy which may arise precipitously with the onset of intractable seizures, coma, cardiorespiratory arrest, and death within 48 hours.

While there is disagreement about what exposure levels are needed to produce the earliest symptoms, most experts agree that symptoms definitely can occur at blood lead levels of 60 µg/dl whole blood and therefore recommend a 40 µg/dl maximum. The central nervous system effects frequently are not reversible following discontinued exposure or chelation therapy and when improvement does occur, it is almost always only partial.

The peripheral neuropathy resulting from lead exposure characteristically involves only motor function with minimal sensory damage and has a marked predilection for the extensor muscles of the most active extremity. The peripheral neuropathy can occur with varying degrees of severity. The earliest and mildest form which can

be detected in workers with blood lead levels as low as 50 µg/dl is manifested by slowing of motor nerve conduction velocity often without clinical symptoms. With progression of the neuropathy there is development of painless extensor muscle weakness usually involving the extensor muscles of the fingers and hand in the most active upper extremity, followed in severe cases by wrist drop or, much less commonly, foot drop.

In addition to slowing of nerve conduction, electromyographical studies in patients with blood lead levels greater than 50 µg/dl have demonstrated a decrease in the number of acting motor unit potentials, an increase in the duration of motor unit potentials, and spontaneous pathological activity including fibrillations and fasciculations. Whether these effects occur at levels of 40 µg/dl is undetermined.

While the peripheral neuropathies can occasionally be reversed with therapy, again such recovery is notassured particularly in the more severe neuropathies and often improvement is only partial. The lack of reversibility is felt to be due in part to segmental demyelination.

3. Gastrointestinal. Lead may also affect the gastrointestinal system producing abdominal colic or diffuse abdominal pain, constipation, obstipation, diarrhea, anorexia, nausea and vomiting. Lead colic rarely develops at blood lead levels below 80 µg/dl.

4. Renal. Renal toxicity represents one of the most serious health effects of lead poisoning. In the early stages of disease nuclear inclusion bodies can frequently be identified in proximal renal tubular cells. Renal function remains normal and the changes in this stage are probably reversible. With more advanced disease there is progressive interstitial fibrosis and impaired renal function. Eventually extensive interstitial fibrosis ensues with sclerotic glomeruli and dilated and atrophied proximal tubules; all represent end stage kidney disease. Azotemia can be progressive, eventually resulting in frank uremia necessitating dialysis. There is occasionallyassociated hypertension and hyperuricemia with or without gout.

Early kidney disease is difficult to detect. The urinalysis is normal in early lead nephropathy and the blood urea nitrogen and serum creatinine increase only when two-thirds of kidney function is lost. Measurement of creatinine clearance can often detect earlier disease as can other methods of measurement of glomerular filtration rate. An abnormal Ca-EDTA mobilization test has been used to differentiate between lead-induced and other nephropathies, but this procedure is not widely accepted. A form of Fanconi syndrome with aminoaciduria, glycosuria, and hyperphosphaturia indicating severe injury to the proximal renal tubules is occasionally seen in children.

5. Reproductive effects. Exposure to lead can have serious effects on reproductive function in both males and females. In male workers exposed to lead there can be a decrease in sexual drive, impotence, decreased ability to produce healthy sperm, and sterility. Malformed sperm (teratospermia), decreased number of sperm (hypospermia), and sperm with decreased motility (asthenospermia) can all occur. Teratospermia has been noted at mean blood lead levels of 53 µg/dl and hypospermia and

asthenospermia at 41 µg/dl. Furthermore, there appears to be a dose-response relationship for teratospermia in lead exposed workers.

Women exposed to lead may experience menstrual disturbances including dysmenorrhea, menorrhagia and amenorrhea. Following exposure to lead, women have a higher frequency of sterility, premature births, spontaneous miscarriages, and stillbirths.

Germ cells can be affected by lead and cause genetic damage in the egg or sperm cells before conception and result in failure to implant, miscarriage, stillbirth, or birth defects.

Infants of mothers with lead poisoning have a higher mortality during the first year and suffer from loweredbirth weights, slower growth, and nervous system disorders.

Lead can pass through the placental barrier and lead levels in the mother's blood are comparable to concentrations of lead in the umbilical cord at birth. Transplacental passage becomes detectable at 12-14 weeks of gestation and increases until birth.

There is little direct data on damage to the fetus from exposure to lead but it is generally assumed that the fetus and newborn would be at least as susceptible to neurological damage as young children. Blood lead levels of 50-60 µg/dl in children can cause significant neurobehavioral impairments and there is evidence of hyperactivity at blood levels as low as 25 µg/dl. Given the overall body of literature concerning the adverse health effects of lead in children, OSHA feels that the blood lead level in children should be maintained below 30 µg/dl with a population mean of 15 µg/dl. Blood lead levels in the fetus and newborn likewise should notexceed 30 µg/dl.

Because of lead's ability to pass through the placental barrier and also because of the demonstrated adverse effects of lead on reproductive function in both the male and female as well as the risk of genetic damage of lead on both the ovum and sperm, OSHA recommends a 30 µg/dl maximum permissible blood lead level in both males and females who wish to bear children.

6. Other toxic effects. Debate and research continue on the effects of lead on the human body. Hypertension has frequently been noted in occupationally exposed individuals although it is difficult to assess whether this is due to lead's adverse effects on the kidney or if some other mechanism is involved. Vascular and electrocardiographic changes have been detected but have not been well characterized. Lead is thought to impair thyroid function and interfere with the pituitary-adrenal axis, but again these effects have not been well defined.

III. Medical Evaluation

The most important principle in evaluating a worker for any occupational disease including lead poisoning is a high index of suspicion on the part of the examining physician. As discussed in Section 2, lead can affect numerous organ systems and produce a wide array of signs and symptoms, most of which are non-specific and subtle in nature at least in the early stages of disease. Unless serious concern for lead toxicity is present, many of the early clues to diagnosis may easily be overlooked.

The crucial initial step in the medical evaluation is recognizing that a worker's employment can result in exposure to lead. The worker will frequently be able to define exposures to lead and lead containing materials but often will not volunteer this information unless specifically asked. In other situations the worker may not know of any exposures to lead but the suspicion might be raised on the part of the physician because of the industry or occupation of the worker. Potential occupational exposure to lead and its compounds occur in many occupations in the construction industry, including demolition and salvaging operations, removal or encapsulation of materials containing lead, construction, alteration, repair or renovation of structures containing lead, transportation, disposal, storage or containment of lead or lead-containing materials on construction sites, and maintenance operations associated with construction activities.

Once the possibility for lead exposure is raised, the focus can then be directed toward eliciting information from the medical history, physical exam, and finally from laboratory data to evaluate the worker for potential lead toxicity.

A complete and detailed work history is important in the initial evaluation. A listing of all previous employment with information on job description, exposure to fumes or dust, known exposures to lead or other toxic substances, a description of any personal protective equipment used, and previous medical surveillance should all be included in the worker's record. Where exposure to lead is suspected, information concerning on-the-job personal hygiene, smoking or eating habits in work areas, laundry procedures, and use of any protective clothing or respiratory protection equipment should be noted. A complete work history is essential in the medical evaluation of a worker with suspected lead toxicity, especially when long term effects such as neurotoxicity and nephrotoxicity are considered.

The medical history is also of fundamental importance and should include a listing of all past and current medical conditions, current medications including proprietary drug intake, previous surgeries and hospitalizations, allergies, smoking history, alcohol consumption, and also non-occupational lead exposures such as hobbies (hunting, riflery). Also known childhood exposures should be elicited. Any previous history of hematological, neurological, gastrointestinal, renal, psychological, gynecological, genetic, or reproductive problems should be specifically noted.

A careful and complete review of systems must be performed to assess both recognized complaints and subtle or slowly acquired symptoms which the worker might not appreciate as being significant. The review of symptoms should include the following:

1. General - weight loss, fatigue, decreased appetite.

2. Head, Eyes, Ears, Nose, Throat (HEENT) - headaches, visual disturbances or decreased visual acuity, hearing deficits or tinnitus, pigmentation of the oral mucosa, or metallic taste in mouth.

3. Cardio-pulmonary - shortness of breath, cough, chest pains, palpitations, or orthopnea.

4. Gastrointestinal - nausea, vomiting, heartburn, abdominal pain, constipation or diarrhea.

5. Neurologic - irritability, insomnia, weakness (fatigue), dizziness, loss of memory, confusion, hallucinations, incoordination, ataxia, decreased strength in hands or feet, disturbances in gait, difficulty in climbing stairs, or seizures.

6. Hematologic - pallor, easy fatigability, abnormal blood loss, melena.

7. Reproductive (male and female and spouse where relevant) - history of infertility, impotence, loss of libido, abnormal menstrual periods, history of miscarriages, stillbirths, or children with birth defects.

8. Musculo-skeletal - muscle and joint pains.

The physical examination should emphasize the neurological, gastrointestinal, and cardiovascular systems. The worker's weight and blood pressure should be recorded and the oral mucosa checked for pigmentation characteristic of a possible Burtonian or lead line on the gingiva. It should be noted, however, that the lead line may not be present even in severe lead poisoning if good oral hygiene is practiced.

The presence of pallor on skin examination may indicate an anemia which, if severe, might also be associated with a tachycardia. If an anemia is suspected, an active search for blood loss should be undertaken including potential blood loss through the gastrointestinal tract.

A complete neurological examination should include an adequate mental status evaluation including a search for behavioral and psychological disturbances, memory testing, evaluation for irritability, insomnia, hallucinations, and mental clouding. Gait and coordination should be examined along with close observation for tremor. A detailed evaluation of peripheral nerve function including careful sensory and motor function testing is warranted. Strength testing particularly of extensor muscle groups of all extremities is of fundamental importance.

Cranial nerve evaluation should also be included in the routine examination.

The abdominal examination should include auscultation for bowel sounds and abdominal bruits and palpationfor organomegaly, masses, and diffuse abdominal tenderness.

Cardiovascular examination should evaluate possible early signs of congestive heart failure. Pulmonary status should be addressed particularly if respirator protection is contemplated.

As part of the medical evaluation, the interim lead standard requires the following laboratory studies:

1. Blood lead level

2. Hemoglobin and hematocrit determinations, red cell indices, and examination of the peripheral blood smearto evaluate red blood cell morphology

3. Blood urea nitrogen

4. Serum creatinine

5. Routine urinalysis with microscopic examination.

6. A zinc protoporphyrin level.

In addition to the above, the physician is authorized to order any further laboratory or other tests which he or she deems necessary in accordance with sound medical practice. The evaluation must also include pregnancytesting or laboratory evaluation of male fertility if requested by the employee. Additional tests which are probably not warranted on a routine basis but may be appropriate when blood lead and ZPP levels are equivocal include delta aminolevulinic acid and coproporphyrin concentrations in the urine, and dark-field illumination for detection of basophilic stippling in red blood cells.

If an anemia is detected further studies including a careful examination of the peripheral smear, reticulocyte count, stool for occult blood, serum iron, total iron binding capacity, bilirubin, and, if appropriate, vitamin B12 and folate may be of value in attempting to identify the cause of the anemia.

If a peripheral neuropathy is suspected, nerve conduction studies are warranted both for diagnosis and as abasis to monitor any therapy.

If renal disease is questioned, a 24 hour urine collection for creatinine clearance, protein, and electrolytes may be indicated. Elevated uric acid levels may result from lead-induced renal disease and a serum uric acid level might be performed.

An electrocardiogram and chest x-ray may be obtained as deemed appropriate.

Sophisticated and highly specialized testing should not be done routinely and where indicated should be under the direction of a specialist.

IV. Laboratory Evaluation

The blood lead level at present remains the single most important test to monitor lead exposure and is the test used in the medical surveillance program under the lead standard to guide employee medical removal. The ZPPhas several advantages over the blood lead level. Because of its relatively recent development and the lack of extensive data concerning its interpretation, the ZPP currently remains an ancillary test.

This section will discuss the blood lead level and ZPP in detail and will outline their relative advantages and disadvantages. Other blood tests currently available to evaluate lead exposure will also be reviewed.

The blood lead level is a good index of current or recent lead absorption when there is no anemia present and when the worker has not taken any chelating agents. However, blood lead levels along with urinary lead levelsdo not necessarily indicate the total body burden of lead and are not adequate measures of past exposure. One reason for this is that lead has a high affinity for bone and up to 90% of the body's total lead is deposited there. A very important component of the total lead body burden is lead in soft tissue (liver, kidney, and brain). This fraction of the lead body burden, the biologically active lead, is not entirely reflected by blood lead levels since it is a function of the dynamics of lead absorption, distribution, deposition in bone and excretion. Following discontinuation of exposure to lead, the excess body burden is only slowly mobilized from bone and other relatively stable body stores and excreted. Consequently, a high blood lead level may only represent recent heavy exposure to lead without a significant total body excess and likewise a low blood lead level does not exclude an elevated total body burden of lead.

Also due to its correlation with recent exposures, the blood lead level may vary considerably over short time intervals.

To minimize laboratory error and erroneous results due to contamination, blood specimens must be carefully collected after thorough cleaning of the skin with appropriate methods using lead-free blood containers and analyzed by a reliable laboratory. Under the standard, samples must be analyzed in laboratories which are approved by OSHA. Analysis is to be made using atomic absorption spectrophotometry, anodic stripping voltammetry or any method which meets the accuracy requirements set forth by the standard.

The determination of lead in urine is generally considered a less reliable monitoring technique than analysis of whole blood primarily due to individual variability in urinary excretion capacity as well as the technical difficulty of obtaining accurate 24 hour urine collections. In addition, workers with renal insufficiency, whether due to lead or some other cause, may have decreased lead clearance and consequently urine lead levels may underestimate the true lead burden. Therefore, urine lead levels should not be used as a routine test.

The zinc protoporphyrin test, unlike the blood lead determination, measures an adverse metabolic effect of lead and as such is a better indicator of lead toxicity than the level of blood lead itself. The level of ZPP reflects lead absorption over the preceding 3 to 4 months, and therefore is a better indicator of lead body burden. TheZPP requires more time than the blood lead to read significantly elevated levels; the return to normal after discontinuing lead exposure is also slower. Furthermore, the ZPP test is simpler, faster, and less expensive toperform and no contamination is possible. Many investigators believe it is the most reliable means of monitoring chronic lead absorption.

Zinc protoporphyrin results from the inhibition of the enzyme ferrochelatase which catalyzes the insertion of an iron molecule into the protoporphyrin molecule, which then becomes heme. If iron is not inserted into the molecule then zinc, having a greater affinity for protoporphyrin, takes the place of the iron, forming ZPP.

An elevation in the level of circulating ZPP may occur at blood lead levels as low as 20-30 µg/dl in some workers. Once the blood lead level has reached 40 µg/dl there is more marked rise in the ZPP value from its normal range of less than 100 µg/dl100 ml. Increases in blood lead levels beyond 40 µg/100 g are associated with exponential increases in ZPP.

Whereas blood lead levels fluctuate over short time spans, ZPP levels remain relatively stable. ZPP is measured directly in red blood cells and is present for the cell's entire 120 day life-span. Therefore, the ZPP level in blood reflects the average ZPP production over the previous 3-4 months and consequently the averagelead exposure during that time interval.

It is recommended that a hematocrit be determined whenever a confirmed ZPP of 50 µg/100 ml whole blood is obtained to rule out a significant underlying anemia. If the ZPP is in excess of 100 µg/100 ml and not associated with abnormal elevations in blood lead levels, the laboratory should be checked to be sure that blood leads were determined using atomic absorption spectrophotometry anodic stripping voltammetry, or any method which meets the accuracy requirements set forth by the standard by an OSHA approved laboratory which is experienced in lead level determinations. Repeat periodic blood lead studies should be obtained in all individuals with elevated ZPP levels to be certain that an associated elevated blood lead level has not been missed due to transient fluctuations in blood leads.

ZPP has a characteristic fluorescence spectrum with a peak at 594 nm which is detectable with a hematofluorimeter. The hematofluorimeter is accurate and portable and can provide on-site, instantaneous results for workers who can be frequently tested via a finger prick.

However, careful attention must be given to calibration and quality control procedures. Limited data on blood lead-ZPP correlations and the ZPP levels which are associated with the adverse health effects discussed in Section 2 are the major limitations of the test. Also it is difficult to correlate ZPP levels with environmental exposure and there is some variation of response with age and sex. Nevertheless, the ZPP promises to be animportant diagnostic test for the early detection of lead toxicity and its value will increase as more data is collected regarding its relationship to other manifestations of lead poisoning.

Levels of delta-aminolevulinic acid (ALA) in the urine are also used as a measure of lead exposure. Increasing concentrations of ALA are believed to result from the inhibition of the enzyme delta-aminolevulinic acid dehydrase (ALA-D). Although the test is relatively easy to perform, inexpensive, and rapid, the disadvantages include variability in results, the necessity to collect a complete 24 hour urine sample which has a specific gravity greater than 1.010, and also the fact that ALA decomposes in the presence of light.

The pattern of porphyrin excretion in the urine can also be helpful in identifying lead intoxication. With lead poisoning, the urine concentrations of coproporphyrins I and II, porphobilinogen and uroporphyrin I rise. The most important increase, however, is that of coproporphyrin III; levels may exceed 5,000 µg/l in the urine in lead poisoned individuals, but its correlation with blood lead levels and ZPP are not as good as those of ALA. Increases in urinary porphyrins are not diagnostic of lead toxicity and may be seen in porphyria, some liver diseases, and in patients with high reticulocyte counts.

Summary. The Occupational Safety and Health Administration's interim standard for inorganic lead in the construction industry places significant emphasis on the medical surveillance of all workers exposed to levels of inorganic lead above 30 µg/m^3 TWA. The physician has a fundamental role in this surveillance program, and in the operation of the medical removal protection program.

Even with adequate worker education on the adverse health effects of lead and appropriate training in work practices, personal hygiene and other control measures, the physician has a primary responsibility for evaluating potential lead toxicity in the worker. It is only through a careful and detailed medical and work history, a complete physical examination and appropriate laboratory testing that an accurate assessment can be made. Many of the adverse health effects of lead toxicity are either irreversible or only partially reversibleand therefore early detection of disease is very important.

This document outlines the medical monitoring program as defined by the occupational safety and health standard for inorganic lead. It reviews the adverse health effects of lead poisoning and describes the important elements of the history and physical examinations as they relate to these adverse effects. Finally, the appropriate laboratory testing for evaluating lead exposure and toxicity is presented.

It is hoped that this review and discussion will give the physician a better understanding of the OSHA standard with the ultimate goal of protecting the health and well-being of the worker exposed to lead under his or her care.

[58 FR 26627, May 4, 1993, as amended at 58 FR 34218, June 24, 1993; 61 FR 5510, Feb. 13, 1996; 63 FR 1296, Jan. 8, 1998; 70 FR 1143, Jan. 5, 2005; 71 FR 16674, Apr. 3, 2006; 71 FR 50191, Aug. 24, 2006; 73 FR 75588, Dec. 12, 2008; 76 FR 33611, June 8, 2011; 76 FR 80741, Dec. 27, 2011; 77 FR 17890, Mar. 26, 2012; 85 FR 8735, Feb. 18, 2020; 87 FR 38986, June 30, 2022]

§ 1926.64 Process safety management of highly hazardous chemicals.

For requirements regarding the process safety management of highly hazardous chemicals as it pertains to construction work, follow the requirements in 29 CFR 1910.119.

[84 FR 21576, May 14, 2019]

§ 1926.65 Hazardous waste operations and emergency response.

(a) *Scope, application, and definitions* -

    (1) **Scope.** This section covers the following operations, unless the employer can demonstrate that the operation does not involve employee exposure or the reasonable possibility for employee exposure to safety or health hazards:

        (i) Clean-up operations required by a governmental body, whether Federal, state, local or other involving hazardous substances that are conducted at uncontrolled hazardous waste sites (including, but not limited to, the EPA's National Priority Site List (NPL), state priority site lists, sites recommended for the EPA NPL, and initial investigations of government identified sites which are conducted before the presence or absence of hazardous substances has been ascertained);

        (ii) Corrective actions involving clean-up operations at sites covered by the Resource Conservation and Recovery Act of 1976 (RCRA) as amended (42 U.S.C. 6901 *et seq.*);

        (iii) Voluntary clean-up operations at sites recognized by Federal, state, local or other governmental bodies as uncontrolled hazardous waste sites;

        (iv) Operations involving hazardous wastes that are conducted at treatment, storage, and disposal (TSD) facilities regulated by 40 CFR parts 264 and 265 pursuant to RCRA; or by agencies under agreement with U.S.E.P.A. to implement RCRA regulations; and

        (v) Emergency response operations for releases of, or substantial threats of releases of, hazardous substances without regard to the location of the hazard.

    (2) **Application.**

        (i) All requirements of 29 CFR parts 1910 and 1926 apply pursuant to their terms to hazardous waste and emergency response operations whether covered by this section or not. If there is a conflict or overlap, the provision more protective of employee safety and health shall apply without regard to 29 CFR 1926.20(e).

        (ii) Hazardous substance clean-up operations within the scope of paragraphs (a)(1)(i) through (a)(1)(iii) of this section must comply with all paragraphs of this section except paragraphs (p) and (q).

        (iii) Operations within the scope of paragraph (a)(1)(iv) of this section must comply only with the requirements of paragraph (p) of this section.

Notes and Exceptions: (A) All provisions of paragraph (p) of this section cover any treatment, storage or disposal (TSD) operation regulated by 40 CFR parts 264 and 265 orby state law authorized under RCRA, and required to have a permit or interim status from EPA pursuant to 40 CFR 270.1 or from a state agency pursuant to RCRA.

(B) Employers who are not required to have a permit or interim status because they are conditionally exempt small quantity generators under 40 CFR 261.5 or are generators who qualify under 40 CFR 262.34 for exemptions from regulation under 40 CFR parts 264, 265 and 270 ("excepted employers") are not covered by paragraphs (p)(1) through (p)(7) of this section. Excepted employers who are required by the EPA or state agency to have their employees engage in emergency response or who direct their employees toengage in emergency response are covered by paragraph (p)(8) of this section, and cannot be exempted by (p)(8)(i) of this section. Excepted employers who are not required to have employees engage in emergency response, who direct their employees

to evacuate in the case of such emergencies and who meet the requirements of paragraph (p)(8)(i) of this section are exempt from the balance of paragraph (p)(8) of this section.

(C) If an area is used primarily for treatment, storage or disposal, any emergency response operations in that area shall comply with paragraph (p)(8) of this section. In other areas not used primarily for treatment, storage, or disposal, any emergency response operations shall comply with paragraph (q) of this section. Compliance with the requirements of paragraph (q) of this section shall be deemed to be in compliance with the requirements of paragraph (p)(8) of this section.

(iv) Emergency response operations for releases of, or substantial threats of releases of, hazardous substances which are not covered by paragraphs (a)(1)(i) through (a)(1)(iv) of this section mustonly comply with the requirements of paragraph (q) of this section.

(3) **Definitions - Buddy system** means a system of organizing employees into work groups in such a manner that each employee of the work group is designated to be observed by at least one otheremployee in the work group. The purpose of the buddy system is to provide rapid assistance toemployees in the event of an emergency.

Clean-up operation means an operation where hazardous substances are removed, contained, incinerated, neutralized, stabilized, cleared-up, or in any other manner processed or handledwith the ultimate goal of making the site safer for people or the environment.

Decontamination means the removal of hazardous substances from employees and their equipmentto the extent necessary to preclude the occurrence of foreseeable adverse health affects.

Emergency response or *responding to emergencies* means a response effort by employees from outside the immediate release area or by other designated responders (i.e., mutual-aid groups, local fire departments, etc.) to an occurrence which results, or is likely to result, in an uncontrolled release of a hazardous substance. Responses to incidental releases of hazardoussubstances where the substance can be absorbed, neutralized, or otherwise controlled at the time of release by employees in the immediate release area, or by maintenance personnel are not considered to be emergency responses within the scope of this standard. Responses to releases of hazardous substances where there is no potential safety or health hazard (i.e., fire, explosion, or chemical exposure) are not considered to be emergency responses.

Facility means

(A) any building, structure, installation, equipment, pipe or pipeline (including any pipe into asewer or publicly owned treatment works), well, pit, pond, lagoon, impoundment, ditch, storage container, motor vehicle, rolling stock, or aircraft, or

(B) any site or area where a hazardous substance has been deposited, stored, disposed of, orplaced, or otherwise come to be located; but does not include any consumer product in consumer use or any water-borne vessel.

Hazardous materials response (HAZMAT) team means an organized group of employees, designatedby the employer, who are expected to perform work to handle and control actual or potential leaks or spills of hazardous substances requiring possible close approach to the substance. The team members perform responses to releases or potential releases of hazardous substances for the purpose of control or stabilization of the incident. A HAZMAT team is not a fire brigade nor is a typical fire brigade a HAZMAT team. A HAZMAT team, however, may be aseparate component of a fire brigade or fire department.

Hazardous substance means any substance designated or listed under paragraphs (A) through (D) of this definition, exposure to which results or may result in adverse affects on the health or safetyof employees:

(A) Any substance defined under section 101(14) of CERCLA;

(B) Any biological agent and other disease-causing agent which after release into the environment and upon exposure, ingestion, inhalation, or assimilation into any person, either directly from the environment or indirectly by ingestion through food chains, will or may reasonably be anticipated to cause death, disease, behavioral abnormalities, cancer,genetic mutation, physiological malfunctions (including malfunctions in reproduction) orphysical deformations in such persons or their offspring;

(C) Any substance listed by the U.S. Department of Transportation as hazardous materialsunder 49 CFR 172.101 and appendices; and

(D) Hazardous waste as herein

Hazardous waste defined.means -

(A) A waste or combination of wastes as defined in 40 CFR

(B) 261.3, orThose substances defined as hazardous wastes

Hazardous waste operation 1.8.

means any operation conducted within the scope of this standard.

Hazardous waste site or *Site* means any facility or location within the scope of this standard at which hazardous waste operations take place.

Health hazard means a chemical or a pathogen where acute or chronic health effects may occur in exposed employees. It also includes stress due to temperature extremes. The term *health hazard* includes chemicals that are classified in accordance with the Hazard Communication Standard, § 1910.1200, as posing one of the following hazardous effects: acute toxicity (any route of exposure); skin corrosion or irritation; serious eye damage or eye irritation; respiratory or skin sensitization; germ cell mutagenicity; carcinogenicity; reproductive toxicity; specific target organ toxicity (single or repeated exposure); aspiration toxicity or simple asphyxiant. (*See* Appendix A to § 1910.1200 - Health Hazard Criteria (Mandatory) for the criteria for determining whether a chemical is classified as a health hazard.)

IDLH or *Immediately dangerous to life or health* means an atmospheric concentration of any toxic, corrosive or asphyxiant substance that poses an immediate threat to life or would cause irreversible or delayed adverse health effects or would interfere with an individual's ability to escape from a dangerous atmosphere.

Oxygen deficiency means that concentration of oxygen by volume below which atmosphere supplying respiratory protection must be provided. It exists in atmospheres where the percentage of oxygen by volume is less than 19.5 percent oxygen.

Permissible exposure limit means the exposure, inhalation or dermal permissible exposure limit specified either in § 1926.55, elsewhere in subpart D, or in other pertinent sections of this part.

Published exposure level means the exposure limits published in "NIOSH Recommendations for Occupational Health Standards" dated 1986 incorporated by reference, or if none is specified, the exposure limits published in the standards specified by the American Conference of Governmental Industrial Hygienists in their publication "Threshold Limit Values and Biological Exposure Indices for 1987-88" dated 1987 incorporated by reference.

Post emergency response means that portion of an emergency response performed after the immediate threat of a release has been stabilized or eliminated and clean-up of the site has begun. If post emergency response is performed by an employer's own employees who were part of the initial emergency response, it is considered to be part of the initial response and not post emergency response. However, if a group of an employer's own employees, separate from the group providing initial response, performs the clean-up operation, then the separate group of employees would be considered to be performing post-emergency response and subject to paragraph (q)(11) of this section.

Qualified person means a person with specific training, knowledge and experience in the area for which the person has the responsibility and the authority to control.

Site safety and health supervisor (or oficial) means the individual located on a hazardous waste site who is responsible to the employer and has the authority and knowledge necessary to implement the site safety and health plan and verify compliance with applicable safety and health requirements.

Small quantity generator means a generator of hazardous wastes who in any calendar month generates no more than 1,000 kilograms (2,205 pounds) of hazardous waste in that month.

Uncontrolled hazardous waste site, means an area identified as an uncontrolled hazardous waste site by a governmental body, whether Federal, state, local or other where an accumulation of hazardous substances creates a threat to the health and safety of individuals or the environment or both. Some sites are found on public lands such as those created by former municipal, county or state landfills where illegal or poorly managed waste disposal has taken place. Other sites are found on private property, often belonging to generators or former generators of hazardous substance wastes. Examples of such sites include, but are not limited to, surface impoundments, landfills, dumps, and tank or drum farms. Normal operations at TSDsites are not covered by this definition.

(b) **Safety and health program.**

Note to (b): Safety and health programs developed and implemented to meet other Federal, state, or local regulations are considered acceptable in meeting this requirement if they cover or are modified to cover the topics required in this paragraph. An additional or separate safety and health program is not required by this paragraph.

> **General.**
>
> (i) Employers shall develop and implement a written safety and health program for their employees involved in hazardous waste operations. The program shall be designed to identify, evaluate, and control safety and health hazards, and provide for emergency response for hazardous waste operations.
>
> (ii) The written safety and health program shall incorporate the
>
> > (A) following: An organizational structure;
> >
> > (B) A comprehensive workplan;
> >
> > (C) A site-specific safety and health plan which need not repeat the employer's standard operating procedures required in paragraph (b)(1)(ii)(F) of this section;
> >
> > (D) The safety and health training
> >
> > (E) program;The medical
> >
> > (F) surveillance program;
> >
> > The employer's standard operating procedures for safety and health; and
> >
> > (G) Any necessary interface between general program and site specific activities.

- (iii) **Site excavation.** Site excavations created during initial site preparation or during hazardous waste operations shall be shored or sloped as appropriate to prevent accidental collapse inaccordance with subpart P of 29 CFR part 1926.

- (iv) **Contractors and sub-contractors.** An employer who retains contractor or sub-contractor services for work in hazardous waste operations shall inform those contractors, sub-contractors, or their representatives of the site emergency response procedures and any potential fire, explosion, health, safety or other hazards of the hazardous waste operation that have been identified by the employer, including those identified in the employer's information program.

- (v) **Program availability.** The written safety and health program shall be made available to any contractor or subcontractor or their representative who will be involved with the hazardous waste operation; to employees; to employee designated representatives; to OSHA personnel, and to personnel of other Federal, state, or local agencies with regulatory authority over the site.

(2) **Organizational structure part of the site program** -

- (i) The organizational structure part of the program shall establish the specific chain of commandand specify the overall responsibilities of supervisors and employees. It shall include, at a minimum, the following elements:

    - (A) A general supervisor who has the responsibility and authority to direct all hazardous wasteoperations.

    - (B) A site safety and health supervisor who has the responsibility and authority to develop andimplement the site safety and health plan and verify compliance.

    - (C) All other personnel needed for hazardous waste site operations and emergency responseand their general functions and responsibilities.

    - (D) The lines of authority, responsibility, and communication.

- (ii) The organizational structure shall be reviewed and updated as necessary to reflect the currentstatus of waste site operations.

(3) **Comprehensive workplan part of the site program.** The comprehensive workplan part of the program shall address the tasks and objectives of the site operations and the logistics and resources requiredto reach those tasks and objectives.

- (i) The comprehensive workplan shall address anticipated clean-up activities as well as normal operating procedures which need not repeat the employer's procedures available elsewhere.

- (ii) The comprehensive workplan shall define work tasks and objectives and identify the methodsfor accomplishing those tasks and objectives.

- (iii) The comprehensive workplan shall establish personnel requirements for implementing the plan.

- (iv) The comprehensive workplan shall provide for the implementation of the training required inparagraph (e) of this section.

- (v) The comprehensive workplan shall provide for the implementation of the required informationalprograms required in paragraph (i) of this section.

(vi) The comprehensive workplan shall provide for the implementation of the medical surveillance program described in paragraph (f) of this section.

(4) *Site-specific safety and health plan part of the program* -

    (i) *General.* The site safety and health plan, which must be kept on site, shall address the safety and health hazards of each phase of site operation and include the requirements and procedures for employee protection.

    (ii) *Elements.* The site safety and health plan, as a minimum, shall address the following:

        (A) A safety and health risk or hazard analysis for each site task and operation found in the workplan.

        (B) Employee training assignments to assure compliance with paragraph (e) of this section.

        (C) Personal protective equipment to be used by employees for each of the site tasks and operations being conducted as required by the personal protective equipment program in paragraph (g)(5) of this section.

        (D) Medical surveillance requirements in accordance with the program in paragraph (f) of this section.

        (E) Frequency and types of air monitoring, personnel monitoring, and environmental sampling techniques and instrumentation to be used, including methods of maintenance and calibration of monitoring and sampling equipment to be used.

        (F) Site control measures in accordance with the site control program required in paragraph (d) of this section.

        (G) Decontamination procedures in accordance with paragraph (k) of this section.

        (H) An emergency response plan meeting the requirements of paragraph (l) of this section for safe and effective responses to emergencies, including the necessary PPE and other equipment.

        (I) Confined space entry procedures.

        (J) A spill containment program meeting the requirements of paragraph (j) of this section.

    (iii) *Pre-entry briefing.* The site specific safety and health plan shall provide for pre-entry briefings to be held prior to initiating any site activity, and at such other times as necessary to ensure that employees are apprised of the site safety and health plan and that this plan is being followed. The information and data obtained from site characterization and analysis work required in paragraph (c) of this section shall be used to prepare and update the site safety and health plan.

    (iv) *Effectiveness of site safety and health plan.* Inspections shall be conducted by the site safety and health supervisor or, in the absence of that individual, another individual who is knowledgeable in occupational safety and health, acting on behalf of the employer as necessary to determine the effectiveness of the site safety and health plan. Any deficiencies in the effectiveness of the site safety and health plan shall be corrected by the employer.

(c) *Site characterization and analysis -*

(1) *General.* Hazardous waste sites shall be evaluated in accordance with this paragraph to identify specific site hazards and to determine the appropriate safety and health control procedures neededto protect employees from the identified hazards.

(2) *Preliminary evaluation.* A preliminary evaluation of a site's characteristics shall be performed prior to site entry by a qualified person in order to aid in the selection of appropriate employee protection methods prior to site entry. Immediately after initial site entry, a more detailed evaluation of the site'sspecific characteristics shall be performed by a qualified person in order to further identify existing site hazards and to further aid in the selection of the appropriate engineering controls and personalprotective equipment for the tasks to be performed.

(3) *Hazard identification.* All suspected conditions that may pose inhalation or skin absorption hazards that are immediately dangerous to life or health (IDLH), or other conditions that may cause death or serious harm, shall be identified during the preliminary survey and evaluated during the detailed survey. Examples of such hazards include, but are not limited to, confined space entry, potentially explosive or flammable situations, visible vapor clouds, or areas where biological indicators such as dead animals or vegetation are located.

(4) *Required information.* The following information to the extent available shall be obtained by the employer prior to allowing employees to enter a site:

(i) Location and approximate size of the site.

(ii) Description of the response activity and/or the job task to be performed.

(iii) Duration of the planned employee activity.

(iv) Site topography and accessibility by air and

(v) roads.Safety and health hazards expected at

(vi) the site.

Pathways for hazardous substance dispersion.

(vii) Present status and capabilities of emergency response teams that would provide assistance tohazardous waste clean-up site employees at the time of an emergency.

(viii) Hazardous substances and health hazards involved or expected at the site, and their chemicaland physical properties.

(5) *Personal protective equipment.* Personal protective equipment (PPE) shall be provided and usedduring initial site entry in accordance with the following requirements:

(i) Based upon the results of the preliminary site evaluation, an ensemble of PPE shall be selectedand used during initial site entry which will provide protection to a level of exposure below permissible exposure limits and published exposure levels for known or suspected hazardous substances and health hazards, and which will provide protection against other known and suspected hazards identified during the preliminary site evaluation. If there is no permissible exposure limit or published exposure level, the employer may use other published studies and information as a guide to appropriate personal protective equipment.

(ii) If positive-pressure self-contained breathing apparatus is not used as part of the entry ensemble, and if respiratory protection is warranted by the potential hazards identified during the preliminary site evaluation, an escape self-contained breathing apparatus of at least five minute's duration shall be carried by employees during initial site entry.

(iii) If the preliminary site evaluation does not produce sufficient information to identify the hazards or suspected hazards of the site, an ensemble providing protection equivalent to Level B PPE shall be provided as minimum protection, and direct reading instruments shall be used as appropriate for identifying IDLH conditions. (See appendix B for a description of Level B hazards and the recommendations for Level B protective equipment.)

(iv) Once the hazards of the site have been identified, the appropriate PPE shall be selected and used in accordance with paragraph (g) of this section.

(6) *Monitoring.* The following monitoring shall be conducted during initial site entry when the site evaluation produces information that shows the potential for ionizing radiation or IDLH conditions, or when the site information is not sufficient reasonably to eliminate these possible conditions:

(i) Monitoring with direct reading instruments for hazardous levels of ionizing radiation.

(ii) Monitoring the air with appropriate direct reading test equipment (i.e., combustible gas meters, detector tubes) for IDLH and other conditions that may cause death or serious harm (combustible or explosive atmospheres, oxygen deficiency, toxic substances).

(iii) Visually observing for signs of actual or potential IDLH or other dangerous conditions.

(iv) An ongoing air monitoring program in accordance with paragraph (h) of this section shall be implemented after site characterization has determined the site is safe for the start-up of operations.

(7) *Risk identification.* Once the presence and concentrations of specific hazardous substances and health hazards have been established, the risks associated with these substances shall be identified. Employees who will be working on the site shall be informed of any risks that have been identified. In situations covered by the Hazard Communication Standard, 29 CFR 1926.59, training required by that standard need not be duplicated.

Note to (c)(7). Risks to consider include, but are not limited to:
(a) Exposures exceeding the permissible exposure limits and published exposure levels.

(b) IDLH concentrations.

(c) Potential skin absorption and irritation sources.

(d) Potential eye irritation sources.

(e) Explosion sensitivity and flammability ranges.

(f) Oxygen deficiency.

(8) **Employee notification.** Any information concerning the chemical, physical, and toxicologic properties of each substance known or expected to be present on site that is available to the employer and relevant to the duties an employee is expected to perform shall be made available to the affected employees prior to the commencement of their work activities. The employer may utilize information developed for the hazard communication standard for this purpose.

(d) **Site control** -

(1) **General.** Appropriate site control procedures shall be implemented to control employee exposure to hazardous substances before clean-up work begins.

(2) **Site control program.** A site control program for protecting employees which is part of the employer's site safety and health program required in paragraph (b) of this section shall be developed during the planning stages of a hazardous waste clean-up operation and modified as necessary as new information becomes available.

(3) **Elements of the site control program.** The site control program shall, as a minimum, include: A site map; site work zones; the use of a "buddy system"; site communications including alerting means for emergencies; the standard operating procedures or safe work practices; and, identification of the nearest medical assistance. Where these requirements are covered elsewhere they need not be repeated.

(e) **Training** -

(1) **General.**

(i) All employees working on site (such as but not limited to equipment operators, general laborers and others) exposed to hazardous substances, health hazards, or safety hazards and their supervisors and management responsible for the site shall receive training meeting the requirements of this paragraph before they are permitted to engage in hazardous waste operations that could expose them to hazardous substances, safety, or health hazards, and they shall receive review training as specified in this paragraph.

(ii) Employees shall not be permitted to participate in or supervise field activities until they have been trained to a level required by their job function and responsibility.

(2) **Elements to be covered.** The training shall thoroughly cover the

(i) following: Names of personnel and alternates responsible for site

(ii) safety and health; Safety, health and other hazards present on the site;

- (iii) Use of personal protective equipment;
- (iv) Work practices by which the employee can minimize risks from
- (v) hazards;Safe use of engineering controls and equipment on the site;
- (vi) Medical surveillance requirements, including recognition of symptoms and signs which mightindicate overexposure to hazards; and
- (vii) The contents of paragraphs (G) through (J) of the site safety and health plan set forth in paragraph (b)(4)(ii) of this section.

(3) *Initial training.*
- (i) General site workers (such as equipment operators, general laborers and supervisory personnel) engaged in hazardous substance removal or other activities which expose or potentially expose workers to hazardous substances and health hazards shall receive a minimum of 40 hours of instruction off the site, and a minimum of three days actual field experience under the direct supervision of a trained, experienced supervisor.
- (ii) Workers on site only occasionally for a specific limited task (such as, but not limited to, ground water monitoring, land surveying, or geo-physical surveying) and who are unlikely to be exposed over permissible exposure limits and published exposure limits shall receive a minimum of 24 hours of instruction off the site, and the minimum of one day actual field experience under the direct supervision of a trained, experienced supervisor.
- (iii) Workers regularly on site who work in areas which have been monitored and fully characterizedindicating that exposures are under permissible exposure limits and published exposure limits where respirators are not necessary, and the characterization indicates that there are no health hazards or the possibility of an emergency developing, shall receive a minimum of 24 hours of instruction off the site and the minimum of one day actual field experience under the direct supervision of a trained, experienced supervisor.
- (iv) Workers with 24 hours of training who are covered by paragraphs (e)(3)(ii) and (e)(3)(iii) of thissection, and who become general site workers or who are required to wear respirators, shall have the additional 16 hours and two days of training necessary to total the training specified inparagraph (e)(3)(i).

(4) *Management and supervisor training.* On-site management and supervisors directly responsible for, or who supervise employees engaged in, hazardous waste operations shall receive 40 hours initialtraining, and three days of supervised field experience (the training may be reduced to 24 hours and one day if the only area of their responsibility is employees covered by paragraphs (e)(3)(ii) and (e)(3)(iii)) and at least eight additional hours of specialized training at the time of job assignment onsuch topics as, but not limited to, the employer's safety and health program and the associated employee training program, personal protective equipment program, spill containment program, and health hazard monitoring procedure and techniques.

(5) *Qualifications for trainers.* Trainers shall be qualified to instruct employees about the subject matter that is being presented in training. Such trainers shall have satisfactorily completed a training program for teaching the subjects they are expected to teach, or they shall have the academic credentials and instructional experience necessary for teaching the subjects.

Instructors shall demonstrate competent instructional skills and knowledge of the applicable subject matter.

(6) *Training certification.* Employees and supervisors that have received and successfully completed the training and field experience specified in paragraphs (e)(1) through (e)(4) of this section shall be certified by their instructor or the head instructor and trained supervisor as having successfully completed the necessary training. A written certificate shall be given to each person so certified. Any person who has not been so certified or who does not meet the requirements of paragraph (e)(9) of this section shall be prohibited from engaging in hazardous waste operations.

(7) *Emergency response.* Employees who are engaged in responding to hazardous emergency situations at hazardous waste clean-up sites that may expose them to hazardous substances shall be trained in how to respond to such expected emergencies.

(8) *Refresher training.* Employees specified in paragraph (e)(1) of this section, and managers and supervisors specified in paragraph (e)(4) of this section, shall receive eight hours of refresher training annually on the items specified in paragraph (e)(2) and/or (e)(4) of this section, any critique of incidents that have occurred in the past year that can serve as training examples of related work, and other relevant topics.

(9) *Equivalent training.* Employers who can show by documentation or certification that an employee's work experience and/or training has resulted in training equivalent to that training required in paragraphs (e)(1) through (e)(4) of this section shall not be required to provide the initial training requirements of those paragraphs to such employees and shall provide a copy of the certification or documentation to the employee upon request. However, certified employees or employees with equivalent training new to a site shall receive appropriate, site specific training before site entry and have appropriate supervised field experience at the new site. Equivalent training includes any academic training or the training that existing employees might have already received from actual hazardous waste site work experience.

(f) **Medical surveillance** -

(1) *General.* Employers engaged in operations specified in paragraphs (a)(1)(i) through (a)(1)(iv) of this section and not covered by (a)(2)(iii) exceptions and employers of employees specified in paragraph (q)(9) shall institute a medical surveillance program in accordance with this paragraph.

(2) *Employees covered.* The medical surveillance program shall be instituted by the employer for the following employees:

(i) All employees who are or may be exposed to hazardous substances or health hazards at or above the permissible exposure limits or, if there is no permissible exposure limit, above the published exposure levels for these substances, without regard to the use of respirators, for 30 days or more a year;

(ii) All employees who wear a respirator for 30 days or more a year or as required by § 1926.103;

(iii) All employees who are injured, become ill or develop signs or symptoms due to possible overexposure involving hazardous substances or health hazards from an emergency response or hazardous waste operation; and

(iv) Members of HAZMAT teams.

(3) ***Frequency of medical examinations and consultations.*** Medical examinations and consultations shall be made available by the employer to each employee covered under paragraph (f)(2) of this section on the following schedules:

- (i) For employees covered under paragraphs (f)(2)(i), (f)(2)(ii), and

    - (A) (f)(2)(iv):Prior to assignment;
    - (B) At least once every twelve months for each employee covered unless the attending physician believes a longer interval (not greater than biennially) is appropriate;
    - (C) At termination of employment or reassignment to an area where the employee would not be covered if the employee has not had an examination within the last six months;
    - (D) As soon as possible upon notification by an employee that the employee has developed signs or symptoms indicating possible overexposure to hazardous substances or health hazards, or that the employee has been injured or exposed above the permissible exposure limits or published exposure levels in an emergency situation;
    - (E) At more frequent times, if the examining physician determines that an increased frequency of examination is medically necessary.

- (ii) For employees covered under paragraph (f)(2)(iii) and for all employees including those of employers covered by paragraph (a)(1)(v) who may have been injured, received a health impairment, developed signs or symptoms which may have resulted from exposure to hazardous substances resulting from an emergency incident, or exposed during an emergency incident to hazardous substances at concentrations above the permissible exposure limits or the published exposure levels without the necessary personal protective equipment being used:

    - (A) As soon as possible following the emergency incident or development of signs or symptoms;
    - (B) At additional times, if the examining physician determines that follow-up examinations or consultations are medically necessary.

(4) ***Content of medical examinations and consultations.***

- (i) Medical examinations required by paragraph (f)(3) of this section shall include a medical and work history (or updated history if one is in the employee's file) with special emphasis on symptoms related to the handling of hazardous substances and health hazards, and to fitness for duty including the ability to wear any required PPE under conditions (i.e., temperature extremes) that may be expected at the work site.

- (ii) The content of medical examinations or consultations made available to employees pursuant to paragraph (f) shall be determined by the attending physician. The guidelines in the *Occupational Safety and Health Guidance Manual for Hazardous Waste Site Activities* (See appendix D, Reference #10) should be consulted.

(5) ***Examination by a physician and costs.*** All medical examinations and procedures shall be performed by or under the supervision of a licensed physician, preferably one knowledgeable in occupational medicine, and shall be provided without cost to the employee, without loss of pay, and at a reasonable time and place.

(6) **Information provided to the physician.** The employer shall provide one copy of this standard and its appendices to the attending physician, and in addition the following for each employee:

   (i) A description of the employee's duties as they relate to the employee's

   (ii) exposures. The employee's exposure levels or anticipated exposure levels.

   (iii) A description of any personal protective equipment used or to be used.

   (iv) Information from previous medical examinations of the employee which is not readily available to the examining physician.

   (v) Information required by § 1926.103.

(7) **Physician's written opinion.**

   (i) The employer shall obtain and furnish the employee with a copy of a written opinion from the attending physician containing the following:

   (A) The physician's opinion as to whether the employee has any detected medical conditions which would place the employee at increased risk of material impairment of the employee's health from work in hazardous waste operations or emergency response, or from respirator use.

   (B) The physician's recommended limitations upon the employee's assigned

   (C) work. The results of the medical examination and tests if requested by the

   (D) employee.

   A statement that the employee has been informed by the physician of the results of the medical examination and any medical conditions which require further examination or treatment.

   (ii) The written opinion obtained by the employer shall not reveal specific findings or diagnoses unrelated to occupational exposures.

(8) **Recordkeeping.**

   (i) An accurate record of the medical surveillance required by paragraph (f) of this section shall be retained. This record shall be retained for the period specified and meet the criteria of 29 CFR 1926.33.

   (ii) The record required in paragraph (f)(8)(i) of this section shall include at least the following information:

   (A) The name of the employee;

   (B) Physician's written opinions, recommended limitations, and results of examinations and tests;

   (C) Any employee medical complaints related to exposure to hazardous substances;

   (D) A copy of the information provided to the examining physician by the employer, with the exception of the standard and its appendices.

(g) **Engineering controls, work practices, and personal protective equipment for employee protection.** Engineering controls, work practices, personal protective equipment, or a combination of these shall beimplemented in accordance with this paragraph to protect employees from exposure to hazardous substances and safety and health hazards.

(1) *Engineering controls, work practices and PPE for substances regulated either in § 1926.55, elsewhere in subpart D, or in other pertinent sections of this part.*

    (i) Engineering controls and work practices shall be instituted to reduce and maintain employeeexposure to or below the permissible exposure limits for substances regulated either in § 1926.55 or other pertinent sections of this part, except to the extent that such controls andpractices are not feasible.

    Note to (g)(1)(i): Engineering controls which may be feasible include the use of pressurized cabs or control booths on equipment, and/or the use of remotely operated material handling equipment. Work practices which may be feasible are removing all non-essential employees from potential exposure during opening of drums, wetting down dusty operations and locating employees upwind of possible hazards.
    (ii) Whenever engineering controls and work practices are not feasible or not required, any reasonable combination of engineering controls, work practices and PPE shall be used to reduce and maintain employee exposures to or below the permissible exposure limits or dose limits for substances regulated either in § 1926.55 or other pertinent sections of this part.

    (iii) The employer shall not implement a schedule of employee rotation as a means of compliance with permissible exposure limits or dose limits except when there is no other feasible way of complying with the airborne or dermal dose limits for ionizing radiation.

    (iv) The provisions of subpart D shall be followed.

(2) *Engineering controls, work practices, and PPE for substances not regulated either in § 1926.55, elsewhere in subpart D, or in other pertinent sections of this part.* An appropriate combination of engineering controls, work practices, and personal protective equipment shall be used to reduce and maintain employee exposure to or below published exposure levels for hazardous substances and health hazards not regulated either in § 1926.55, elsewhere in subpart D, or in other pertinent sections of this part. The employer may use the published literature and Safety Data Sheets (SDS) asa guide in making the employer's determination as to what level of protection the employer believes is appropriate for hazardous substances and health hazards for which there is no permissible exposure limit or published exposure limit.

(3) *Personal protective equipment selection.*

    (i) Personal protective equipment (PPE) shall be selected and used which will protect employees from the hazards and potential hazards they are likely to encounter as identified during the sitecharacterization and analysis.

    (ii) Personal protective equipment selection shall be based on an evaluation of the performance characteristics of the PPE relative to the requirements and limitations of the site, the task- specific conditions and duration, and the hazards and potential hazards

(iii) Positive pressure self-contained breathing apparatus, or positive pressure air-line respirators equipped with an escape air supply, shall be used when chemical exposure levels present will create a substantial possibility of immediate death, immediate serious illness or injury, or impair the ability to escape.

(iv) Totally-encapsulating chemical protective suits (protection equivalent to Level A protection as recommended in appendix B) shall be used in conditions where skin absorption of a hazardous substance may result in a substantial possibility of immediate death, immediate serious illness or injury, or impair the ability to escape.

(v) The level of protection provided by PPE selection shall be increased when additional information on site conditions indicates that increased protection is necessary to reduce employee exposures below permissible exposure limits and published exposure levels for hazardous substances and health hazards. (See appendix B for guidance on selecting PPE ensembles.)

Note to (g)(3): The level of employee protection provided may be decreased when additional information or site conditions show that decreased protection will not result in hazardous exposures to employees.

(vi) Personal protective equipment shall be selected and used to meet the requirements of subpart E of this part and additional requirements specified in this section.

(4) *Totally-encapsulating chemical protective suits.*

(i) Totally-encapsulating suits shall protect employees from the particular hazards which are identified during site characterization and analysis.

(ii) Totally-encapsulating suits shall be capable of maintaining positive air pressure. (See appendix A for a test method which may be used to evaluate this requirement.)

(iii) Totally-encapsulating suits shall be capable of preventing inward test gas leakage of more than
0.5 percent. (See appendix A for a test method which may be used to evaluate this requirement.)

(5) *Personal protective equipment (PPE) program.* A written personal protective equipment program, which is part of the employer's safety and health program required in paragraph (b) of this section or required in paragraph (p)(1) of this section and which is also a part of the site-specific safety and health plan shall be established. The PPE program shall address the elements listed below. When elements, such as donning and doffing procedures, are provided by the manufacturer of a piece of equipment and are attached to the plan, they need not be rewritten into the plan as long as they adequately address the procedure or element.

(i) PPE selection based upon site

(ii) hazards, PPE use and limitations of

(iii) the equipment, Work mission

(iv) duration,

PPE maintenance and storage,

(v) PPE decontamination and

(vi) disposal, PPE training and

(vii) proper fitting,

PPE donning and doffing procedures,

(viii) PPE inspection procedures prior to, during, and after use,

(ix) Evaluation of the effectiveness of the PPE program, and

(x) Limitations during temperature extremes, heat stress, and other appropriate medical considerations.

(h) *Monitoring* -

  (1) *General.*

    (i) Monitoring shall be performed in accordance with this paragraph where there may be a question of employee exposure to hazardous concentrations of hazardous substances in order to assure proper selection of engineering controls, work practices and personal protective equipment so that employees are not exposed to levels which exceed permissible exposure limits, or published exposure levels if there are no permissible exposure limits, for hazardous substances.

    (ii) Air monitoring shall be used to identify and quantify airborne levels of hazardous substances and safety and health hazards in order to determine the appropriate level of employee protection needed on site.

  (2) *Initial entry.* Upon initial entry, representative air monitoring shall be conducted to identify any IDLH condition, exposure over permissible exposure limits or published exposure levels, exposure over a radioactive material's dose limits or other dangerous condition such as the presence of flammable atmospheres or oxygen-deficient environments.

  (3) *Periodic monitoring.* Periodic monitoring shall be conducted when the possibility of an IDLH condition or flammable atmosphere has developed or when there is indication that exposures may have risen over permissible exposure limits or published exposure levels since prior monitoring.
Situations where it shall be considered whether the possibility that exposures have risen are as follows:

    (i) When work begins on a different portion of the site.

    (ii) When contaminants other than those previously identified are being handled.

    (iii) When a different type of operation is initiated (e.g., drum opening as opposed to exploratory well drilling).

    (iv) When employees are handling leaking drums or containers or working in areas with obvious liquid contamination (e.g., a spill or lagoon).

  (4) *Monitoring of high-risk employees.* After the actual clean-up phase of any hazardous waste operation commences; for example, when soil, surface water or containers are moved or disturbed; the employer shall monitor those employees likely to have the highest exposures to

hazardous substances and health hazards likely to be present above permissible exposure limits or published exposure levels by using personal sampling frequently enough to characterize employee exposures.If the employees likely to have the highest exposure are over permissible exposure limits or published exposure limits, then monitoring shall continue to determine all employees likely to be above those limits. The employer may utilize a representative sampling approach by documenting that the employees and chemicals chosen for monitoring are based on the criteria stated above.

Note to (h): It is not required to monitor employees engaged in site characterization operations covered by paragraph (c) of this section.

(i) **Informational programs.** Employers shall develop and implement a program, which is part of the employer's safety and health program required in paragraph (b) of this section, to inform employees, contractors, and subcontractors (or their representative) actually engaged in hazardous waste operations of the nature, level and degree of exposure likely as a result of participation in such hazardous waste operations. Employees, contractors and subcontractors working outside of the operations part of a site are not covered by this standard.

(j) **Handling drums and containers -**

   (1) **General.**

      (i) Hazardous substances and contaminated soils, liquids, and other residues shall be handled, transported, labeled, and disposed of in accordance with this paragraph.

      (ii) Drums and containers used during the clean-up shall meet the appropriate DOT, OSHA, and EPA regulations for the wastes that they contain.

      (iii) When practical, drums and containers shall be inspected and their integrity shall be assured prior to being moved. Drums or containers that cannot be inspected before being moved because of storage conditions (i.e., buried beneath the earth, stacked behind other drums, stacked several tiers high in a pile, etc.) shall be moved to an accessible location and inspectedprior to further handling.

      (iv) Unlabelled drums and containers shall be considered to contain hazardous substances and handled accordingly until the contents are positively identified and labeled.

      (v) Site operations shall be organized to minimize the amount of drum or container movement.

      (vi) Prior to movement of drums or containers, all employees exposed to the transfer operationshall be warned of the potential hazards associated with the contents of the drums or containers.

      (vii) U.S. Department of Transportation specified salvage drums or containers and suitable quantities of proper absorbent shall be kept available and used in areas where spills, leaks, orruptures may occur.

      (viii) Where major spills may occur, a spill containment program, which is part of the employer'ssafety and health program required in paragraph (b) of this section, shall be implemented tocontain and isolate the entire volume of the hazardous substance being transferred.

(ix) Drums and containers that cannot be moved without rupture, leakage, or spillage shall be emptied into a sound container using a device classified for the material being transferred.

(x) A ground-penetrating system or other type of detection system or device shall be used to estimate the location and depth of buried drums or containers.

(xi) Soil or covering material shall be removed with caution to prevent drum or container rupture.

(xii) Fire extinguishing equipment meeting the requirements of subpart F of this part shall be onhand and ready for use to control incipient fires.

(2) ***Opening drums and containers.*** The following procedures shall be followed in areas where drums orcontainers are being opened:

(i) Where an airline respirator system is used, connections to the source of air supply shall be protected from contamination and the entire system shall be protected from physical damage.

(ii) Employees not actually involved in opening drums or containers shall be kept a safe distancefrom the drums or containers being opened.

(iii) If employees must work near or adjacent to drums or containers being opened, a suitable shieldthat does not interfere with the work operation shall be placed between the employee and the drums or containers being opened to protect the employee in case of accidental explosion.

(iv) Controls for drum or container opening equipment, monitoring equipment, and fire suppressionequipment shall be located behind the explosion-resistant barrier.

(v) When there is a reasonable possibility of flammable atmospheres being present, material handling equipment and hand tools shall be of the type to prevent sources of ignition.

(vi) Drums and containers shall be opened in such a manner that excess interior pressure will besafely relieved. If pressure can not be relieved from a remote location, appropriate shielding shall be placed between the employee and the drums or containers to reduce the risk of employee injury.

(vii) Employees shall not stand upon or work from drums or containers.

(3) ***Material handling equipment.*** Material handiing equipment used to transfer drums and containers shall be selected, positioned and operated to minimize sources of ignition related to the equipmentfrom igniting vapors released from ruptured drums or containers.

(4) ***Radioactive wastes.*** Drums and containers containing radioactive wastes shall not be handled until such time as their hazard to employees is properly assessed.

(5) ***Shock sensitive wastes.*** As a minimum, the following special precautions shall be taken when drumsand containers containing or suspected of containing shock-sensitive wastes are handled:

(i) All non-essential employees shall be evacuated from the area of transfer.

(ii) Material handling equipment shall be provided with explosive containment devices or protectiveshields to protect equipment operators from exploding containers.

(iii) An employee alarm system capable of being perceived above surrounding light and noise conditions shall be used to signal the commencement and completion of explosive waste

(iv) Continuous communications (i.e., portable radios, hand signals, telephones, as appropriate) shall be maintained between the employee-in-charge of the immediate handling area and both the site safety and health supervisor and the command post until such time as the handling operation is completed. Communication equipment or methods that could cause shock sensitive materials to explode shall not be used.

(v) Drums and containers under pressure, as evidenced by bulging or swelling, shall not be moveduntil such time as the cause for excess pressure is determined and appropriate containmentprocedures have been implemented to protect employees from explosive relief of the drum.

(vi) Drums and containers containing packaged laboratory wastes shall be considered to containshock-sensitive or explosive materials until they have been characterized.

CAUTION: Shipping of shock sensitive wastes may be prohibited under U.S. Department of Transportation regulations. Employers and their shippers should refer to 49 CFR 173.21 and 173.50.

(6) *Laboratory waste packs.* In addition to the requirements of paragraph (j)(5) of this section, the following precautions shall be taken, as a minimum, in handling laboratory waste packs (lab packs):

(i) Lab packs shall be opened only when necessary and then only by an individual knowledgeable in the inspection, classification, and segregation of the containers within the pack according tothe hazards of the wastes.

(ii) If crystalline material is noted on any container, the contents shall be handled as a shock-sensitive waste until the contents are identified.

(7) *Sampling of drum and container contents.* Sampling of containers and drums shall be done in accordance with a sampling procedure which is part of the site safety and health plan developed forand available to employees and others at the specific worksite.

(8) *Shipping and transport.*

(i) Drums and containers shall be identified and classified prior to packaging for shipment.

(ii) Drum or container staging areas shall be kept to the minimum number necessary to identify andclassify materials safely and prepare them for transport.

(iii) Staging areas shall be provided with adequate access and egress routes.

(iv) Bulking of hazardous wastes shall be permitted only after a thorough characterization of the materials has been completed.

(9) *Tank and vault procedures.*

(i) Tanks and vaults containing hazardous substances shall be handled in a manner similar to thatfor drums and containers, taking into consideration the size of the tank or vault.

(ii) Appropriate tank or vault entry procedures as described in the employer's safety and health planshall be followed whenever employees must enter a tank or vault.

(k) *Decontamination* -

(1) ***General.*** Procedures for all phases of decontamination shall be developed and implemented in accordance with this paragraph.

(2) ***Decontamination procedures.***

    (i) A decontamination procedure shall be developed, communicated to employees and implemented before any employees or equipment may enter areas on site where potential forexposure to hazardous substances exists.

    (ii) Standard operating procedures shall be developed to minimize employee contact with hazardous substances or with equipment that has contacted hazardous substances.

    (iii) All employees leaving a contaminated area shall be appropriately decontaminated; all contaminated clothing and equipment leaving a contaminated area shall be appropriately disposed of or decontaminated.

    (iv) Decontamination procedures shall be monitored by the site safety and health supervisor to determine their effectiveness. When such procedures are found to be ineffective, appropriatesteps shall be taken to correct any deficiencies.

(3) ***Location.*** Decontamination shall be performed in geographical areas that will minimize the exposureof uncontaminated employees or equipment to contaminated employees or equipment.

(4) ***Equipment and solvents.*** All equipment and solvents used for decontamination shall be decontaminated or disposed of properly.

(5) ***Personal protective clothing and equipment.***

    (i) Protective clothing and equipment shall be decontaminated, cleaned, laundered, maintained orreplaced as needed to maintain their effectiveness.

    (ii) Employees whose non-impermeable clothing becomes wetted with hazardous substances shall immediately remove that clothing and proceed to shower. The clothing shall be disposed of ordecontaminated before it is removed from the work zone.

(6) ***Unauthorized employees.*** Unauthorized employees shall not remove protective clothing or equipment from change rooms.

(7) ***Commercial laundries or cleaning establishments.*** Commercial laundries or cleaning establishments that decontaminate protective clothing or equipment shall be informed of the potentially harmful effects of exposures to hazardous substances.

(8) ***Showers and change rooms.*** Where the decontamination procedure indicates a need for regular showers and change rooms outside of a contaminated area, they shall be provided and meet the requirements of 29 CFR 1910.141. If temperature conditions prevent the effective use of water, thenother effective means for cleansing shall be provided and used.

(l) ***Emergency response by employees at uncontrolled hazardous waste sites -***

(1) ***Emergency response plan.***

    (i) An emergency response plan shall be developed and implemented by all employers within the scope of paragraphs (a)(1) (i)-(ii) of this section to handle anticipated emergencies prior to thecommencement of hazardous waste operations. The plan shall be in writing and available forinspection and copying by employees, their representatives, OSHA personnel and other governmental agencies with relevant responsibilities.

(ii) Employers who will evacuate their employees from the danger area when an emergency occurs, and who do not permit any of their employees to assist in handling the emergency, are exempt from the requirements of this paragraph if they provide an emergency action plan complying with § 1926.35 of this part.

(2) **Elements of an emergency response plan.** The employer shall develop an emergency response plan for emergencies which shall address, as a minimum, the following:

(i) Pre-emergency planning.

(ii) Personnel roles, lines of authority, and communication.

(iii) Emergency recognition and prevention.

(iv) Safe distances and places of

(v) refuge. Site security and control.

(vi) Evacuation routes and procedures.

(vii) Decontamination procedures which are not covered by the site safety and health plan.

(viii) Emergency medical treatment and first aid.

(ix) Emergency alerting and response procedures.

(x) Critique of response and follow-up.

(xi) PPE and emergency equipment.

(3) **Procedures for handling emergency incidents.**

(i) In addition to the elements for the emergency response plan required in paragraph (1)(2) of this section, the following elements shall be included for emergency response plans:

(A) Site topography, layout, and prevailing weather conditions.

(B) Procedures for reporting incidents to local, state, and federal governmental

(ii) agencies. The emergency response plan shall be a separate section of the Site Safety

(iii) and Health Plan.

The emergency response plan shall be compatible and integrated with the disaster, fire and/or emergency response plans of local, state, and federal agencies.

(iv) The emergency response plan shall be rehearsed regularly as part of the overall training program for site operations.

(v) The site emergency response plan shall be reviewed periodically and, as necessary, be amended to keep it current with new or changing site conditions or information.

(vi) An employee alarm system shall be installed to notify employees of an emergency situation; to stop work activities if necessary; to lower background noise in order to speed communication; and to begin emergency procedures.

(vii) Based upon the information available at time of the emergency, the employer shall evaluate the incident and the site response capabilities and proceed with the appropriate steps to implement the site emergency response plan.

(m) **Illumination.** Areas accessible to employees shall be lighted to not less than the minimum illumination intensities listed in the following Table D-65.1 while any work is in progress:

Table D-65.1 - Minimum Illumination Intensities in Foot-Candles

Foot-candles	Area or operations
5	General site areas.

(n)

Foot-candles	Area or operations
3	Excavation and waste areas, accessways, active storage areas, loading platforms, refueling, and field maintenance areas.
5	Indoors: Warehouses, corridors, hallways, and exitways.
5	Tunnels, shafts, and general underground work areas. (Exception: Minimum of 10 foot-candles is required at tunnel and shaft heading during drilling mucking, and scaling. Mine Safety and Health Administration approved cap lights shall be acceptable for use in the tunnel heading.)
10	General shops (e.g., mechanical and electrical equipment rooms, active storerooms, barracks or living quarters, locker or dressing rooms, dining areas, and indoor toilets and workrooms.)
30	First aid stations, infirmaries, and offices.

Sanitation at temporary workplaces -

(1) **Potable water.**

    (i) An adequate supply of potable water shall be provided on the site.

    (ii) Portable containers used to dispense drinking water shall be capable of being tightly closed, and equipped with a tap. Water shall not be dipped from containers.

    (iii) Any container used to distribute drinking water shall be clearly marked as to the nature of its contents and not used for any other purpose.

    (iv) Where single service cups (to be used but once) are supplied, both a sanitary container for the unused cups and a receptacle for disposing of the used cups shall be provided.

(2) **Nonpotable water.**

    (i) Outlets for nonpotable water, such as water for firefighting purposes, shall be identified to indicate clearly that the water is unsafe and is not to be used for drinking, washing, or cooking purposes.

    (ii) There shall be no cross-connection, open or potential, between a system furnishing potable water and a system furnishing nonpotable water.

(3) **Toilet facilities.**

    (i) Toilets shall be provided for employees according to the following Table D-65.2.

Table D-65.2 - Toilet Facilities

Number of employees	Minimum number of facilities
20 or fewer	One.
More than 20, fewer than 200	One toilet seat and one urinal per 40 employees.
More than 200	One toilet seat and one urinal per 50 employees.

- (ii) Under temporary field conditions, provisions shall be made to assure that at least one toilet facility is available.

- (iii) Hazardous waste sites not provided with a sanitary sewer shall be provided with the following toilet facilities unless prohibited by local codes:

    - (A) Chemical
    - (B) toilets;
    - (C) Recirculating
    - (D) toilets;

- (iv) Combustion

    toilets; or Flush

    toilets.

    The requirements of this paragraph for sanitation facilities shall not apply to mobile crews having transportation readily available to nearby toilet facilities.

- (v) Doors entering toilet facilities shall be provided with entrance locks controlled from inside the facility.

(4) **Food handling.** All food service facilities and operations for employees shall meet the applicable laws, ordinances, and regulations of the jurisdictions in which they are located.

(5) **Temporary sleeping quarters.** When temporary sleeping quarters are provided, they shall be heated, ventilated, and lighted.

(6) **Washing facilities.** The employer shall provide adequate washing facilities for employees engaged in operations where hazardous substances may be harmful to employees. Such facilities shall be in near proximity to the worksite; in areas where exposures are below permissible exposure limits and published exposure levels and which are under the controls of the employer; and shall be so equipped as to enable employees to remove hazardous substances from themselves.

(7) **Showers and change rooms.** When hazardous waste clean-up or removal operations commence on a site and the duration of the work will require six months or greater time to complete, the employer shall provide showers and change rooms for all employees exposed to hazardous substances and health hazards involved in hazardous waste clean-up or removal operations.

- (i) Showers shall be provided and shall meet the requirements of 29 CFR 1926.51(f)(4).

- (ii) Change rooms shall be provided and shall meet the requirements of 29 CFR 1926.51(i). Change rooms shall consist of two separate change areas separated by the shower area required in paragraph (n)(7)(i) of this section. One change area, with an exit leading off the worksite, shallprovide employees with a clean area where they can remove, store, and put on street clothing.The second area, with an exit to the worksite, shall provide employees with an area where they can put on, remove and store work clothing and personal protective equipment.

    (iii)   Showers and change rooms shall be located in areas where exposures are below the permissible exposure limits and published exposure levels. If this cannot be accomplished,then a ventilation system shall be provided that will supply air that is below the permissibleexposure limits and published exposure levels.

    (iv)   Employers shall assure that employees shower at the end of their work shift and when leavingthe hazardous waste site.

(o) *New technology programs.*

    (1)   The employer shall develop and implement procedures for the introduction of effective new technologies and equipment developed for the improved protection of employees working with hazardous waste clean-up operations, and the same shall be implemented as part of the site safetyand health program to assure that employee protection is being maintained.

    (2)   New technologies, equipment or control measures available to the industry, such as the use of foams, absorbents, adsorbents, neutralizers, or other means to suppress the level of air contaminates while excavating the site or for spill control, shall be evaluated by employers or theirrepresentatives. Such an evaluation shall be done to determine the effectiveness of the new methods, materials, or equipment before implementing their use on a large scale for enhancing employee protection. Information and data from manufacturers or suppliers may be used as part ofthe employer's evaluation effort. Such evaluations shall be made available to OSHA upon request.

(p) *Certain operations conducted under the Resource Conservation and Recovery Act of 1976 (RCRA).* Employers conducting operations at treatment, storage and disposal (TSD) facilities specified in paragraph (a)(1)(iv) of this section shall provide and implement the programs specified in this paragraph.See the "Notes and Exceptions" to paragraph (a)(2)(iii) of this section for employers not covered.)".

    (1)   *Safety and health program.* The employer shall develop and implement a written safety and health program for employees involved in hazardous waste operations that shall be available for inspectionby employees, their representatives and OSHA personnel. The program shall be designed to identify, evaluate and control safety and health hazards in their facilities for the purpose of employee protection, to provide for emergency response meeting the requirements of paragraph (p)(8) of thissection and to address as appropriate site analysis, engineering controls, maximum exposure limits, hazardous waste handling procedures and uses of new technologies.

    (2)   *Hazard communication program.* The employer shall implement a hazard communication programmeeting the requirements of 29 CFR 1926.59 as part of the employer's safety and program.

    Note to 1926.65: The exemption for hazardous waste provided in § 1926.59 is applicable to this section.

    Medical surveillance program. The employer shall develop and implement a medical surveillanceprogram meeting the requirements of paragraph (f) of this section.

(3)

(4) **Decontamination program.** The employer shall develop and implement a decontamination proceduremeeting the requirements of paragraph (k) of this section.

(5) **New technology program.** The employer shall develop and implement procedures meeting the requirements of paragraph (o) of this section for introducing new and innovative equipment into theworkplace.

(6) **Material handling program.** Where employees will be handling drums or containers, the employer shall develop and implement procedures meeting the requirements of paragraphs (j)(1) (ii) through

(viii) and (xi) of this section, as well as (j)(3) and (j)(8) of this section prior to starting such work.

(7) **Training program** -

    (i) **New employees.** The employer shall develop and implement a training program, which is part ofthe employer's safety and health program, for employees exposed to health hazards or hazardous substances at TSD operations to enable the employees to perform their assigned

    duties and functions in a safe and healthful manner so as not endanger themselves or other employees. The initial training shall be for 24 hours and refresher training shall be for eight hours annually. Employees who have received the initial training required by this paragraph shallbe given a written certificate attesting that they have successfully completed the necessary training.

    (ii) **Current employees.** Employers who can show by an employee's previous work experience and/ or training that the employee has had training equivalent to the initial training required by thisparagraph, shall be considered as meeting the initial training requirements of this paragraph asto that employee. Equivalent training includes the training that existing employees might havealready received from actual site work experience. Current employees shall receive eight hours of refresher training annually.

    (iii) **Trainers.** Trainers who teach initial training shall have satisfactorily completed a training course for teaching the subjects they are expected to teach or they shall have the academic credentials and instruction experience necessary to demonstrate a good command of the subject matter of the courses and competent instructional skills.

(8) **Emergency response program** -

    (i) **Emergency response plan.** An emergency response plan shall be developed and implemented by all employers. Such plans need not duplicate any of the subjects fully addressed in the employer's contingency planning required by permits, such as those issued by the U.S. Environmental Protection Agency, provided that the contingency plan is made part of the emergency response plan. The emergency response plan shall be a written portion of the employers safety and health program required in paragraph (p)(1) of this section. Employers who will evacuate their employees from the worksite location when an emergency occurs and who do not permit any of their employees to assist in handling the emergency are exempt from the requirements of paragraph (p)(8) if they provide an emergency action plan complying with §1926.35 of this part.

    (ii) **Elements of an emergency response plan.** The employer shall develop an emergency response plan for emergencies which shall address, as a minimum, the following areas to the extent thatthey are not addressed in any specific program required in this paragraph:

        (A) Pre-emergency planning and coordination with outside

        (B)

        (C)

parties. Personnel roles, lines of authority, and communication.

Emergency recognition and
- (D) prevention. Safe distances and
- (E) places of refuge.

Site security and control.
- (F) Evacuation routes and
- (G) procedures.
- (H) Decontamination procedures.

Emergency medical treatment and first
- (I) aid. Emergency alerting and response
- (J) procedures. Critique of response and follow-up.
- (K) PPE and emergency equipment.

(iii) **Training.**

(A) Training for emergency response employees shall be completed before they are called upon to perform in real emergencies. Such training shall include the elements of the emergency response plan, standard operating procedures the employer has established for the job, the personal protective equipment to be worn and procedures for handling emergency incidents.

Exception #1: An employer need not train all employees to the degree specified if the employer divides the work force in a manner such that a sufficient number of employees who have responsibility to control emergencies have the training specified, and all other employees, who may first respond to an emergency incident, have sufficient awareness training to recognize that an emergency response situation exists and that they are instructed in that case to summon the fully trained employees and not attempt control activities for which they are not trained.

Exception #2: An employer need not train all employees to the degree specified if arrangements have been made in advance for an outside fully-trained emergency response team to respond in a reasonable period and all employees, who may come to the incident first, have sufficient awareness training to recognize that an emergency response situation exists and they have been instructed to call the designated outside fully-trained emergency response team for assistance.

(B) Employee members of TSD facility emergency response organizations shall be trained to a level of competence in the recognition of health and safety hazards to protect themselves and other employees. This would include training in the methods used to minimize the risk from safety and health hazards; in the safe use of control equipment; in the selection anduse of appropriate personal protective equipment; in the safe operating procedures to be used at the incident scene; in the techniques of coordination with other employees to minimize risks; in the appropriate response to over exposure from health hazards or injury to themselves and other employees; and in the recognition of subsequent symptoms which may result from over exposures.

    (C)   The employer shall certify that each covered employee has attended and successfully completed the training required in paragraph (p)(8)(iii) of this section, or shall certify theemployee's competency at least yearly. The method used to demonstrate competency forcertification of training shall be recorded and maintained by the employer.

  (iv)   ***Procedures for handling emergency incidents.***

    (A)   In addition to the elements for the emergency response plan required in paragraph (p)(8)(ii) of this section, the following elements shall be included for emergency responseplans to the extent that they do not repeat any information already contained in the emergency response plan:

        (1)   Site topography, layout, and prevailing weather conditions.

        (2)   Procedures for reporting incidents to local, state, and federal governmental agencies.

    (B)   The emergency response plan shall be compatible and integrated with the disaster, fireand/or emergency response plans of local, state, and federal agencies.

    (C)   The emergency response plan shall be rehearsed regularly as part of the overall trainingprogram for site operations.

    (D)   The site emergency response plan shall be reviewed periodically and, as necessary, beamended to keep it current with new or changing site conditions or information.

    (E)   An employee alarm system shall be installed to notify employees of an emergency situation; to stop work activities if necessary; to lower background noise in order to speedcommunication; and to begin emergency procedures.

    (F)   Based upon the information available at time of the emergency, the employer shall evaluate the incident and the site response capabilities and proceed with the appropriatesteps to implement the site emergency response plan.

(q)   ***Emerqency response to hazardous substance releases.*** This paragraph covers employers whose employees are engaged in emergency response no matter where it occurs except that it does not cover employees engaged in operations specified in paragraphs (a)(1)(i) through (a)(1)(iv) of this section. Those emergency response organizations who have developed and implemented programs equivalent to this paragraph for handling releases of hazardous substances pursuant to section 303 of the Superfund Amendments and Reauthorization Act of 1986 (Emergency Planning and Community Right-to-Know Act of1986, 42 U.S.C. 11003) shall be deemed to have met the requirements of this paragraph.

  (1)   ***Emergency response plan.*** An emergency response plan shall be developed and implemented to handle anticipated emergencies prior to the commencement of emergency

response operations. The plan shall be in writing and available for inspection and copying by employees, their representatives and OSHA personnel. Employers who will evacuate their employees from the danger area when an emergency occurs, and who do not permit any of their employees to assist in handling the emergency, are exempt from the requirements of this paragraph if they provide an emergency action plan in accordance with § 1926.35 of this part.

(2) **Elements of an emergency response plan.** The employer shall develop an emergency response plan for emergencies which shall address, as a minimum, the following to the extent that they are not addressed elsewhere:

(i) Pre-emergency planning and coordination with outside

(ii) parties. Personnel roles, lines of authority, training, and

(iii) communication. Emergency recognition and prevention.

(iv) Safe distances and places of

(v) refuge. Site security and control.

(vi) Evacuation routes and

(vii) procedures.

(viii) Decontamination.

Emergency medical treatment and first aid.

(ix) Emergency alerting and response (x)

procedures. Critique of response and follow-

up.

(xi) PPE and emergency equipment.

(xii) Emergency response organizations may use the local emergency response plan or the state emergency response plan or both, as part of their emergency response plan to avoid duplication. Those items of the emergency response plan that are being properly addressed by the SARA Title III plans may be substituted into their emergency plan or otherwise kept together for the employer and employee's use.

(3) **Procedures for handling emergency response.**

(i) The senior emergency response official responding to an emergency shall become the individual in charge of a site-specific Incident Command System (ICS). All emergency responders and their communications shall be coordinated and controlled through the individual in charge of the ICS assisted by the senior official present for each employer.

Note to (g)(3)(i): The *senior oficial* at an emergency response is the most senior official on the site who has the responsibility for controlling the operations at the site. Initially it is the senior officer on the first-due piece of responding emergency apparatus to arrive on the incident scene. As more senior officers arrive (i.e., battalion chief, fire chief, state law enforcement official, site coordinator, etc.) the position is passed up the line of authority which has been previously established.

(ii) The individual in charge of the ICS shall identify, to the extent possible, all hazardous substances or conditions present and shall address as appropriate site analysis, use of engineering controls, maximum exposure limits, hazardous substance handling procedures,and use of any new technologies.

(iii) Based on the hazardous substances and/or conditions present, the individual in charge of the ICS shall implement appropriate emergency operations, and assure that the personal protectiveequipment worn is appropriate for the hazards to be encountered.

(iv) Employees engaged in emergency response and exposed to hazardous substances presentingan inhalation hazard or potential inhalation hazard shall wear positive pressure self-contained breathing apparatus while engaged in emergency response, until such time that the individual incharge of the ICS determines through the use of air monitoring that a decreased level of respiratory protection will not result in hazardous exposures to employees.

(v) The individual in charge of the ICS shall limit the number of emergency response personnel atthe emergency site, in those areas of potential or actual exposure to incident or site hazards, tothose who are actively performing emergency operations. However, operations in hazardous areas shall be performed using the buddy system in groups of two or more.

(vi) Back-up personnel shall stand by with equipment ready to provide assistance or rescue. Advance first aid support personnel, as a minimum, shall also stand by with medical equipmentand transportation capability.

(vii) The individual in charge of the ICS shall designate a safety official, who is knowledgable in theoperations being implemented at the emergency response site, with specific responsibility to identify and evaluate hazards and to provide direction with respect to the safety of operationsfor the emergency at hand.

(viii) When activities are judged by the safety official to be an IDLH condition and/or to involve an imminent danger condition, the safety official shall have the authority to alter, suspend, or terminate those activities. The safety official shall immediately inform the individual in charge of the ICS of any actions needed to be taken to correct these hazards at the emergency scene.

(ix) After emergency operations have terminated, the individual in charge of the ICS shall implementappropriate decontamination procedures.

(x) When deemed necessary for meeting the tasks at hand, approved self-contained compressed air breathing apparatus may be used with approved cylinders from other approved self- contained compressed air breathing apparatus provided that such cylinders are of the same capacity and pressure rating. All compressed air cylinders used with self-contained breathing apparatus shall meet U.S. Department of Transportation and National Institute for OccupationalSafety and Health criteria.

(4) **Skilled support personnel.** Personnel, not necessarily an employer's own employees, who are skilled in the operation of certain equipment, such as mechanized earth moving or digging equipment or crane and hoisting equipment, and who are needed temporarily to perform immediate emergency support work that cannot reasonably be performed in a timely fashion by an employer's own employees, and who will be or may be exposed to the hazards at an emergency response scene, arenot required to meet the training required in this paragraph for the employer's regular employees.

However, these personnel shall be given an initial briefing at the site prior to their participation in anyemergency response. The initial briefing shall include instruction in the wearing of appropriate personal protective equipment, what chemical hazards are involved, and what duties are to be performed. All other appropriate safety and health precautions provided to the employer's own employees shall be used to assure the safety and health of these personnel.

(5) **Specialist employees.** Employees who, in the course of their regular job duties, work with and are trained in the hazards of specific hazardous substances, and who will be called upon to provide technical advice or assistance at a hazardous substance release incident to the individual in charge,shall receive training or demonstrate competency in the area of their specialization annually.

(6) **Training.** Training shall be based on the duties and function to be performed by each responder of anemergency response organization. The skill and knowledge levels required for all new responders, those hired after the effective date of this standard, shall be conveyed to them through training before they are permitted to take part in actual emergency operations on an incident. Employees who participate, or are expected to participate, in emergency response, shall be given training in accordance with the following paragraphs:

(i) **First responder awareness level.** First responders at the awareness level are individuals who arelikely to witness or discover a hazardous substance release and who have been trained to initiate an emergency response sequence by notifying the proper authorities of the release.

They would take no further action beyond notifying the authorities of the release. First responders at the awareness level shall have sufficient training or have had sufficient experience to objectively demonstrate competency in the following areas:

(A) An understanding of what hazardous substances are, and the risks associated with themin an incident.

(B) An understanding of the potential outcomes associated with an emergency created whenhazardous substances are present.

(C) The ability to recognize the presence of hazardous substances in an

(D) emergency.The ability to identify the hazardous substances, if possible.

(E) An understanding of the role of the first responder awareness individual in the employer'semergency response plan including site security and control and the U.S. Department of Transportation's Emergency Response Guidebook.

(F) The ability to realize the need for additional resources, and to make appropriate notifications to the communication center.

(ii) **First responder operations level.** First responders at the operations level are individuals who respond to releases or potential releases of hazardous substances as part of the initial response to the site for the purpose of protecting nearby persons, property, or the environment from the effects of the release. They are trained to respond in a defensive fashion without actually trying to stop the release. Their function is to contain the release from a safe distance, keep it from spreading, and prevent exposures. First responders at the operational level shall have received at least eight hours of training or have had sufficient experience to objectively demonstrate competency in the following areas in addition to those listed for the awareness level and the employer shall so certify:

(A) Knowledge of the basic hazard and risk assessment techniques.

(B) Know how to select and use proper personal protective equipment provided to the first responder operational level.

(C) An understanding of basic hazardous materials terms.

(D) Know how to perform basic control, containment and/or confinement operations within the capabilities of the resources and personal protective equipment available with their unit.

(E) Know how to implement basic decontamination procedures.

(F) An understanding of the relevant standard operating procedures and termination procedures.

(iii) **Hazardous materials technician.** Hazardous materials technicians are individuals who respond to releases or potential releases for the purpose of stopping the release. They assume a more aggressive role than a first responder at the operations level in that they will approach the point of release in order to plug, patch or otherwise stop the release of a hazardous substance.
Hazardous materials technicians shall have received at least 24 hours of training equal to the first responder operations level and in addition have competency in the following areas and the employer shall so certify:

(A) Know how to implement the employer's emergency response plan.

(B) Know the classification, identification and verification of known and unknown materials by using field survey instruments and equipment.

(C) Be able to function within an assigned role in the Incident Command System.

(D) Know how to select and use proper specialized chemical personal protective equipment provided to the hazardous materials technician.

(E) Understand hazard and risk assessment techniques.

(F) Be able to perform advance control, containment, and/or confinement operations within the capabilities of the resources and personal protective equipment available with the unit.

- (G) Understand and implement decontamination procedures.
- (H) Understand termination procedures.
- (I) Understand basic chemical and toxicological terminology and behavior.

(iv) *Hazardous materials specialist.* Hazardous materials specialists are individuals who respond with and provide support to hazardous materials technicians. Their duties parallel those of the hazardous materials technician, however, those duties require a more directed or specific knowledge of the various substances they may be called upon to contain. The hazardous materials specialist would also act as the site liaison with Federal, state, local and other government authorities in regards to site activities. Hazardous materials specialists shall have received at least 24 hours of training equal to the technician level and in addition have competency in the following areas and the employer shall so certify:

- (A) Know how to implement the local emergency response plan.
- (B) Understand classification, identification and verification of known and unknown materials by using advanced survey instruments and equipment.
- (C) Know of the state emergency response plan.
- (D) Be able to select and use proper specialized chemical personal protective equipment provided to the hazardous materials specialist.
- (E) Understand in-depth hazard and risk techniques.
- (F) Be able to perform specialized control, containment, and/or confinement operations within the capabilities of the resources and personal protective equipment available.
- (G) Be able to determine and implement decontamination procedures.
- (H) Have the ability to develop a site safety and control plan.
- (I) Understand chemical, radiological and toxicological terminology and behavior.

(v) *On scene incident commander.* Incident commanders, who will assume control of the incident scene beyond the first responder awareness level, shall receive at least 24 hours of training equal to the first responder operations level and in addition have competency in the following areas and the employer shall so certify:

- (A) Know and be able to implement the employer's incident command system.
- (B) Know how to implement the employer's emergency response plan.
- (C) Know and understand the hazards and risks associated with employees working in chemical protective clothing.
- (D) Know how to implement the local emergency response plan.
- (E) Know of the state emergency response plan and of the Federal Regional Response Team.
- (F) Know and understand the importance of decontamination procedures.

(7) **Trainers.** Trainers who teach any of the above training subjects shall have satisfactorily completed a training course for teaching the subjects they are expected to teach, such as the courses offered by the U.S. National Fire Academy, or they shall have the training and/or academic credentials and instructional experience necessary to demonstrate competent instructional skills and a good command of the subject matter of the courses they are to teach.

(8) **Refresher training.**

   (i) Those employees who are trained in accordance with paragraph (q)(6) of this section shall receive annual refresher training of sufficient content and duration to maintain their competencies, or shall demonstrate competency in those areas at least yearly.

   (ii) A statement shall be made of the training or competency, and if a statement of competency is made, the employer shall keep a record of the methodology used to demonstrate competency.

(9) **Medical surveillance and consultation.**

   (i) Members of an organized and designated HAZMAT team and hazardous materials specialists shall receive a baseline physical examination and be provided with medical surveillance as required in paragraph (f) of this section.

   (ii) Any emergency response employees who exhibits signs or symptoms which may have resulted from exposure to hazardous substances during the course of an emergency incident, either immediately or subsequently, shall be provided with medical consultation as required in paragraph (f)(3)(ii) of this section.

(10) **Chemical protective clothing.** Chemical protective clothing and equipment to be used by organized and designated HAZMAT team members, or to be used by hazardous materials specialists, shall meet the requirements of paragraphs (g) (3) through (5) of this section.

(11) **Post-emergency response operations.** Upon completion of the emergency response, if it is determined that it is necessary to remove hazardous substances, health hazards, and materials contaminated with them (such as contaminated soil or other elements of the natural environment) from the site of the incident, the employer conducting the clean-up shall comply with one of the following:

   (i) Meet all of the requirements of paragraphs (b) through (o) of this section; or

   (ii) Where the clean-up is done on plant property using plant or workplace employees, such employees shall have completed the training requirements of the following: 29 CFR 1926.35, 1926.59, and 1926.103, and other appropriate safety and health training made necessary by the tasks that they are expected to be performed such as personal protective equipment and decontamination procedures. All equipment to be used in the performance of the clean-up work shall be in serviceable condition and shall have been inspected prior to use.

Appendices to § 1926.65 - Hazardous Waste Operations and EmergencyResponse

Note: The following appendices serve as non-mandatory guidelines to assist employees and employers in complying with the appropriate requirements of this section. However § 1926.65(g)

makes mandatory in certain circumstances the use of Level A and Level B PPE protection.

Appendix A to § 1926.65 - Personal Protective Equipment Test Methods

This appendix sets forth the non-mandatory examples of tests which may be used to evaluate compliance with
§ 1926.65(g)(4) (ii) and (iii). Other tests and other challenge agents may be used to evaluate compliance.

A. Totally-encapsulating chemical protective suit pressure test

1.0 - Scope

1.1 This practice measures the ability of a gas tight totally-encapsulating chemical protective suit material, seams, and closures to maintain a fixed positive pressure. The results of this practice allow the gas tight integrity of a totally-encapsulating chemical protective suit to be evaluated.

1.2 Resistance of the suit materials to permeation, penetration, and degradation by specific hazardoussubstances is not determined by this test method.

2.0 - Definition of terms

2.1 *Totally-encapsulated chemical protective suit (TECP suit)* means a full body garment which is constructed of protective clothing materials; covers the wearer's torso, head, arms, legs and respirator; may cover the wearer's hands and feet with tightly attached gloves and boots; completely encloses the wearer and respirator by itselfor in combination with the wearer's gloves and boots.

2.2 *Protective clothing material* means any material or combination of materials used in an item of clothing for the purpose of isolating parts of the body from direct contact with a potentially hazardous liquid or gaseous chemicals.

2.3 *Gas tight* means, for the purpose of this test method, the limited flow of a gas under pressure from the inside of a TECP suit to atmosphere at a prescribed pressure and time interval.

3.0 - Summary of test method

3.1 The TECP suit is visually inspected and modified for the test. The test apparatus is attached to the suit to permit inflation to the pre-test suit expansion pressure for removal of suit wrinkles and creases. The pressure is lowered to the test pressure and monitored for three minutes. If the pressure drop is excessive, the TECP suit fails the test and is removed from service. The test is repeated after leak location and repair.

4.0 - Required Supplies

4.1 Source of compressed air.

4.2 Test apparatus for suit testing, including a pressure measurement device with a sensitivity of at least $1/4$ inch water gauge.

4.3 Vent valve closure plugs or sealing tape.

4.4 Soapy water solution and soft brush.

4.5 Stop watch or appropriate timing device.

5.0 - Safety Precautions

5.1 Care shall be taken to provide the correct pressure safety devices required for the source of compressed air used.

6.0 - Test Procedure

6.1 Prior to each test, the tester shall perform a visual inspection of the suit. Check the suit for seam integrity by visually examining the seams and gently pulling on the seams. Ensure that all air supply lines, fittings, visor, zippers, and valves are secure and show no signs of deterioration.

6.1.1 Seal off the vent valves along with any other normal inlet or exhaust points (such as umbilical air line fittings or face piece opening) with tape or other appropriate means (caps, plugs, fixture, etc.). Care should be exercised in the sealing process not to damage any of the suit components.

6.1.2 Close all closure assemblies.

6.1.3 Prepare the suit for inflation by providing an improvised connection point on the suit for connecting an airline. Attach the pressure test apparatus to the suit to permit suit inflation from a compressed air source equipped with a pressure indicating regulator. The leak tightness of the pressure test apparatus should be tested before and after each test by closing off the end of the tubing attached to the suit and assuring a pressure of three inches water gauge for three minutes can be maintained. If a component is removed for the test, that component shall be replaced and a second test conducted with another component removed to permit a complete test of the ensemble.

6.1.4 The pre-test expansion pressure (A) and the suit test pressure (B) shall be supplied by the suit manufacturer, but in no case shall they be less than: (A) = three inches water gauge; and (B) = two inches water gauge. The ending suit pressure (C) shall be no less than 80 percent of the test pressure (B); i.e., the pressuredrop shall not exceed 20 percent of the test pressure (B).

6.1.5 Inflate the suit until the pressure inside is equal to pressure (A), the pre-test expansion suit pressure. Allow at least one minute to fill out the wrinkles in the suit. Release sufficient air to reduce the suit pressure to pressure (B), the suit test pressure. Begin timing. At the end of three minutes, record the suit pressure as pressure (C), the ending suit pressure. The difference between the suit test pressure and the ending suit testpressure (B-C) shall be defined as the suit pressure drop.

6.1.6 If the suit pressure drop is more than 20 percent of the suit test pressure (B) during the three-minute testperiod, the suit fails the test and shall be removed from service.

7.0 - Retest Procedure

7.1 If the suit fails the test check for leaks by inflating the suit to pressure (A) and brushing or wiping the entire suit (including seams, closures, lens gaskets, glove-to-sleeve joints, etc.) with a mild soap and water solution. Observe the suit for the formation of soap bubbles, which is an indication of a leak. Repair all identified leaks.

7.2 Retest the TECP suit as outlined in Test procedure 6.0.

8.0 - Report

8.1 Each TECP suit tested by this practice shall have the following information recorded:

8.1.1 Unique identification number, identifying brand name, date of purchase, material of construction, andunique fit features, e.g., special breathing apparatus.

8.1.2 The actual values for test pressures (A), (B), and (C) shall be recorded along with the specific observationtimes. If the ending pressure (C) is less than 80 percent of the test pressure (B), the suit shall be identified as failing the test. When possible, the specific leak location shall be identified in the test records. Retest pressure data shall be recorded as an additional test.

8.1.3 The source of the test apparatus used shall be identified and the sensitivity of the pressure gauge shall be recorded.

8.1.4 Records shall be kept for each pressure test even if repairs are being made at the test location.

CAUTION

Visually inspect all parts of the suit to be sure they are positioned correctly and secured tightly before puttingthe suit back into service. Special care should be taken to examine each exhaust valve to make sure it is not blocked.

Care should also be exercised to assure that the inside and outside of the suit is completely dry before it is putinto storage.

B. Totally-encapsulating chemical protective suit qualitative leak test

1.0 - Scope

1.1 This practice semi-qualitatively tests gas tight totally-encapsulating chemical protective suit integrity by detecting inward leakage of ammonia vapor. Since no modifications are made to the suit to carry out this test, the results from this practice provide a realistic test for the integrity of the entire suit.

1.2 Resistance of the suit materials to permeation, penetration, and degradation is not determined by this testmethod. ASTM test methods are available to test suit materials for these characteristics and the tests are usually conducted by the manufacturers of the suits.

2.0 - Definition of terms

2.1 *Totally-encapsulated chemical protective suit (TECP suit)* means a full body garment which is constructed of protective clothing materials; covers the wearer's torso, head, arms, legs and respirator; may cover the wearer's hands and feet with tightly attached gloves and boots; completely encloses the wearer and respirator by itself or in combination with the wearer's gloves, and boots.

2.2 *Protective clothing material* means any material or combination of materials used in an item of clothing for the purpose of isolating parts of the body from direct contact with a potentially hazardous liquid or gaseous chemicals.

2.3 *Gas tight* means, for the purpose of this test method, the limited flow of a gas under pressure from the inside of a TECP suit to atmosphere at a prescribed pressure and time interval.

2.4 *Intrusion Coeficient* means a number expressing the level of protection provided by a gas tight totally- encapsulating chemical protective suit. The intrusion coefficient is calculated by dividing the test room challenge agent concentration by the concentration of challenge agent found inside the suit. The accuracy of the intrusion coefficient is dependent on the challenge agent monitoring methods. The larger the intrusion coefficient the greater the protection provided by the TECP suit.

3.0 - Summary of recommended practice

3.1 The volume of concentrated aqueous ammonia solution (ammonia hydroxide NH_4OH) required to generate the test atmosphere is determined using the directions outlined in 6.1. The suit is donned by a person wearing the appropriate respiratory equipment (either a positive pressure self-contained breathing apparatus or a positive pressure supplied air respirator) and worn inside the enclosed test room. The concentrated aqueous ammonia solution is taken by the suited individual into the test room and poured into an open plastic pan. A two-minute evaporation period is observed before the test room concentration is measured, using a high range ammonia length of stain detector tube. When the ammonia vapor reaches a concentration of between 1000 and 1200 ppm, the suited individual starts a standardized exercise protocol to stress and flex the suit. After this protocol is completed, the test room concentration is measured again. The suited individual exits the test room and his stand-by person measures the ammonia concentration inside the suit using a low range ammonia length of stain detector tube or other more sensitive ammonia detector. A stand-by person is required to observe the test individual during the test procedure; aid the person in donning and doffing the TECP suit; and monitor the suit interior. The intrusion coefficient of the suit can be calculated by dividing the average test area concentration by the interior suit concentration. A colorimetric ammonia indicator strip of bromophenol blue or equivalent is placed on the inside of the suit face piece lens so that the suited individual is able to detect a color change and know if the suit has a significant leak. If a color change is observed the individual shall leave the test room immediately.

4.0 - Required supplies

4.1 A supply of concentrated aqueous ammonium hydroxide (58% by weight).

4.2 A supply of bromophenol/blue indicating paper or equivalent, sensitive to 5-10 ppm ammonia or greaterover a two-minute period of exposure. [pH 3.0 (yellow) to pH 4.6 (blue)]

4.3 A supply of high range (0.5-10 volume percent) and low range (5-700 ppm) detector tubes for ammonia andthe corresponding sampling pump. More sensitive ammonia detectors can be substituted for the low range detector tubes to improve the sensitivity of this practice.

4.4 A shallow plastic pan (PVC) at least 12":14":1" and a half pint plastic container (PVC) with tightly closing lid.

4.5 A graduated cylinder or other volumetric measuring device of at least 50 milliliters in volume with an accuracy of at least ±1 milliliters.

5.0 - Safety precautions

5.1 Concentrated aqueous ammonium hydroxide, NH_4OH, is a corrosive volatile liquid requiring eye, skin, andrespiratory protection. The person conducting the test shall review the Safety Data Sheet (SDS) for aqueousammonia.

5.2 Since the established permissible exposure limit for ammonia is 35 ppm as a 15 minute STEL, only personswearing a positive pressure self-contained breathing apparatus or a positive pressure supplied air respirator shall be in the chamber. Normally only the person wearing the totally-encapsulating suit will be inside the chamber. A stand-by person shall have a positive pressure self-contained breathing apparatus, or a positive pressure supplied air respirator available to enter the test area should the suited individual need assistance.

5.3 A method to monitor the suited individual must be used during this test. Visual contact is the simplest butother methods using communication devices are acceptable.

5.4 The test room shall be large enough to allow the exercise protocol to be carried out and then to beventilated to allow for easy exhaust of the ammonia test atmosphere after the test(s) are completed.

5.5 Individuals shall be medically screened for the use of respiratory protection and checked for allergies to ammonia before participating in this test procedure.

6.0 - Test procedure

6.1.1 Measure the test area to the nearest foot and calculate its volume in cubic feet. Multiply the test area volume by 0.2 milliliters of concentrated aqueous ammonia solution per cubic foot of test area volume to determine the approximate volume of concentrated aqueous ammonia required to generate 1000 ppm in thetest area.

6.1.2 Measure this volume from the supply of concentrated aqueous ammonia and place it into a closed plasticcontainer.

6.1.3 Place the container, several high range ammonia detector tubes, and the pump in the clean test pan andlocate it near the test area entry door so that the suited individual has easy access to these supplies.

6.2.1 In a non-contaminated atmosphere, open a pre-sealed ammonia indicator strip and fasten one end of thestrip to the inside of the suit face shield lens where it can be seen by the wearer. Moisten the indicator strip with distilled water. Care shall be taken not to contaminate the detector part of the indicator paper by touching it. A small piece of masking tape or equivalent should be used to attach the indicator strip to the interior of the suit face shield.

6.2.2 If problems are encountered with this method of attachment, the indicator strip can be attached to theoutside of the respirator face piece lens being used during the test.

6.3 Don the respiratory protective device normally used with the suit, and then don the TECP suit to be tested. Check to be sure all openings which are intended to be sealed (zippers, gloves, etc.) are completely sealed. DO NOT, however, plug off any venting valves.

6.4 Step into the enclosed test room such as a closet, bathroom, or test booth, equipped with an exhaust fan.No air should be exhausted from the chamber during the test because this will dilute the ammonia challenge concentrations.

6.5 Open the container with the pre-measured volume of concentrated aqueous ammonia within the enclosed test room, and pour the liquid into the empty plastic test pan. Wait two minutes to allow for adequate volatilization of the concentrated aqueous ammonia. A small mixing fan can be used near the evaporation panto increase the evaporation rate of the ammonia solution.

6.6 After two minutes a determination of the ammonia concentration within the chamber should be made usingthe high range colorimetric detector tube. A concentration of 1000 ppm ammonia or greater shall be generated before the exercises are started.

6.7 To test the integrity of the suit the following four minute exercise protocol should be followed:

6.7.1 Raising the arms above the head with at least 15 raising motions completed in one minute.

6.7.2 Walking in place for one minute with at least 15 raising motions of each leg in a one-minute period.

6.7.3 Touching the toes with a least 10 complete motions of the arms from above the head to touching of thetoes in a one-minute period.

6.7.4 Knee bends with at least 10 complete standing and squatting motions in a one-minute period.

6.8 If at any time during the test the colorimetric indicating paper should change colors, the test should bestopped and section 6.10 and 6.12 initiated (See ¶ 4.2).

6.9 After completion of the test exercise, the test area concentration should be measured again using the high range colorimetric detector tube.

6.10 Exit the test area.

6.11 The opening created by the suit zipper or other appropriate suit penetration should be used to determine the ammonia concentration in the suit with the low range length of stain detector tube or other ammonia monitor. The internal TECP suit air should be sampled far enough from the enclosed test area to prevent a false ammonia reading.

6.12 After completion of the measurement of the suit interior ammonia concentration the test is concluded and the suit is doffed and the respirator removed.

6.13 The ventilating fan for the test room should be turned on and allowed to run for enough time to removethe ammonia gas. The fan shall be vented to the outside of the building.

6.14 Any detectable ammonia in the suit interior (five ppm ammonia (NH_3) or more for the length of stain detector tube) indicates that the suit has failed the test. When other ammonia detectors are used a lower level of detection is possible, and it should be specified as the pass/fail criteria.

6.15 By following this test method, an intrusion coefficient of approximately 200 or more can be measured with the suit in a completely operational condition. If the intrusion coefficient is 200 or more, then the suit is suitable for emergency response and field use.

6.16

7.0 - Retest procedures

7.1 If the suit fails this test, check for leaks by following the pressure test in test A above.

7.2 Retest the TECP suit as outlined in the test procedure 6.0.

7.3

8.0 - Report

8.1 Each gas tight totally-encapsulating chemical protective suit tested by this practice shall have the following information recorded.

8.1.1 Unique identification number, identifying brand name, date of purchase, material of construction, andunique suit features; e.g., special breathing apparatus.

8.1.2 General description of test room used for test.

8.1.3 Brand name and purchase date of ammonia detector strips and color change data.

8.1.4 Brand name, sampling range, and expiration date of the length of stain ammonia detector tubes. The brand name and model of the sampling pump should also be recorded. If another type of ammonia detector isused, it should be identified along with its minimum detection limit for ammonia.

8.1.5 Actual test results shall list the two test area concentrations, their average, the interior suit concentration, and the calculated intrusion coefficient. Retest data shall be recorded as an additional test

8.2 The evaluation of the data shall be specified as "suit passed" or "suit failed," and the date of the test. Any detectable ammonia (five ppm or greater for the length of stain detector tube) in the suit interior indicates the suit has failed this test. When other ammonia detectors are used, a lower level of detection is possible and it should be specified as the pass fail criteria.

CAUTION

Visually inspect all parts of the suit to be sure they are positioned correctly and secured tightly before puttingthe suit back into service. Special care should be taken to examine each exhaust valve to make sure it is not blocked.

Care should also be exercised to assure that the inside and outside of the suit is completely dry before it is putinto storage.

Appendix B to § 1926.65 - General Description and Discussion of theLevels of Protection and Protective Gear

This appendix sets forth information about personal protective equipment (PPE) protection levels which maybe used to assist employers in complying with the PPE requirements of this section.

As required by the standard, PPE must be selected which will protect employees from the specific hazards which they are likely to encounter during their work on-site.

Selection of the appropriate PPE is a complex process which should take into consideration a variety of factors. Key factors involved in this process are identification of the hazards, or suspected hazards; their routes of potential hazard to employees (inhalation, skin absorption, ingestion, and eye or skin contact); and the performance of the PPE *materials* (and seams) in providing a barrier to these hazards. The amount of protection provided by PPE is material-hazard specific. That is, protective equipment materials will protect well against some hazardous substances and poorly, or not at all, against others. In many instances, protective equipment materials cannot be found which will provide continuous protection from the particular hazardous substance. In these cases the breakthrough time of the protective material should exceed the work durations.

Other factors in this selection process to be considered are matching the PPE to the employee's work requirements and task-specific conditions. The durability of PPE materials, such as tear strength and seam strength, should be considered in relation to the employee's tasks. The effects of PPE in relation to heat stress and task duration are a factor in selecting and using PPE. In some cases layers of PPE may be necessary to provide sufficient protection, or to protect expensive PPE inner garments, suits or equipment.

The more that is known about the hazards at the site, the easier the job of PPE selection becomes. As more information about the hazards and conditions at the site becomes available, the site supervisor can make decisions to up-grade or down-grade the level of PPE protection to match the tasks at hand.

The following are guidelines which an employer can use to begin the selection of the appropriate PPE. As noted above, the site information may suggest the use of combinations of PPE selected from the different protection levels (i.e., A, B, C, or D) as being more suitable to the hazards of the work. It should be cautioned that the listing below does not fully address the performance of the specific PPE material in relation to the specific hazards at the job site, and that PPE selection, evaluation and re-selection is an ongoing process until sufficient information about the hazards and PPE performance is obtained.

Part A. Personal protective equipment is divided into four categories based on the degree of protection afforded. (See part B of this appendix for further explanation of Levels A, B, C, and D hazards.)

I. *Level A* - To be selected when the greatest level of skin, respiratory, and eye protection is required.

The following constitute Level A equipment; it may be used as appropriate;

1. Positive pressure, full face-piece self-contained breathing apparatus (SCBA), or positive pressure supplied air respirator with escape SCBA, approved by the National Institute for Occupational Safety and Health (NIOSH).

2. Totally-encapsulating chemical-protective suit.

3. Coveralls.[1]

4. Long underwear.[1]

5. Gloves, outer, chemical-resistant.

6. Gloves, inner, chemical-resistant.

7. Boots, chemical-resistant, steel toe and shank.

8. Hard hat (under suit).[1]

9. Disposable protective suit, gloves and boots (depending on suit construction, may be worn over totally-encapsulating suit).

II. *Level B* - The highest level of respiratory protection is necessary but a lesser level of skin protection is needed.

The following constitute Level B equipment; it may be used as appropriate.

1. Positive pressure, full-facepiece self-contained breathing apparatus (SCBA), or positive pressure supplied air respirator with escape SCBA (NIOSH approved).

2. Hooded chemical-resistant clothing (overalls and long-sleeved jacket; coveralls; one or two-piece chemical- splash suit; disposable chemical-resistant overalls).

3. Coveralls.[1]

4. Gloves, outer, chemical-resistant.

5. Gloves, inner, chemical-resistant.

6. Boots, outer, chemical-resistant steel toe and shank.

7. Boot-covers, outer, chemical-resistant (disposable).[1]

8. Hard hat.[1]

9. [Reserved]

10. Face shield.[1]

III. *Level C* - The concentration(s) and type(s) of airborne substance(s) is known and the criteria for using airpurifying respirators are met.

The following constitute Level C equipment; it may be used as appropriate.

1. Full-face or half-mask, air purifying respirators (NIOSH approved).

2. Hooded chemical-resistant clothing (overalls; two-piece chemical-splash suit; disposable chemical-resistant overalls).

3. Coveralls.[1]

4. Gloves, outer, chemical-resistant.

5. Gloves, inner, chemical-resistant.

6. Boots (outer), chemical-resistant steel toe and shank.[1]

7. Boot-covers, outer, chemical-resistant (disposable)[1].

8. Hard hat.[1]

9. Escape mask.[1]

10. Face shield.[1]

IV. *Level D* - A work uniform affording minimal protection, used for nuisance contamination only.

The following constitute Level D equipment; it may be used as appropriate:

1. Coveralls.

2. Gloves.[1]

[1] Optional, as applicable.

3. Boots/shoes, chemical-resistant steel toe and shank.

4. Boots, outer, chemical-resistant (disposable).[1]

5. Safety glasses or chemical splash goggles *.

6. Hard hat.[1]

7. Escape mask.[1]

8. Face shield.[1]

Part B. The types of hazards for which levels A, B, C, and D protection are appropriate are described below:

I. *Level A* - Level A protection should be used when:

1. The hazardous substance has been identified and requires the highest level of protection for skin, eyes, and the respiratory system based on either the measured (or potential for) high concentration of atmospheric vapors, gases, or particulates; or the site operations and work functions involve a high potential for splash, immersion, or exposure to unexpected vapors, gases, or particulates of materials that are harmful to skin or capable of being absorbed through the skin;

2. Substances with a high degree of hazard to the skin are known or suspected to be present, and skin contact is possible; or

3. Operations are being conducted in confined, poorly ventilated areas, and the absence of conditions requiring Level A have not yet been determined.

II. *Level B* - Level B protection should be used when:

1. The type and atmospheric concentration of substances have been identified and require a high level of respiratory protection, but less skin protection;

2. The atmosphere contains less than 19.5 percent oxygen; or

3. The presence of incompletely identified vapors or gases is indicated by a direct-reading organic vapor detection instrument, but vapors and gases are not suspected of containing high levels of chemicals harmful to skin or capable of being absorbed through the skin.

> Note: This involves atmospheres with IDLH concentrations of specific substances that present severe inhalation hazards and that do not represent a severe skin hazard; or that do not meet the criteria for use of air-purifying respirators.

III. *Level C* - Level C protection should be used when:

1. The atmospheric contaminants, liquid splashes, or other direct contact will not adversely affect or be absorbed through any exposed skin;

2. The types of air contaminants have been identified, concentrations measured, and an air-purifying respirator is available that can remove the contaminants; and

3. All criteria for the use of air-purifying respirators are met.

IV. *Level D* - Level D protection should be used when:

1. The atmosphere contains no known hazard; and

2. Work functions preclude splashes, immersion, or the potential for unexpected inhalation of or contact with hazardous levels of any chemicals.

> Note: As stated before, combinations of personal protective equipment other than those described for Levels A, B, C, and D protection may be more appropriate and may be used to provide the proper level of protection.
>
> As an aid in selecting suitable chemical protective clothing, it should be noted that the National Fire Protection Association (NFPA) has developed standards on chemical protectiveclothing. The standards that have been adopted by include:
>
> NFPA 1991 - Standard on Vapor-Protective Suits for Hazardous Chemical Emergencies (EPALevel A Protective Clothing).
>
> NFPA 1992 - Standard on Liquid Splash-Protective Suits for Hazardous Chemical Emergencies (EPA Level B Protective Clothing).
>
> NFPA 1993 - Standard on Liquid Splash-Protective Suits for Non-emergency, Non-flammable Hazardous Chemical Situations (EPA Level B Protective Clothing).
>
> These standards apply documentation and performance requirements to the manufacture of chemical protective suits. Chemical protective suits meeting these requirements are labelled as compliant with the appropriate standard. It is recommended that chemical protective suitsthat meet these standards be used.

Appendix C to § 1926.65 - Compliance Guidelines

1. *Occupational Safety and Health Program.* Each hazardous waste site clean-up effort will require an occupational safety and health program headed by the site coordinator or the employer's representative. The purpose of the program will be the protection of employees at the site and will be an extension of the employer's overall safety and health program. The program will need to be developed before work begins on thesite and implemented as work proceeds as stated in paragraph (b). The program is to facilitate coordination and communication of safety and health issues among personnel responsible for the various activities which

will take place at the site. It will provide the overall means for planning and implementing the needed safety and health training and job orientation of employees who will be working at the site. The program will provide the means for identifying and controlling worksite hazards and the means for monitoring program effectiveness.

The program will need to cover the responsibilities and authority of the site coordinator or the employer's manager on the site for the safety and health of employees at the site, and the relationships with contractors or support services as to what each employer's safety and health responsibilities are for their employees on the site. Each contractor on the site needs to have its own safety and health program so structured that it will smoothly interface with the program of the site coordinator or principal contractor.

Also those employers involved with treating, storing or disposal of hazardous waste as covered in paragraph
(p) must have implemented a safety and health program for their employees. This program is to include the hazard communication program required in paragraph (p)(1) and the training required in paragraphs (p)(7) and(p)(8) as parts of the employers comprehensive overall safety and health program. This program is to be in writing.

Each site or workplace safety and health program will need to include the following: (1) Policy statements of the line of authority and accountability for implementing the program, the objectives of the program and the role of the site safety and health supervisor or manager and staff; (2) means or methods for the development of procedures for identifying and controlling workplace hazards at the site; (3) means or methods for the development and communication to employees of the various plans, work rules, standard operating procedures and practices that pertain to individual employees and supervisors; (4) means for the training of supervisors and employees to develop the needed skills and knowledge to perform their work in a safe and healthful manner; (5) means to anticipate and prepare for emergency situations; and (6) means for obtaining information feedback to aid in evaluating the program and for improving the effectiveness of the program. The management and employees should be trying continually to improve the effectiveness of the program therebyenhancing the protection being afforded those working on the site.

Accidents on the site or workplace should be investigated to provide information on how such occurrences can be avoided in the future. When injuries or illnesses occur on the site or workplace, they will need to be investigated to determine what needs to be done to prevent this incident from occurring again. Such information will need to be used as feedback on the effectiveness of the program and the information turned into positive steps to prevent any reoccurrence. Receipt of employee suggestions or complaints relating to safety and health issues involved with site or workplace activities is also a feedback mechanism that can be used effectively to improve the program and may serve in part as an evaluative tool(s).

For the development and implementation of the program to be the most effective, professional safety andhealth personnel should be used. Certified Safety Professionals, Board Certified Industrial Hygienists or Registered Professional Safety Engineers are good examples of professional stature for safety and health managers who will administer the employer's program.

2. *Training.* The training programs for employees subject to the requirements of paragraph (e) of this standard should address: the safety and health hazards employees should expect to find on hazardous waste clean-up sites; what control measures or techniques are effective for those hazards; what monitoring procedures are effective in characterizing exposure levels; what makes an effective

employer's safety and health program; what a site safety and health plan should include; hands on training with personal protective equipment and clothing they may be expected to use; the contents of the OSHA standard relevant to the employee's duties and function; and, employee's responsibilities under OSHA and other regulations. Supervisors will need training in

their responsibilities under the safety and health program and its subject areas such as the spill containment program, the personal protective equipment program, the medical surveillance program, the emergency response plan and other areas.

The training programs for employees subject to the requirements of paragraph (p) of this standard should address: the employers safety and health program elements impacting employees; the hazard communication program; the medical surveillance program; the hazards and the controls for such hazards that employees need to know for their job duties and functions. All require annual refresher training.

The training programs for employees covered by the requirements of paragraph (q) of this standard should address those competencies required for the various levels of response such as: the hazards associated with hazardous substances; hazard identification and awareness; notification of appropriate persons; the need for and use of personal protective equipment including respirators; the decontamination procedures to be used; preplanning activities for hazardous substance incidents including the emergency reponse plan; company standard operating procedures for hazardous substance emergency responses; the use of the incident command system and other subjects. Hands-on training should be stressed whenever possible. Critiques done after an incident which include an evaluation of what worked and what did not and how could the incident be better handled the next time may be counted as training time.

For hazardous materials specialists (usually members of hazardous materials teams), the training should address the care, use and/or testing of chemical protective clothing including totally encapsulating suits, the medical surveillance program, the standard operating procedures for the hazardous materials team including the use of plugging and patching equipment and other subject areas.

Officers and leaders who may be expected to be in charge at an incident should be fully knowledgeable of their company's incident command system. They should know where and how to obtain additional assistance and be familiar with the local district's emergency response plan and the state emergency response plan.

Specialist employees such as technical experts, medical experts or environmental experts that work with hazardous materials in their regular jobs, who may be sent to the incident scene by the shipper, manufacturer or governmental agency to advise and assist the person in charge of the incident should have training on an annual basis. Their training should include the care and use of personal protective equipment including respirators; knowledge of the incident command system and how they are to relate to it; and those areas needed to keep them current in their respective field as it relates to safety and health involving specific hazardous substances.

Those skilled support personnel, such as employees who work for public works departments or equipment operators who operate bulldozers, sand trucks, backhoes, etc., who may be called to the incident scene to provide emergency support assistance, should have at least a safety and health briefing before entering the area of potential or actual exposure. These skilled support personnel, who have not been a part of the emergency response plan and do not meet the training requirements, should be made aware of the hazards they face and should be provided all necessary protective clothing and equipment required for their tasks.

There are two National Fire Protection Association standards, NFPA 472 - "Standard for Professional Competence of Responders to Hazardous Material Incidents" and NFPA 471 - "Recommended Practice for Responding to Hazardous Material Incidents", which are excellent resource documents to aid fire departments and other emergency response organizations in developing their training program materials. NFPA 472

provides guidance on the skills and knowledge needed for first responder awareness level, first responder operations level, hazmat technicians, and hazmat specialist. It also offers guidance for the officer corp who will be in charge of hazardous substance incidents.

3. *Decontamination.* Decontamination procedures should be tailored to the specific hazards of the site, and may vary in complexity and number of steps, depending on the level of hazard and the employee's exposure to the hazard. Decontamination procedures and PPE decontamination methods will vary depending upon the specific substance, since one procedure or method may not work for all substances. Evaluation of decontamination methods and procedures should be performed, as necessary, to assure that employees are not exposed to hazards by re-using PPE. References in appendix D may be used for guidance in establishing an effective decontamination program. In addition, the U.S. Coast Guard's Manual, "Policy Guidance for Response to Hazardous Chemical Releases," U.S. Department of Transportation, Washington, DC (COMDTINST M16465.30) is a good reference for establishing an effective decontamination program.

4. *Emergency response plans.* States, along with designated districts within the states, will be developing or have developed local emergency response plans. These state and district plans should be utilized in the emergency response plans called for in the standard. Each employer should assure that its emergency response plan is compatible with the local plan. The major reference being used to aid in developing the state and local district plans is the *Hazardous Materials Emergency Planning Guide,* NRT-1. The current Emergency Response Guidebook from the U.S. Department of Transportation, CMA's CHEMTREC and the Fire Service Emergency Management Handbook may also be used as resources.

Employers involved with treatment, storage, and disposal facilities for hazardous waste, which have the required contingency plan called for by their permit, would not need to duplicate the same planning elements. Those items of the emergency response plan that are properly addressed in the contingency plan may be substituted into the emergency response plan required in 1926.65 or otherwise kept together for employer and employee use.

5. *Personal protective equipment programs.* The purpose of personal protective clothing and equipment (PPE) is to shield or isolate individuals from the chemical, physical, and biologic hazards that may be encountered at a hazardous substance site.

As discussed in appendix B, no single combination of protective equipment and clothing is capable of protecting against all hazards. Thus PPE should be used in conjunction with other protective methods and its effectiveness evaluated periodically.

The use of PPE can itself create significant worker hazards, such as heat stress, physical and psychological stress, and impaired vision, mobility, and communication. For any given situation, equipment and clothing should be selected that provide an adequate level of protection. However, over-protection, as well as under-protection, can be hazardous and should be avoided where possible.

Two basic objectives of any PPE program should be to protect the wearer from safety and health hazards, and to prevent injury to the wearer from incorrect use and/or malfunction of the PPE. To accomplish these goals, acomprehensive PPE program should include hazard identification, medical monitoring, environmental surveillance, selection, use, maintenance, and decontamination of PPE and its associated training.

The written PPE program should include policy statements, procedures, and guidelines. Copies should be made available to all employees, and a reference copy should be made available at the worksite. Technical data on equipment, maintenance manuals, relevant regulations, and other essential information should also be collected and maintained.

6. *Incident command system (ICS)*. Paragraph 1926.65(q)(3)(ii) requires the implementation of an ICS. The ICSis an organized approach to effectively control and *manage* operations at an emergency incident. The individual in charge of the ICS is the senior official responding to the incident. The ICS is not much different than the "command post" approach used for many years by the fire service. During large complex fires involving several companies and many pieces of apparatus, a command post would be established. This enabled *one* individual to be in charge of managing the incident, rather than having several officers from different companies makingseparate, and sometimes conflicting, decisions. The individual in charge of the command post would delegateresponsibility for performing various tasks to subordinate officers. Additionally, all communications were routed through the command post to reduce the number of radio transmissions and eliminate confusion. However, strategy, tactics, and all decisions were made by one individual.

The ICS is a very similar system, except it is implemented for emergency response to all incidents, both largeand small, that involve hazardous substances.

For a small incident, the individual in charge of the ICS may perform many tasks of the ICS. There may not beany, or little, delegation of tasks to subordinates. For example, in response to a small incident, the individual in charge of the ICS, in addition to normal command activities, may become the safety officer and may designate only one employee (with proper equipment) as a back-up to provide assistance if needed. OSHA does recommend, however, that at least two employees be designated as back-up personnel since the assistance needed may include rescue.

To illustrate the operation of the ICS, the following scenario might develop during a small incident, such as an overturned tank truck with a small leak of flammable liquid.

The first responding senior officer would implement and take command of the ICS. That person would size-up the incident and determine if additional personnel and apparatus were necessary; would determine what actions to take to control the leak; and, determine the proper level of personal protective equipment. If additional assistance is not needed, the individual in charge of the ICS would implement actions to stop and control the leak using the fewest number of personnel that can effectively accomplish the tasks. The individual in charge of the ICS then would designate himself as the safety officer and two other employees as a back-up in case rescue may become necessary. In this scenario, decontamination procedures would not be necessary.

A large complex incident may require many employees and difficult, time-consuming efforts to control. In these situations, the individual in charge of the ICS will want to delegate different tasks to subordinates in order to maintain a span of control that will keep the number of subordinates, that are reporting, to a manageable level.

Delegation of task at large incidents may be by location, where the incident scene is divided into sectors, and subordinate officers coordinate activities within the sector that they have been assigned.

Delegation of tasks can also be by function. Some of the functions that the individual in charge of the ICS may want to delegate at a large incident are: medical services; evacuation; water supply; resources (equipment, apparatus); media relations; safety; and, site control (integrate activities with police for crowd and traffic control). Also for a large incident, the individual in charge of the ICS will designate several employees as back- up personnel; and a number of safety officers to monitor conditions and recommend safety precautions.

Therefore, no matter what size or complexity an incident may be, by implementing an ICS there will be *one individual in charge* who makes the decisions and gives directions; and, all actions, and communications are coordinated through one central point of command. Such a system should reduce confusion, improve safety, organize and coordinate actions, and should facilitate effective management of the incident.

7. *Site Safety and Control Plans.* The safety and security of response personnel and others in the area of an emergeny response incident site should be of primary concern to the incident commander. The use of a site safety and control plan could greatly assist those in charge of assuring the safety and health of employees on the site.

A comprehensive site safety and control plan should include the following: summary analysis of hazards on the site and a risk analysis of those hazards; site map or sketch; site work zones (clean zone, transition or decontamination zone, work or hot zone); use of the buddy system; site communications; command post or command center; standard operating procedures and safe work practices; medical assistance and triage area; hazard monitoring plan (air contaminate monitoring, etc.); decontamination procedures and area; and other relevant areas. This plan should be a part of the employer's emergency response plan or an extension of it to the specific site.

8. *Medical surveillance programs.* Workers handling hazardous substances may be exposed to toxic chemicals, safety hazards, biologic hazards, and radiation. Therefore, a medical surveillance program is essential to assess and monitor workers' health and fitness for employment in hazardous waste operations and during the course of work; to provide emergency and other treatment as needed; and to keep accurate records for futurereference.

The *Occupational Safety and Health Guidance Manual for Hazardous Waste Site Activities* developed by the National Institute for Occupational Safety and Health (NIOSH), the Occupational Safety and Health Administration (OSHA), the U.S. Coast Guard (USCG), and the Environmental Protection Agency (EPA); October 1985 provides an excellent example of the types of medical testing that should be done as part of a medical surveillance program.

9. *New Technology and Spill Containment Programs.* Where hazardous substances may be released by spilling from a container that will expose employees to the hazards of the materials, the employer will need to implement a program to contain and control the spilled material. Diking and ditching, as well as use of absorbents like diatomaceous earth, are traditional techniques which have proven to be effective over the years. However, in recent years new products have come into the marketplace, the

use of which complement and increase the effectiveness of these traditional methods. These new products also provide emergency responders and others with additional tools or agents to use to reduce the hazards of spilled materials.

These agents can be rapidly applied over a large area and can be uniformly applied or otherwise can be used to build a small dam, thus improving the workers' ability to control spilled material. These application techniques enhance the intimate contact between the agent and the spilled material allowing for the quickest effect by the

agent or quickest control of the spilled material. Agents are available to solidify liquid spilled materials, to suppress vapor generation from spilled materials, and to do both. Some special agents, which when applied as recommended by the manufacturer, will react in a controlled manner with the spilled material to neutralize acids or caustics, or greatly reduce the level of hazard of the spilled material.

There are several modern methods and devices for use by emergency response personnel or others involved with spill control efforts to safely apply spill control agents to control spilled material hazards. These include portable pressurized applicators similar to hand-held portable fire extinguishing devices, and nozzle and hose systems similar to portable fire fighting foam systems which allow the operator to apply the agent without having to come into contact with the spilled material. The operator is able to apply the agent to the spilled material from a remote position.

The solidification of liquids provides for rapid containment and isolation of hazardous substance spills. By directing the agent at run-off points or at the edges of the spill, the reactant solid will automatically create a barrier to slow or stop the spread of the material. Clean-up of hazardous substances is greatly improved when solidifying agents, acid or caustic neutralizers, or activated carbon adsorbents are used. Properly applied, these agents can totally solidify liquid hazardous substances or neutralize or absorb them, which results in materials which are less hazardous and easier to handle, transport, and dispose of. The concept of spill treatment, to create less hazardous substances, will improve the safety and level of protection of employees working at spill clean-up operations or emergency response operations to spills of hazardous substances.

The use of vapor suppression agents for volatile hazardous substances, such as flammable liquids and those substances which present an inhalation hazard, is important for protecting workers. The rapid and uniform distribution of the agent over the surface of the spilled material can provide quick vapor knockdown. There are temporary and long-term foam-type agents which are effective on vapors and dusts, and activated carbon adsorption agents which are effective for vapor control and soaking-up of the liquid. The proper use of hose lines or hand-held portable pressurized applicators provides good mobility and permits the worker to deliver the agent from a safe distance without having to step into the untreated spilled material. Some of these systems can be recharged in the field to provide coverage of larger spill areas than the design limits of a single charged applicator unit. Some of the more effective agents can solidify the liquid flammable hazardous substances and at the same time elevate the flashpoint above 140 °F so the resulting substance may be handled as a nonhazardous waste material if it meets the U.S. Environmental Protection Agency's 40 CFR part 261 requirements (See particularly § 261.21).

All workers performing hazardous substance spill control work are expected to wear the proper protective clothing and equipment for the materials present and to follow the employer's established standard operating procedures for spill control. All involved workers need to be trained in the established operating procedures; in the use and care of spill control equipment; and in the associated hazards and control of such hazards of spill containment work.

These new tools and agents are the things that employers will want to evaluate as part of their new technologyprogram. The treatment of spills of hazardous substances or wastes at an emergency incident as part of the immediate spill containment and control efforts is sometimes acceptable to EPA and a permit exception is described in 40 CFR 264.1(g)(8) and 265.1(c)(11).

Appendix D to § 1926.65 - References

The following references may be consulted for further information on the subject of this standard:

1. OSHA Instruction DFO CPL 2.70 - January 29, 1986, *Special Emphasis Program: Hazardous Waste Sites.*

2. OSHA Instruction DFO CPL 2-2.37A - January 29, 1986, *Technical Assistance and Guidelines for Superfund and Other Hazardous Waste Site Activities.*

3. OSHA Instruction DTS CPL 2.74 - January 29, 1986, *Hazardous Waste Activity Form, OSHA 175.*

4. *Hazardous Waste Inspections Reference Manual,* U.S. Department of Labor, Occupational Safety and Health Administration, 1986.

5. Memorandum of Understanding Among the National Institute for Occupational Safety and Health, the Occupational Safety and Health Administration, the United States Coast Guard, and the United States Environmental Protection Agency, *Guidance for Worker Protection During Hazardous Waste Site Investigations and Clean-up and Hazardous Substance Emergencies.* December 18, 1980.

6. *National Priorities List,* 1st Edition, October 1984; U.S. Environmental Protection Agency, Revised periodically.

7. *The Decontamination of Response Personnel,* Field Standard Operating Procedures (F.S.O.P.) 7; U.S. Environmental Protection Agency, Office of Emergency and Remedial Response, Hazardous Response Support Division, December 1984.

8. *Preparation of a Site Safety Plan,* Field Standard Operating Procedures (F.S.O.P.) 9; U.S. Environmental Protection Agency, Office of Emergency and Remedial Response, Hazardous Response Support Division, April 1985.

9. *Standard Operating Safety Guidelines;* U.S. Environmental Protection Agency, Office of Emergency and Remedial Response, Hazardous Response Support Division, Environmental Response Team; November 1984.

10. *Occupational Safety and Health Guidance Manual for Hazardous Waste Site Activities,* National Institute for Occupational Safety and Health (NIOSH), Occupational Safety and Health Administration (OSHA), U.S. CoastGuard (USCG), and Environmental Protection Agency (EPA); October 1985.

11. *Protecting Health and Safety at Hazardous Waste Sites: An Overview,* U.S. Environmental Protection Agency, EPA/625/9-85/006; September 1985.

12. *Hazardous Waste Sites and Hazardous Substance Emergencies,* NIOSH Worker Bulletin, U.S. Department of Health and Human Services, Public Health Service, Centers for Disease Control, National Institute for Occupational Safety and Health; December 1982.

13. *Personal Protective Equipment for Hazardous Materials Incidents: A Selection Guide;* U.S. Department of Health and Human Services, Public Health Service, Centers for Disease Control, National Institute for Occupational Safety and Health; October 1984.

14. *Fire Service Emergency Management Handbook,* International Association of Fire Chiefs Foundation, 101 East Holly Avenue, Unit 10B, Sterling, VA 22170, January 1985.

15. *Emergency Response Guidebook,* U.S Department of Transportation, Washington, DC, 1987.

16. *Report to the Congress on Hazardous Materials Training, Planning and Preparedness,* Federal Emergency Management Agency, Washington, DC, July 1986.

17. *Workbook for Fire Command,* Alan V. Brunacini and J. David Beageron, National Fire Protection Association, Batterymarch Park, Quincy, MA 02269, 1985.

18. *Fire Command,* Alan V. Brunacini, National Fire Protection Association, Batterymarch Park,, Quincy, MA 02269, 1985.

19. *Incident Command System,* Fire Protection Publications, Oklahoma State University, Stillwater, OK 74078, 1983.

20. *Site Emergency Response Planning,* Chemical Manufacturers Association, Washington, DC 20037, 1986.

21. *Hazardous Materials Emergency Planning Guide,* NRT-1, Environmental Protection Agency, Washington, DC, March 1987.

22. *Community Teamwork: Working Together to Promote Hazardous Materials Transportation Safety.* U.S. Department of Transportation, Washington, DC, May 1983.

23. *Disaster Planning Guide for Business and Industry,* Federal Emergency Management Agency, Publication No. FEMA 141, August 1987.

Appendix E to § 1926.65 - Training Curriculum Guidelines

The following non-mandatory general criteria may be used for assistance in developing site-specific training curriculum used to meet the training requirements of 29 CFR 1926.65(e); 29 CFR 1926.65(p)(7), (p)(8)(iii); and 29 CFR 1926.65(q)(6), (q)(7), and (q)(8). These are generic guidelines and they are not presented as a complete training curriculum for any specific employer. Site-specific training programs must be developed on the basis of a needs assessment of the hazardous waste site, RCRA/TSDF, or emergency response operation in accordance with 29 CFR 1926.65.

It is noted that the legal requirements are set forth in the regulatory text of § 1926.65. The guidance set forth here presents a highly effective program that in the areas covered would meet or exceed the regulatory requirements. In addition, other approaches could meet the regulatory requirements.

Suggested General Criteria

Definitions:

Competent means possessing the skills, knowledge, experience, and judgment to perform assigned tasks or activities satisfactorily as determined by the employer.

Demonstration means the showing by actual use of equipment or procedures.

Hands-on training means training in a simulated work environment that permits each student to have experience performing tasks, making decisions, or using equipment appropriate to the job assignment for which the training is being conducted.

Initial training means training required prior to beginning work.

Lecture means an interactive discourse with a class lead by an instructor.

Proficient means meeting a stated level of achievement.

Site-specific means individual training directed to the operations of a specific job site.

Training hours means the number of hours devoted to lecture, learning activities, small group work sessions, demonstration, evaluations, or hands-on experience.

Suggested Core Criteria:

1. *Training facility.* The training facility should have available sufficient resources, equipment, and site locations to perform didactic and hands-on training when appropriate. Training facilities should have sufficient organization, support staff, and services to conduct training in each of the courses offered.

2. *Training Director.* Each training program should be under the direction of a training director who is responsible for the program. The Training Director should have a minimum of two years of employee education experience.

3. *Instructors.* Instructors should be deem competent on the basis of previous documented experience in their area of instruction, successful completion of a "train-the-trainer" program specific to the topics they will teach, and an evaluation of instructional competence by the Training Director.

Instructors should be required to maintain professional competency by participating in continuing education or professional development programs or by completing successfully an annual refresher course and having an annual review by the Training Director.

The annual review by the Training Director should include observation of an instructor's delivery, a review of those observations with the trainer, and an analysis of any instructor or class evaluations completed by the students during the previous year.

4. *Course materials.* The Training Director should approve all course materials to be used by the

training provider. Course materials should be reviewed and updated at least annually. Materials and equipment should be in good working order and maintained properly.

All written and audio-visual materials in training curricula should be peer reviewed by technically competent outside reviewers or by a standing advisory committee.

Reviews should possess expertise in the following disciplines were applicable: occupational health, industrial hygiene and safety, chemical/environmental engineering, employee education, or emergency response. One or more of the peer reviewers should be a employee experienced in the work activities to which the training is directed.

5. *Students.* The program for accepting students should include:

a. Assurance that the student is or will be involved in work where chemical exposures are likely and that thestudent possesses the skills necessary to perform the work.

b. A policy on the necessary medical clearance.

6. *Ratios.* Student-instructor ratios should not exceed 30 students per instructor. Hands-on activity requiring the use of personal protective equipment should have the following student-instructor ratios. For Level C or Level D personal protective equipment the ratio should be 10 students per instructor. For Level A or Level B personal protective equipment the ratio should be 5 students per instructor.

7. *Proficiency assessment.* Proficiency should be evaluated and documented by the use of a written assessment and a skill demonstration selected and developed by the Training Director and training staff. Theassessment and demonstration should evaluate the knowledge and individual skills developed in the course oftraining. The level of minimum achievement necessary for proficiency shall be specified in writing by the Training Director.

If a written test is used, there should be a minimum of 50 questions. If a written test is used in combinationwith a skills demonstration, a minimum of 25 questions should be used. If a skills demonstration is used, the tasks chosen and the means to rate successful completion should be fully documented by the Training Director.

The content of the written test or of the skill demonstration shall be relevant to the objectives of the course. The written test and skill demonstration should be updated as necessary to reflect changes in the curriculum and any update should be approved by the Training Director.

The proficiency assessment methods, regardless of the approach or combination of approaches used, should be justified, document and approved by the Training Director.

The proficiency of those taking the additional courses for supervisors should be evaluated and document byusing proficiency assessment methods acceptable to the Training Director. These proficiency assessment methods must reflect the additional responsibilities borne by supervisory personnel in hazardous waste operations or emergency response.

8. *Course certificate.* Written documentation should be provided to each student who satisfactorily completes the training course. The documentation should include:

a. Student's name.

b. Course title.

c. Course date.

d. Statement that the student has successfully completed the course.

e. Name and address of the training provider.

f. An individual identification number for the certificate.

g. List of the levels of personal protective equipment used by the student to complete the course.

This documentation may include a certificate and an appropriate wallet-sized laminated card with a photograph of the student and the above information. When such course certificate cards are used, the individual identification number for the training certificate should be shown on the card.

9. *Recordkeeping.* Training providers should maintain records listing the dates courses were presented, the names of the individual course attenders, the names of those students successfully completing each course, and the number of training certificates issued to each successful student. These records should be maintained for a minimum of five years after the date an individual participated in a training program offered by the trainingprovider. These records should be available and provided upon the student's request or as mandated by law.

10. *Program quality control.* The Training Director should conduct or direct an annual written audit of the training program. Program modifications to address deficiencies, if any, should be documented, approved, andimplemented by the training provider. The audit and the program modification documents should be maintained at the training facility.

Suggested Program Quality Control Criteria

Factors listed here are suggested criteria for determining the quality and appropriateness of employee health and safety training for hazardous waste operations and emergency response.

A. Training Plan.

Adequacy and appropriateness of the training program's curriculum development, instructor training, distribution of course materials, and direct student training should be considered, including

1. The duration of training, course content, and course schedules/agendas;

2. The different training requirements of the various target populations, as specified in the appropriate generic training curriculum;

3. The process for the development of curriculum, which includes appropriate technical input, outside review, evaluation, program pretesting.

4. The adequate and appropriate inclusion of hands-on, demonstration, and instruction methods;

5. Adequate monitoring of student safety, progress, and performance during the training.

B. Program management, Training Director, staff, and consultants.

Adequacy and appropriateness of staff performance and delivering an effective training program should be considered, including

1. Demonstration of the training director's leadership in assuring quality of health and safety training.

2. Demonstration of the competency of the staff to meet the demands of delivering high quality hazardouswaste employee health and safety training.

3. Organization charts establishing clear lines of authority.

4. Clearly defined staff duties including the relationship of the training staff to the overall program.

5. Evidence that the training organizational structure suits the needs of the training program.

6. Appropriateness and adequacy of the training methods used by the instructors.

7. Sufficiency of the time committed by the training director and staff to the training program.

8. Adequacy of the ratio of training staff to students.

9. Availability and commitment of the training program of adequate human and equipment resources in theareas of

a. Health effects,

b. Safety,

c. Personal protective equipment (PPE),

d. Operational procedures,

e. Employee protection practices/procedures.

10. Appropriateness of management controls.

11. Adequacy of the organization and appropriate resources assigned to assure appropriate training.

12. In the case of multiple-site training programs, adequacy of satellite centers management.

C. Training facilities and resources.

Adequacy and appropriateness of the facilities and resources for supporting the training program should be considered, including,

1. Space and equipment to conduct the training.

2. Facilities for representative hands-on training.

3. In the case of multiple-site programs, equipment and facilities at the satellite centers.

4. Adequacy and appropriateness of the quality control and evaluations program to account for instructor performance.

5. Adequacy and appropriateness of the quality control and evaluation program to ensure appropriate course evaluation, feedback, updating, and corrective action.

6. Adequacy and appropriateness of disciplines and expertise being used within the quality control and evaluation program.

7. Adequacy and appropriateness of the role of student evaluations to provide feedback for training program improvement.

D. Quality control and evaluation.

Adequacy and appropriateness of quality control and evaluation plans for training programs should be considered, including:

1. A balanced advisory committee and/or competent outside reviewers to give overall policy guidance;

2. Clear and adequate definition of the composition and active programmatic role of the advisory committee or outside reviewers.

3. Adequacy of the minutes or reports of the advisory committee or outside reviewers' meetings or written communication.

4. Adequacy and appropriateness of the quality control and evaluations program to account for instructor performance.

5. Adequacy and appropriateness of the quality control and evaluation program to ensure appropriate course evaluation, feedback, updating, and corrective action.

6. Adequacy and appropriateness of disciplines and expertise being used within the quality control and evaluation program.

7. Adequacy and appropriateness of the role of student evaluations to provide feedback for training program improvement.

E. Students

Adequacy and appropriateness of the program for accepting students should be considered, including

1. Assurance that the student already possess the necessary skills for their job, including necessarydocumentation.

2. Appropriateness of methods the program uses to ensure that recruits are capable of satisfactorilycompleting training.

3. Review and compliance with any medical clearance policy.

F. Institutional Environment and Administrative Support. The adequacy and appropriateness of the institutional environment and administrative support system for the training program should be considered, including

1. Adequacy of the institutional commitment to the employee training program.

2. Adequacy and appropriateness of the administrative structure and administrative support.

G. Summary of Evaluation Questions Key questions for evaluating the quality and appropriateness of an overall training program should include the following:

1. Are the program objectives clearly stated?

2. Is the program accomplishing its objectives?

3. Are appropriate facilities and staff available?

4. Is there an appropriate mix of classroom, demonstration, and hands-on training?

5. Is the program providing quality employee health and safety training that fully meets the intent of regulatory requirements?

6. What are the program's main strengths?

7. What are the program's main weaknesses?

8. What is recommended to improve the program?

9. Are instructors instructing according to their training outlines?

10. Is the evaluation tool current and appropriate for the program content?

11. Is the course material current and relevant to the target group?

Suggested Training Curriculum Guidelines

The following training curriculum guidelines are for those operations specifically identified in 29 CFR 1926.65 as requiring training. Issues such as qualifications of instructors, training certification, and similar criteria appropriate to all categories of operations addressed in 1926.65 have been covered in the preceding sectionand are not re-addressed in each of the generic guidelines. Basic core requirements for training programs that are addressed include

1. General Hazardous Waste Operations

2. RCRA operations - Treatment, storage, and disposal facilities.

3. Emergency Response.

A. General Hazardous Waste Operations and Site-specific Training 1. *Off-site training.*

Minimum training course content for hazardous waste operations, required by 29 CFR 1926.65(e), should include the following topics or procedures:

a. Regulatory knowledge.

(1) A review of 29 CFR 1926.65 and the core elements of an occupational safety and health program.

(2) The content of a medical surveillance program as outlined in 29 CFR 1926.65(f).

(3) The content of an effective site safety and health plan consistent with the requirements of 29 CFR1926.65(b)(4)(ii).

(4) Emergency response plan and procedures as outlined in 29 CFR 1910.38 and 29 CFR 1926.65(l).

(5) Adequate illumination.

(6) Sanitation recommendation and equipment.

(7) Review and explanation of OSHA's hazard-communication standard (29 CFR 1910.1200) and lock-out-tag- out standard (29 CFR 1910.147).

(8) Review of other applicable standards including but not limited to those in the construction standards (29CFR part 1926).

(9) Rights and responsibilities of employers and employees under applicable OSHA and EPA laws.

b. *Technical knowledge.* (1) Type of potential exposures to chemical, biological, and radiological hazards; types of human responses to these hazards and recognition of those responses; principles of toxicology and information about acute and chronic hazards; health and safety considerations of new technology.

(2) Fundamentals of chemical hazards including but not limited to vapor pressure, boiling points, flash points, ph, other physical and chemical properties.

(3) Fire and explosion hazards of chemicals.

(4) General safety hazards such as but not limited to electrical hazards, powered equipment hazards, motor vehicle hazards, walking-working surface hazards, excavation hazards, and hazards associated with working in hot and cold temperature extremes.

(5) Review and knowledge of confined space entry procedures in 29 CFR 1910.146.

(6) Work practices to minimize employee risk from site hazards.

(7) Safe use of engineering controls, equipment, and any new relevant safety technology or safety procedures.

(8) Review and demonstration of competency with air sampling and monitoring equipment that may be used ina site monitoring program.

(9) Container sampling procedures and safeguarding; general drum and container handling procedures including special requirement for laboratory waste packs, shock-sensitive wastes, and radioactive wastes.

(10) The elements of a spill control program.

(11) Proper use and limitations of material handling equipment.

(12) Procedures for safe and healthful preparation of containers for shipping and transport.

(13) Methods of communication including those used while wearing respiratory protection.

c. *Technical skills.* (1) Selection, use maintenance, and limitations of personal protective equipment including the components and procedures for carrying out a respirator program to comply with 29 CFR 1910.134.

(2) Instruction in decontamination programs including personnel, equipment, and hardware; hands-on training including level A, B, and C ensembles and appropriate decontamination lines; field activities including the donning and doffing of protective equipment to a level commensurate with the employee's anticipated job function and responsibility and to the degree required by potential hazards.

(3) Sources for additional hazard information; exercises using relevant manuals and hazard coding systems.

d. *Additional suggested items.* (1) A laminated, dated card or certificate with photo, denoting limitations andlevel of protection for which the employee is trained should be issued to those students successfully completing a course.

(2) Attendance should be required at all training modules, with successful completion of exercises and a finalwritten or oral examination with at least 50 questions.

(3) A minimum of one-third of the program should be devoted to hands-on exercises.

(4) A curriculum should be established for the 8-hour refresher training required by 29 CFR 1926.65(e)(8),

with delivery of such courses directed toward those areas of previous training that need improvement or reemphasis.

(5) A curriculum should be established for the required 8-hour training for supervisors. Demonstrated competency in the skills and knowledge provided in a 40-hour course should be a prerequisite for supervisor training.

2. *Refresher training.* The 8-hour annual refresher training required in 29 CFR 1926.65(e)(8) should be conducted by qualified training providers. Refresher training should include at a minimum the following topics and procedures:

(a) Review of and retraining on relevant topics covered in the 40-hour program, as appropriate, using reports by the students on their work experiences.

(b) Update on developments with respect to material covered in the 40-hour course.

(c) Review of changes to pertinent provisions of EPA or OSHA standards or laws.

(d) Introduction of additional subject areas as appropriate.

(e) Hands-on review of new or altered PPE or decontamination equipment or procedures. Review of new developments in personal protective equipment.

(f) Review of newly developed air and contaminant monitoring equipment.

3. *On-site training.* a. The employer should provide employees engaged in hazardous waste site activities with information and training prior to initial assignment into their work area, as follows:

(1) The requirements of the hazard communication program including the location and availability of the written program, required lists of hazardous chemicals, and safety data sheets.

(2) Activities and locations in their work area where hazardous substance may be present.

(3) Methods and observations that may be used to detect the present or release of a hazardous chemical in the work area (such as monitoring conducted by the employer, continuous monitoring devices, visual appearances, or other evidence (sight, sound or smell) of hazardous chemicals being released, and applicable alarms from monitoring devices that record chemical releases.

(4) The physical and health hazards of substances known or potentially present in the work area.

(5) The measures employees can take to help protect themselves from work-site hazards, including specific procedures the employer has implemented.

(6) An explanation of the labeling system and safety data sheets and how employees can obtain and use appropriate hazard information.

(7) The elements of the confined space program including special PPE, permits, monitoring requirements, communication procedures, emergency response, and applicable lock-out

procedures.

b. The employer should provide hazardous waste employees information and training and should provide a review and access to the site safety and plan as follows:

(1) Names of personnel and alternate responsible for site safety and health.

(2) Safety and health hazards present on the site.

(3) Selection, use, maintenance, and limitations of personal protective equipment specific to the site.

(4) Work practices by which the employee can minimize risks from hazards.

(5) Safe use of engineering controls and equipment available on site.

(6) Safe decontamination procedures established to minimize employee contact with hazardous substances, including:

(A) Employee decontamination,

(B) Clothing decontamination, and

(C) Equipment decontamination.

(7) Elements of the site emergency response plan, including:

(A) Pre-emergency planning.

(B) Personnel roles and lines of authority and communication.

(C) Emergency recognition and prevention.

(D) Safe distances and places of refuge.

(E) Site security and control.

(F) Evacuation routes and procedures.

(G) Decontamination procedures not covered by the site safety and health plan.

(H) Emergency medical treatment and first aid.

(I) Emergency equipment and procedures for handling emergency incidents.

c. The employer should provide hazardous waste employees information and training on personal protective equipment used at the site, such as the following:

(1) PPE to be used based upon known or anticipated site hazards.

(2) PPE limitations of materials and construction; limitations during temperature extremes, heat stress, and other appropriate medical considerations; use and limitations of respirator equipment as well as documentation procedures as outlined in 29 CFR 1910.134.

(3) PPE inspection procedures prior to, during, and after use.

(4) PPE donning and doffing procedures.

(5) PPE decontamination and disposal procedures.

(6) PPE maintenance and storage.

(7) Task duration as related to PPE limitations.

d. The employer should instruct the employee about the site medical surveillance program relative to theparticular site, including

(1) Specific medical surveillance programs that have been adapted for the site.

(2) Specific signs and symptoms related to exposure to hazardous materials on the site.

(3) The frequency and extent of periodic medical examinations that will be used on the site.

(4) Maintenance and availability of records.

(5) Personnel to be contacted and procedures to be followed when signs and symptoms of exposures arerecognized.

e. The employees will review and discuss the site safety plan as part of the training program. The location of the site safety plan and all written programs should be discussed with employees including a discussion of themechanisms for access, review, and references described.

B. RCRA Operations Training for Treatment, Storage and Disposal Facilities.

1. As a minimum, the training course required in 29 CFR 1926.65 (p) should include the following topics:

(a) Review of the applicable paragraphs of 29 CFR 1926.65 and the elements of the employer's occupational safety and health plan.

(b) Review of relevant hazards such as, but not limited to, chemical, biological, and radiological exposures; fire and explosion hazards; thermal extremes; and physical hazards.

(c) General safety hazards including those associated with electrical hazards, powered equipment hazards,lock-out-tag-out procedures, motor vehicle hazards and walking-working surface hazards.

(d) Confined-space hazards and procedures.

(e) Work practices to minimize employee risk from workplace hazards.

(f) Emergency response plan and procedures including first aid meeting the requirements of paragraph (p)(8).

(g) A review of procedures to minimize exposure to hazardous waste and various type of waste streams, including the materials handling program and spill containment program.

(h) A review of hazard communication programs meeting the requirements of 29 CFR 1910.1200.

(i) A review of medical surveillance programs meeting the requirements of 29 CFR 1926.65(p)(3) including the recognition of signs and symptoms of overexposure to hazardous substance including known synergistic interactions.

(j) A review of decontamination programs and procedures meeting the requirements of 29 CFR 1926.65(p)(4).

(k) A review of an employer's requirements to implement a training program and its elements.

(l) A review of the criteria and programs for proper selection and use of personal protective equipment, including respirators.

(m) A review of the applicable appendices to 29 CFR 1926.65.

(n) Principles of toxicology and biological monitoring as they pertain to occupational health.

(o) Rights and responsibilities of employees and employers under applicable OSHA and EPA laws.

(p) Hands-on exercises and demonstrations of competency with equipment to illustrate the basic equipment principles that may be used during the performance of work duties, including the donning and doffing of PPE.

(q) Sources of reference, efficient use of relevant manuals, and knowledge of hazard coding systems to include information contained in hazardous waste manifests.

(r) At least 8 hours of hands-on training.

(s) Training in the job skills required for an employee's job function and responsibility before they are permitted to participate in or supervise field activities.

2. The individual employer should provide hazardous waste employees with information and training prior to an employee's initial assignment into a work area. The training and information should cover the following topics:

(a) The Emergency response plan and procedures including first aid.

(b) A review of the employer's hazardous waste handling procedures including the materials handling program and elements of the spill containment program, location of spill response kits or equipment, and the names of those trained to respond to releases.

(c) The hazardous communication program meeting the requirements of 29 CFR 1910.1200.

(d) A review of the employer's medical surveillance program including the recognition of signs and symptomsof exposure to relevant hazardous substance including known synergistic interactions.

(e) A review of the employer's decontamination program and procedures.

(f) An review of the employer's training program and the parties responsible for that program.

(g) A review of the employer's personal protective equipment program including the proper selection and use ofPPE based upon specific site hazards.

(h) All relevant site-specific procedures addressing potential safety and health hazards. This may include, as appropriate, biological and radiological exposures, fire and explosion hazards, thermal hazards, and physicalhazards such as electrical hazards, powered equipment hazards, lock-out-tag-out hazards, motor vehicle hazards, and walking-working surface hazards.

(i) Safe use engineering controls and equipment on site.

(j) Names of personnel and alternates responsible for safety and health.

C. Emergency response training.

Federal OSHA standards in 29 CFR 1926.65(q) are directed toward private sector emergency responders. Therefore, the guidelines provided in this portion of the appendix are directed toward that employee population.However, they also impact indirectly through State OSHA or USEPA regulations some public sector emergency responders. Therefore, the guidelines provided in this portion of the appendix may be applied to both employee populations.

States with OSHA state plans must cover their employees with regulations at least as effective as the FederalOSHA standards. Public employees in states without approved state OSHA programs covering hazardous waste operations and emergency response are covered by the U.S. EPA under 40 CFR 311, a regulation virtuallyidentical to § 1926.65.

Since this is a non-mandatory appendix and therefore not an enforceable standard, OSHA recommends that those employers, employees or volunteers in public sector emergency response organizations outside Federal OSHA jurisdiction consider the following criteria in developing their own training programs. A unified approach to training at the community level between emergency response organizations covered by Federal OSHA andthose not covered directly by Federal OSHA can help ensure an effective community response to the release orpotential release of hazardous substances in the community.

a. General considerations.

Emergency response organizations are required to consider the topics listed in § 1926.65(q)(6). Emergency response organizations may use some or all of the following topics to supplement those mandatory topics when developing their response training programs. Many of the topics would require an interaction between the response provider and the individuals responsible for the site where the response would be expected.

(1) Hazard recognition, including:

(A) Nature of hazardous substances present,

(B) Practical applications of hazard recognition, including presentations on biology, chemistry, and physics.

(2) Principles of toxicology, biological monitoring, and risk assessment.

(3) Safe work practices and general site safety.

(4) Engineering controls and hazardous waste operations.

(5) Site safety plans and standard operating procedures.

(6) Decontamination procedures and practices.

(7) Emergency procedures, first aid, and self-rescue.

(8) Safe use of field equipment.

(9) Storage, handling, use and transportation of hazardous substances.

(10) Use, care, and limitations of personal protective equipment.

(11) Safe sampling techniques.

(12) Rights and responsibilities of employees under OSHA and other related laws concerning right-to-know, safety and health, compensations and liability.

(13) Medical monitoring requirements.

(14) Community relations.

b. Suggested criteria for specific courses.

(1) First responder awareness level.

(A) Review of and demonstration of competency in performing the applicable skills of 29 CFR 1926.65(q).

(B) Hands-on experience with the U.S. Department of Transportation's *Emergency Response Guidebook* (ERG) and familiarization with OSHA standard 29 CFR 1926.60.

(C) Review of the principles and practices for analyzing an incident to determine both the hazardous substances present and the basic hazard and response information for each hazardous substance present.

(D) Review of procedures for implementing actions consistent with the local emergency response plan, theorganization's standard operating procedures, and the current edition of DOT's ERG including emergency notification procedures and follow-up communications.

(E) Review of the expected hazards including fire and explosions hazards, confined space hazards, electrical hazards, powered equipment hazards, motor vehicle hazards, and walking-working surface hazards.

(F) Awareness and knowledge of the competencies for the First Responder at the Awareness Level covered inthe National Fire Protection Association's Standard No. 472, *Professional Competence of Responders to Hazardous Materials Incidents.*

(2) First responder operations level.

(A) Review of and demonstration of competency in performing the applicable skills of 29 CFR 1926.65(q).

(B) Hands-on experience with the U.S. Department of Transportation's *Emergency Response Guidebook* (ERG), manufacturer safety data sheets, CHEMTREC/CANUTEC, shipper or manufacturer contacts and other relevant sources of information addressing hazardous substance releases. Familiarization with OSHA standard 29 CFR 1926.60.

(C) Review of the principles and practices for analyzing an incident to determine the hazardous substances present, the likely behavior of the hazardous substance and its container, the types of hazardous substancetransportation containers and vehicles, the types and selection of the appropriate defensive strategy for containing the release.

(D) Review of procedures for implementing continuing response actions consistent with the local emergency response plan, the organization's standard operating procedures, and the current edition of DOT's ERG includingextended emergency notification procedures and follow-up communications.

(E) Review of the principles and practice for proper selection and use of personal protective equipment.

(F) Review of the principles and practice of personnel and equipment decontamination.

(G) Review of the expected hazards including fire and explosions hazards, confined space hazards, electrical hazards, powered equipment hazards, motor vehicle hazards, and walking-working surface hazards.

(H) Awareness and knowledge of the competencies for the First Responder at the Operations Level covered inthe National Fire Protection Association's Standard No. 472, *Professional Competence of Responders to Hazardous Materials Incidents.*

(3) Hazardous materials technician.

(A) Review of and demonstration of competency in performing the applicable skills of 29 CFR 1926.65(q).

(B) Hands-on experience with written and electronic information relative to response decision making including but not limited to the U.S. Department of Transportation's *Emergency Response Guidebook* (ERG), manufacturer safety data sheets, CHEMTREC/CANUTEC, shipper or manufacturer contacts, computer data bases and response models, and other relevant sources of information addressing hazardous substance releases. Familiarization with 29 CFR 1926.60.

(C) Review of the principles and practices for analyzing an incident to determine the hazardous substances present, their physical and chemical properties, the likely behavior of the hazardous substance and its container, the types of hazardous substance transportation containers and vehicles involved in the release, the appropriate strategy for approaching release sites and containing the release.

(D) Review of procedures for implementing continuing response actions consistent with the local emergency response plan, the organization's standard operating procedures, and the current edition of DOT's ERG includingextended emergency notification procedures and follow-up communications.

(E) Review of the principles and practice for proper selection and use of personal protective equipment.

(F) Review of the principles and practices of establishing exposure zones, proper decontamination and medical surveillance stations and procedures.

(G) Review of the expected hazards including fire and explosions hazards, confined space hazards, electrical hazards, powered equipment hazards, motor vehicle hazards, and walking-working surface hazards.

(H) Awareness and knowledge of the competencies for the Hazardous Materials Technician covered in the National Fire Protection Association's Standard No. 472, *Professional Competence of Responders to Hazardous Materials Incidents.*

(4) Hazardous materials specialist.

(A) Review of and demonstration of competency in performing the applicable skills of 29 CFR 1926.65(q).

(B) Hands-on experience with retrieval and use of written and electronic information relative to response decision making including but not limited to the U.S. Department of Transportation's *Emergency Response Guidebook* (ERG), manufacturer safety data sheets, CHEMTREC/CANUTEC, shipper or manufacturer contacts, computer data bases and response models, and other relevant sources of information addressing hazardous substance releases. Familiarization with 29 CFR 1926.60.

(C) Review of the principles and practices for analyzing an incident to determine the hazardous substancespresent, their physical and chemical properties, and the likely behavior of the hazardous substance and its container, vessel, or vehicle.

(D) Review of the principles and practices for identification of the types of hazardous substance transportation containers, vessels and vehicles involved in the release; selecting and using the various types of equipment available for plugging or patching transportation containers, vessels or vehicles; organizing and directing the use of multiple teams of hazardous material technicians and selecting the appropriate strategy for approaching release sites and containing or stopping the release.

(E) Review of procedures for implementing continuing response actions consistent with the local emergency response plan, the organization's standard operating procedures, including knowledge of the available public and private response resources, establishment of an incident command post, direction of hazardous materialtechnician teams, and extended emergency notification procedures and follow-up communications.

(F) Review of the principles and practice for proper selection and use of personal protective equipment.

(G) Review of the principles and practices of establishing exposure zones and proper decontamination,monitoring and medical surveillance stations and procedures.

(H) Review of the expected hazards including fire and explosions hazards, confined space hazards, electrical hazards, powered equipment hazards, motor vehicle hazards, and walking-working surface hazards.

(I) Awareness and knowledge of the competencies for the Off-site Specialist Employee covered in the National Fire Protection Association's Standard No. 472, *Professional Competence of Responders to Hazardous Materials Incidents.*

(5) *Incident commander.* The incident commander is the individual who, at any one time, is responsible for and in control of the response effort. This individual is the person responsible for the direction and coordination ofthe response effort. An incident commander's position should be occupied by the most senior, appropriately trained individual present at the response site. Yet, as necessary and appropriate by the level of response provided, the position may be occupied by many individuals during a particular response as the need for greater authority, responsibility, or training increases. It is possible for the first responder at the awareness level to assume the duties of incident commander until a more senior and appropriately trained individual arrives at the response site.

Therefore, any emergency responder expected to perform as an incident commander should be trained to fulfill the obligations of the position at the level of response they will be providing including the following:

(A) Ability to analyze a hazardous substance incident to determine the magnitude of the response problem.

(B) Ability to plan and implement an appropriate response plan within the capabilities of available personneland equipment.

(C) Ability to implement a response to favorably change the outcome of the incident in a manner consistentwith the local emergency response plan and the organization's standard operating procedures.

(D) Ability to evaluate the progress of the emergency response to ensure that the response objectives are beingmet safely, effectively, and efficiently.

(E) Ability to adjust the response plan to the conditions of the response and to notify higher levels of response when required by the changes to the response plan.

[58 FR 35129, June 30, 1993, as amended at 59 FR 43275, Aug. 22, 1994; 61 FR 5510, Feb. 13, 1996; 77 FR 17890, Mar. 26, 2012;
78 FR 9315, Feb. 8, 2013; 85 FR 8736, Feb. 18, 2020]

§ 1926.66 Criteria for design and construction of spray booths.

(a) *Definitions applicable to this section* -

(1) *Aerated solid powders.* Aerated powders shall mean any powdered material used as a coating material which shall be fluidized within a container by passing air uniformly from below. It is common practice to fluidize such materials to form a fluidized powder bed and then dip the part tobe coated into the bed in a manner similar to that used in liquid dipping. Such beds are also used assources for powder spray operations.

(2) *Spraying area.* Any area in which dangerous quantities of flammable vapors or mists, or combustible residues, dusts, or deposits are present due to the operation of spraying processes.

(3) *Spray booth.* A power-ventilated structure provided to enclose or accommodate a spraying operationto confine and limit the escape of spray, vapor, and residue, and to safely conduct or direct them toan exhaust system.

(4) *Waterwash spray booth.* A spray booth equipped with a water washing system designed to minimizedusts or residues entering exhaust ducts and to permit the recovery of overspray finishing material.

(5) *Dry spray booth.* A spray booth not equipped with a water washing system as described in paragraph(a)(4) of this section. A dry spray booth may be equipped with

(i) Distribution or baffle plates to promote an even flow of air through the booth or cause thedeposit of overspray before it enters the exhaust duct; or

(ii) Overspray dry filters to minimize dusts; or

(iii) Overspray dry filters to minimize dusts or residues entering exhaust ducts; or

(iv) Overspray dry filter rolls designed to minimize dusts or residues entering exhaust ducts; or

(v) Where dry powders are being sprayed, with powder collection systems so arranged in theexhaust to capture oversprayed material.

(6) *Fluidized bed.* A container holding powder coating material which is aerated from below so as to form an air-supported expanded cloud of such material through which the preheated object to becoated is immersed and transported.

(7) *Electrostatic fluidized bed.* A container holding powder coating material which is aerated from below so as to form an air-supported expanded cloud of such material which is electrically charged with a charge opposite to the charge of the object to be coated; such object is transported, through the container immediately above the charged and aerated materials in order to be coated.

(8) *Approved.* Shall mean approved and listed by a nationally recognized testing laboratory.

(9) *Listed.* See "approved" in paragraph (a)(8) of this section.

(b) *Spray booths* -

(1) **Construction.** Spray booths shall be substantially constructed of steel, securely and rigidly supported, or of concrete or masonry except that aluminum or other substantial noncombustible material may be used for intermittent or low volume spraying. Spray booths shall be designed to sweep air currents toward the exhaust outlet.

(2) **Interiors.** The interior surfaces of spray booths shall be smooth and continuous without edges and otherwise designed to prevent pocketing of residues and facilitate cleaning and washing without injury.

(3) **Floors.** The floor surface of a spray booth and operator's working area, if combustible, shall be covered with noncombustible material of such character as to facilitate the safe cleaning and removal of residues.

(4) **Distribution or baffle plates.** Distribution or baffle plates, if installed to promote an even flow of air through the booth or cause the deposit of overspray before it enters the exhaust duct, shall be of noncombustible material and readily removable or accessible on both sides for cleaning. Such plates shall not be located in exhaust ducts.

(5) **Dry type overspray collectors - (exhaust air filters).** In conventional dry type spray booths, overspray dry filters or filter rolls, if installed, shall conform to the following:

   (i) The spraying operations except electrostatic spraying operations shall be so designed, installed and maintained that the average air velocity over the open face of the booth (or booth cross section during spraying operations) shall be not less than 100 linear feet per minute.
   Electrostatic spraying operations may be conducted with an air velocity over the open face of the booth of not less than 60 linear feet per minute, or more, depending on the volume of the finishing material being applied and its flammability and explosion characteristics. Visible gauges or audible alarm or pressure activated devices shall be installed to indicate or insure that the required air velocity is maintained. Filter rolls shall be inspected to insure proper replacement of filter media.

   (ii) All discarded filter pads and filter rolls shall be immediately removed to a safe, well-detached location or placed in a water-filled metal container and disposed of at the close of the day's operation unless maintained completely in water.

   (iii) The location of filters in a spray booth shall be so as to not reduce the effective booth enclosure of the articles being sprayed.

   (iv) Space within the spray booth on the downstream and upstream sides of filters shall be protected with approved automatic sprinklers.

   (v) Filters or filter rolls shall not be used when applying a spray material known to be highly susceptible to spontaneous heating and ignition.

   (vi) Clean filters or filter rolls shall be noncombustible or of a type having a combustibility not in excess of class 2 filters as listed by Underwriters' Laboratories, Inc. Filters and filter rolls shall not be alternately used for different types of coating materials, where the combination of materials may be conducive to spontaneous ignition.

(6) **Frontal area.** Each spray booth having a frontal area larger than 9 square feet shall have a metal deflector or curtain not less than $2^1/_2$ inches (5.35 cm) deep installed at the upper outer edge of the booth over the opening.

(7) **Conveyors.** Where conveyors are arranged to carry work into or out of spray booths, the openings therefor shall be as small as practical.

(8) **Separation of operations.** Each spray booth shall be separated from other operations by not less than 3 feet (0.912 m), or by a greater distance, or by such partition or wall as to reduce the danger from juxtaposition of hazardous operations. See also paragraph (c)(1) of this section.

(9) **Cleaning.** Spray booths shall be so installed that all portions are readily accessible for cleaning. A clear space of not less than 3 feet (0.912 m) on all sides shall be kept free from storage or combustible construction.

(10) **Illumination.** When spraying areas are illuminated through glass panels or other transparent materials, only fixed lighting units shall be used as a source of illumination. Panels shall effectively isolate the spraying area from the area in which the lighting unit is located, and shall be of a noncombustible material of such a nature or so protected that breakage will be unlikely. Panels shall be so arranged that normal accumulations of residue on the exposed surface of the panel will not be raised to a dangerous temperature by radiation or conduction from the source of illumination.

(c) **Electrical and other sources of ignition** -

(1) **Conformance.** All electrical equipment, open flames and other sources of ignition shall conform to the requirements of this paragraph, except as follows:

(i) Electrostatic apparatus shall conform to the requirements of paragraphs (e) and (f) of this section;

(ii) Drying, curing, and fusion apparatus shall conform to the requirements of paragraph (g) of this section;

(iii) [Reserved]

(iv) Powder coating equipment shall conform to the requirements of paragraph (c)(1) of this section.

(2) **Minimum separation.** There shall be no open flame or spark producing equipment in any spraying area nor within 20 feet (6.08 m) thereof, unless separated by a partition.

(3) **Hot surfaces.** Space-heating appliances, steampipes, or hot surfaces shall not be located in a spraying area where deposits of combustible residues may readily accumulate.

(4) **Wiring conformance.** Electrical wiring and equipment shall conform to the provisions of this paragraph and shall otherwise be in accordance with subpart S of this part.

(5) **Combustible residues, areas.** Unless specifically approved for locations containing both deposits of readily ignitable residue and explosive vapors, there shall be no electrical equipment in any spraying area, whereon deposits of combustible residues may readily accumulate, except wiring in rigid conduit or in boxes or fittings containing no taps, splices, or terminal connections.

(6) **Wiring type approved.** Electrical wiring and equipment not subject to deposits of combustible residues but located in a spraying area as herein defined shall be of explosion-proof type approved for Class I, group D locations and shall otherwise conform to the provisions of subpart S of this part, for Class I, Division 1, Hazardous Locations. Electrical wiring, motors, and other equipment outside of but within 20 feet (6.08 m) of any spraying

area, and not separated therefrom by partitions, shall not produce sparks under normal operating conditions and shall otherwise conform to the provisions of subpart S of this part for Class I, Division 2 Hazardous Locations.

(7) **Lamps.** Electric lamps outside of, but within 20 feet (6.08 m) of any spraying area, and not separated therefrom by a partition, shall be totally enclosed to prevent the falling of hot particles and shall be protected from mechanical injury by suitable guards or by location.

(8) **Portable lamps.** Portable electric lamps shall not be used in any spraying area during spraying operations. Portable electric lamps, if used during cleaning or repairing operations, shall be of the type approved for hazardous Class I locations.

(9) **Grounding.**

(i) All metal parts of spray booths, exhaust ducts, and piping systems conveying flammable or combustible liquids or aerated solids shall be properly electrically grounded in an effective and permanent manner.

(d) *Ventilation* -

(1) **Conformance.** Ventilating and exhaust systems shall be in accordance with the Standard for Blower and Exhaust Systems for Vapor Removal, NFPA No. 91-1961, where applicable and shall also conform to the provisions of this section.

(2) **General.** All spraying areas shall be provided with mechanical ventilation adequate to remove flammable vapors, mists, or powders to a safe location and to confine and control combustible residues so that life is not endangered. Mechanical ventilation shall be kept in operation at all times while spraying operations are being conducted and for a sufficient time thereafter to allow vapors from drying coated articles and drying finishing material residue to be exhausted.

(3) **Independent exhaust.** Each spray booth shall have an independent exhaust duct system discharging to the exterior of the building, except that multiple cabinet spray booths in which identical spray finishing material is used with a combined frontal area of not more than 18 square feet may have a common exhaust. If more than one fan serves one booth, all fans shall be so interconnected that one fan cannot operate without all fans being operated.

(4) **Fan-rotating element.** The fan-rotating element shall be nonferrous or nonsparking or the casing shall consist of or be lined with such material. There shall be ample clearance between the fan-rotating element and the fan casing to avoid a fire by friction, necessary allowance being made for ordinary expansion and loading to prevent contact between moving parts and the duct or fan housing. Fan blades shall be mounted on a shaft sufficiently heavy to maintain perfect alignment even when the blades of the fan are heavily loaded, the shaft preferably to have bearings outside the duct and booth. All bearings shall be of the self-lubricating type, or lubricated from the outside duct.

(5) **Electric motors.** Electric motors driving exhaust fans shall not be placed inside booths or ducts. See also paragraph (c) of this section.

(6) **Belts.** Belts shall not enter the duct or booth unless the belt and pulley within the duct or booth are thoroughly enclosed.

(7) **Exhaust ducts.** Exhaust ducts shall be constructed of steel and shall be substantially supported. Exhaust ducts without dampers are preferred; however, if dampers are installed, they shall be maintained so that they will be in a full open position at all times the ventilating system is in operation.

    (i) Exhaust ducts shall be protected against mechanical damage and have a clearance from unprotected combustible construction or other combustible material of not less than 18 inches (45.72 cm).

    (ii) If combustible construction is provided with the following protection applied to all surfaces within 18 inches (45.72 cm), clearances may be reduced to the distances indicated:

(*a*) 28-gage sheet metal on 1/4-inch asbestos mill board	12 inches (30.48 cm).
(*b*) 28-gage sheet metal on 1/8-inch asbestos mill board spaced out 1 inch (2.54 cm) on noncombustible spacers	9 inches (22.86 cm).
(*c*) 22-gage sheet metal on 1-inch rockwool batts reinforced with wire mesh or the equivalent	3 inches (7.62 cm).
(*d*) Where ducts are protected with an approved automatic sprinkler system, properly maintained, the clearance required in paragraph (d)(7)(i) of this section may be reduced to 6 inches (15.24 cm)	

(8) **Discharge clearance.** Unless the spray booth exhaust duct terminal is from a water-wash spray booth, the terminal discharge point shall be not less than 6 feet from any combustible exterior wall or roof nor discharge in the direction of any combustible construction or unprotected opening in any noncombustible exterior wall within 25 feet (7.6 m).

(9) **Air exhaust.** Air exhaust from spray operations shall not be directed so that it will contaminate makeup air being introduced into the spraying area or other ventilating intakes, nor directed so as to create a nuisance. Air exhausted from spray operations shall not be recirculated.

(10) **Access doors.** When necessary to facilitate cleaning, exhaust ducts shall be provided with an ample number of access doors.

(11) **Room intakes.** Air intake openings to rooms containing spray finishing operations shall be adequate for the efficient operation of exhaust fans and shall be so located as to minimize the creation of dead air pockets.

(12) **Drying spaces.** Freshly sprayed articles shall be dried only in spaces provided with adequate ventilation to prevent the formation of explosive vapors. In the event adequate and reliable ventilation is not provided such drying spaces shall be considered a spraying area.

(e) *Fixed electrostatic apparatus* -

    (1) **Conformance.** Where installation and use of electrostatic spraying equipment is used, such installation and use shall conform to all other paragraphs of this section, and shall also

conform to the requirements of this paragraph.

(2) **Type approval.** Electrostatic apparatus and devices used in connection with coating operations shall be of approved types.

(3) **Location.** Transformers, power packs, control apparatus, and all other electrical portions of the equipment, with the exception of high-voltage grids, electrodes, and electrostatic atomizing heads and their connections, shall be located outside of the spraying area, or shall otherwise conform to the requirements of paragraph (c) of this section.

(4) **Support.** Electrodes and electrostatic atomizing heads shall be adequately supported in permanent locations and shall be effectively insulated from the ground. Electrodes and electrostatic atomizing heads which are permanently attached to their bases, supports, or reciprocators, shall be deemed to comply with this section. Insulators shall be nonporous and noncombustible.

(5) **Insulators, grounding.** High-voltage leads to electrodes shall be properly insulated and protected from mechanical injury or exposure to destructive chemicals. Electrostatic atomizing heads shall be effectively and permanently supported on suitable insulators and shall be effectively guarded against accidental contact or grounding. An automatic means shall be provided for grounding the electrode system when it is electrically deenergized for any reason. All insulators shall be kept clean and dry.

(6) **Safe distance.** A safe distance shall be maintained between goods being painted and electrodes or electrostatic atomizing heads or conductors of at least twice the sparking distance. A suitable sign indicating this safe distance shall be conspicuously posted near the assembly.

(7) **Conveyors required.** Goods being painted using this process are to be supported on conveyors. The conveyors shall be so arranged as to maintain safe distances between the goods and the electrodes or electrostatic atomizing heads at all times. Any irregularly shaped or other goods subject to possible swinging or movement shall be rigidly supported to prevent such swinging or movement which would reduce the clearance to less than that specified in paragraph (e)(6) of this section.

(8) **Prohibition.** This process is not acceptable where goods being coated are manipulated by hand. When finishing materials are applied by electrostatic equipment which is manipulated by hand, see paragraph (f) of this section for applicable requirements.

(9) **Fail-safe controls.** Electrostatic apparatus shall be equipped with automatic controls which will operate without time delay to disconnect the power supply to the high voltage transformer and to signal the operator under any of the following conditions:

(i) Stoppage of ventilating fans or failure of ventilating equipment from any cause.

(ii) Stoppage of the conveyor carrying goods through the high voltage field.

(iii) Occurrence of a ground or of an imminent ground at any point on the high voltage system.

(iv) Reduction of clearance below that specified in paragraph (e)(6) of this section.

(10) **Guarding.** Adequate booths, fencing, railings, or guards shall be so placed about the equipment that they, either by their location or character or both, assure that a safe isolation of the process is maintained from plant storage or personnel. Such railings, fencing, and guards shall be of conducting material, adequately grounded.

(11) **Ventilation.** Where electrostatic atomization is used the spraying area shall be so ventilated as to insure safe conditions from a fire and health standpoint.

(12) **Fire protection.** All areas used for spraying, including the interior of the booth, shall be protected by automatic sprinklers where this protection is available. Where this protection is not available, otherapproved automatic extinguishing equipment shall be provided.

(f) *Electrostatic hand spraying equipment -*

(1) **Application.** This paragraph shall apply to any equipment using electrostatically charged elements for the atomization and/or, precipitation of materials for coatings on articles, or for other similar purposes in which the atomizing device is hand held and manipulated during the spraying operation.

(2) **Conformance.** Electrostatic hand spraying equipment shall conform with the other provisions of this section.

(3) **Equipment approval and specifications.** Electrostatic hand spray apparatus and devices used in connection with coating operations shall be of approved types. The high voltage circuits shall be designed so as to not produce a spark of sufficient intensity to ignite any vapor-air mixtures nor result in appreciable shock hazard upon coming in contact with a grounded object under all normaloperating conditions. The electrostatically charged exposed elements of the handgun shall be capable of being energized only by a switch which also controls the coating material supply.

(4) **Electrical support equipment.** Transformers, powerpacks, control apparatus, and all other electrical portions of the equipment, with the exception of the handgun itself and its connections to the powersupply shall be located outside of the spraying area or shall otherwise conform to the requirements of paragraph (c) of this section.

(5) **Spray gun ground.** The handle of the spraying gun shall be electrically connected to ground by a metallic connection and to be so constructed that the operator in normal operating position is inintimate electrical contact with the grounded handle.

(6) **Grounding-general.** All electrically conductive objects in the spraying area shall be adequately grounded. This requirement shall apply to paint containers, wash cans, and any other objects or devices in the area. The equipment shall carry a prominent permanently installed warning regardingthe necessity for this grounding feature.

(7) **Maintenance of grounds.** Objects being painted or coated shall be maintained in metallic contact with the conveyor or other grounded support. Hooks shall be regularly cleaned to insure this contact and areas of contact shall be sharp points or knife edges where possible. Points of support of the object shall be concealed from random spray where feasible and where the objects being sprayed are supported from a conveyor, the point of attachment to the conveyor shall be so located as to notcollect spray material during normal operation.

(8) **Interlocks.** The electrical equipment shall be so interlocked with the ventilation of the spraying areathat the equipment cannot be operated unless the ventilation fans are in operation.

(9) **Ventilation.** The spraying operation shall take place within a spray area which is adequately ventilated to remove solvent vapors released from the operation.

(g) *Drying, curing, or fusion apparatus -*

(1) ***Conformance.*** Drying, curing, or fusion apparatus in connection with spray application of flammable and combustible finishes shall conform to the Standard for Ovens and Furnaces, NFPA 86A-1969, where applicable and shall also conform with the following requirements of this paragraph.

(2) ***Alternate use prohibited.*** Spray booths, rooms, or other enclosures used for spraying operations shall not alternately be used for the purpose of drying by any arrangement which will cause a material increase in the surface temperature of the spray booth, room, or enclosure.

(3) ***Adjacent system interlocked.*** Except as specifically provided in paragraph (g)(4) of this section, drying, curing, or fusion units utilizing a heating system having open flames or which may produce sparks shall not be installed in a spraying area, but may be installed adjacent thereto when equipped with an interlocked ventilating system arranged to:

  (i) Thoroughly ventilate the drying space before the heating system can be started;

  (ii) Maintain a safe atmosphere at any source of ignition;

  (iii) Automatically shut down the heating system in the event of failure of the ventilating system.

(4) ***Alternate use permitted.*** Automobile refinishing spray booths or enclosures, otherwise installed and maintained in full conformity with this section, may alternately be used for drying with portable electrical infrared drying apparatus when conforming with the following:

  (i) Interior (especially floors) of spray enclosures shall be kept free of overspray deposits.

  (ii) During spray operations, the drying apparatus and electrical connections and wiring thereto shall not be located within spray enclosure nor in any other location where spray residues maybe deposited thereon.

  (iii) The spraying apparatus, the drying apparatus, and the ventilating system of the spray enclosure shall be equipped with suitable interlocks so arranged that:

   (*a*) The spraying apparatus cannot be operated while the drying apparatus is inside the spray enclosure.

   (*b*) The spray enclosure will be purged of spray vapors for a period of not less than 3 minutes before the drying apparatus can be energized.

   (*c*) The ventilating system will maintain a safe atmosphere within the enclosure during the drying process and the drying apparatus will automatically shut off in the event of failure of the ventilating system.

  (iv) All electrical wiring and equipment of the drying apparatus shall conform with the applicable sections of subpart S of this part. Only equipment of a type approved for Class I, Division 2 hazardous locations shall be located within 18 inches (45.72 cm) of floor level. All metallic parts of the drying apparatus shall be properly electrically bonded and grounded.

  (v) The drying apparatus shall contain a prominently located, permanently attached warning sign indicating that ventilation should be maintained during the drying period and that spraying should not be conducted in the vicinity that spray will deposit on apparatus.

[58 FR 35149, June 30, 1993]

Subpart E - Personal Protective and Life Saving Equipment

Authority: 40 U.S.C. 3701 *et seq.;* 29 U.S.C. 653, 655, 657; Secretary of Labor's Order No. 12-71 (36 FR 8754), 8-76

(41 FR 25059), 9-83 (48 FR 35736), 1-90 (55 FR 9033), 6-96 (62 FR 111), 5-2002 (67 FR 65008), 5-2007 (72 FR

31160), 4-2010 (75 FR 55355), or 1-2012 (77 FR 3912), as applicable; and 29 CFR part 1911.

§ 1926.95 Criteria for personal protective equipment.

(a) *Application.* Protective equipment, including personal protective equipment for eyes, face, head, and extremities, protective clothing, respiratory devices, and protective shields and barriers, shall be provided, used, and maintained in a sanitary and reliable condition wherever it is necessary by reason of hazards of processes or environment, chemical hazards, radiological hazards, or mechanical irritants encountered in a manner capable of causing injury or impairment in the function of any part of the body through absorption, inhalation or physical contact.

(b) *Employee-owned equipment.* Where employees provide their own protective equipment, the employer shall be responsible to assure its adequacy, including proper maintenance, and sanitation of such equipment.

(c) *Design.* All personal protective equipment shall be of safe design and construction for the work to be performed.

(d) *Payment for protective equipment.*

    (1) Except as provided by paragraphs (d)(2) through (d)(6) of this section, the protective equipment, including personal protective equipment (PPE), used to comply with this part, shall be provided by the employer at no cost to employees.

    (2) The employer is not required to pay for non-specialty safety-toe protective footwear (including steel-toe shoes or steel-toe boots) and non-specialty prescription safety eyewear, provided that the employer permits such items to be worn off the job-site.

    (3) When the employer provides metatarsal guards and allows the employee, at his or her request, to use shoes or boots with built-in metatarsal protection, the employer is not required to reimburse the employee for the shoes or boots.

    (4) The employer is not required to pay for:

        (i) Everyday clothing, such as long-sleeve shirts, long pants, street shoes, and normal work boots; or

        (ii) Ordinary clothing, skin creams, or other items, used solely for protection from weather, such as winter coats, jackets, gloves, parkas, rubber boots, hats, raincoats, ordinary sunglasses, and sunscreen.

    (5) The employer must pay for replacement PPE, except when the employee has lost or intentionally damaged the PPE.

    (6) Where an employee provides adequate protective equipment he or she owns pursuant to paragraph

(b) of this section, the employer may allow the employee to use it and is not required to reimburse the employee for that equipment. The employer shall not require an employee to provide or pay for his or her own PPE, unless the PPE is excepted by paragraphs (d)(2) through (d)(5) of this section.

(7) This section shall become effective on February 13, 2008. Employers must implement the PPE payment requirements no later than May 15, 2008.

Note to § 1926.95(d): When the provisions of another OSHA standard specify whether or not the

employer must pay for specific equipment, the payment provisions of that standard shall prevail.

[58 FR 35152, June 30, 1993, as amended at 72 FR 64429, Nov. 15, 2007]

§ 1926.96 Occupational foot protection.

Safety-toe footwear for employees shall meet the requirements and specifications in American National Standard for Men's Safety-Toe Footwear, Z41.1-1967.

[58 FR 35152, June 30, 1993]

§ 1926.97 Electrical protective equipment.

(a) ***Design requirements for specific types of electrical protective equipment.*** Rubber insulating blankets, rubber insulating matting, rubber insulating covers, rubber insulating line hose, rubber insulating gloves, and rubber insulating sleeves shall meet the following requirements:

    (1) ***Manufacture and marking of rubber insulating equipment.***

        (i) Blankets, gloves, and sleeves shall be produced by a seamless

        (ii) process. Each item shall be clearly marked as follows:

            (A) Class 00 equipment shall be marked

            (B) Class 00. Class 0 equipment shall be

            (C) marked Class 0.

            Class 1 equipment shall be marked

            (D) Class 1. Class 2 equipment shall be

            (E) marked Class 2. Class 3 equipment

            (F) shall be marked Class 3. Class 4

            (G) equipment shall be marked Class 4.

            Nonozone-resistant equipment shall be marked

            (H) Type I. Ozone-resistant equipment shall be

            (I) marked Type II.

Other relevant markings, such as the manufacturer's identification and the size of the equipment, may also be provided.

(iii) Markings shall be nonconducting and shall be applied in such a manner as not to impair the insulating qualities of the equipment.

(iv) Markings on gloves shall be confined to the cuff portion of the glove.

(2) **Electrical requirements.**

(i) Equipment shall be capable of withstanding the ac proof-test voltage specified in Table E-1 or the dc proof-test voltage specified in Table E-2.

(A) The proof test shall reliably indicate that the equipment can withstand the voltage involved.

(B) The test voltage shall be applied continuously for 3 minutes for equipment other than matting and shall be applied continuously for 1 minute for matting.

(C) Gloves shall also be capable of separately withstanding the ac proof-test voltage specified in Table E-1 after a 16-hour water soak. (See the note following paragraph (a)(3)(ii)(B) of this section.)

(ii) When the ac proof test is used on gloves, the 60-hertz proof-test current may not exceed the values specified in Table E-1 at any time during the test period.

(A) If the ac proof test is made at a frequency other than 60 hertz, the permissible proof-test current shall be computed from the direct ratio of the frequencies.

(B) For the test, gloves (right side out) shall be filled with tap water and immersed in water to a depth that is in accordance with Table E-3. Water shall be added to or removed from the glove, as necessary, so that the water level is the same inside and outside the glove.

(C) After the 16-hour water soak specified in paragraph (a)(2)(i)(C) of this section, the 60-hertz proof-test current may not exceed the values given in Table E-1 by more than 2 milliamperes.

(iii) Equipment that has been subjected to a minimum breakdown voltage test may not be used for electrical protection. (See the note following paragraph (a)(3)(ii)(B) of this section.)

(iv) Material used for Type II insulating equipment shall be capable of withstanding an ozone test, with no visible effects. The ozone test shall reliably indicate that the material will resist ozone exposure in actual use. Any visible signs of ozone deterioration of the material, such as checking, cracking, breaks, or pitting, is evidence of failure to meet the requirements for ozone-resistant material. (See the note following paragraph (a)(3)(ii)(B) of this section.)

(3) **Workmanship and finish.**

(i) Equipment shall be free of physical irregularities that can adversely affect the insulating properties of the equipment and that can be detected by the tests or inspections required under this section.

(ii) Surface irregularities that may be present on all rubber goods (because of imperfections on forms or molds or because of inherent difficulties in the manufacturing process) and that may appear as indentations, protuberances, or imbedded foreign material are acceptable under the following conditions:

(A) The indentation or protuberance blends into a smooth slope when the material is stretched.

(B) Foreign material remains in place when the insulating material is folded and stretches with the insulating material surrounding it.

Note to paragraph (a): Rubber insulating equipment meeting the following national consensus standards is deemed to be in compliance with the performance requirements of paragraph (a) of this section:

American Society for Testing and Materials (ASTM) D120-09, *Standard Specification for Rubber Insulating Gloves.*

ASTM D178-01 (2010), *Standard Specification for Rubber Insulating Matting.*

ASTM D1048-12, *Standard Specification for Rubber Insulating Blankets.*

ASTM D1049-98 (2010), *Standard Specification for Rubber Insulating Covers.* ASTM D1050-05 (2011), *Standard Specification for Rubber Insulating Line Hose.* ASTM D1051-08, *Standard Specification for Rubber Insulating Sleeves.*

The preceding standards also contain specifications for conducting the various tests required in paragraph (a) of this section. For example, the ac and dc proof tests, the breakdown test, the water-soak procedure, and the ozone test mentioned in this paragraph are described in detail in these ASTM standards.

ASTM F1236-96 (2012), *Standard Guide for Visual Inspection of Electrical Protective Rubber Products,* presents methods and techniques for the visual inspection of electrical protective equipment made of rubber. This guide also contains descriptions and photographs of irregularities that can be found in this equipment.

ASTM F819-10, *Standard Terminology Relating to Electrical Protective Equipment for Workers,* includes definitions of terms relating to the electrical protective equipment covered under this section.

(2)

(b) ***Design requirements for other types of electrical protective equipment.*** The following requirements apply to the design and manufacture of electrical protective equipment that is not covered by paragraph (a) of this section:

    (1) ***Voltage withstand.*** Insulating equipment used for the protection of employees shall be capable of withstanding, without failure, the voltages that may be imposed upon it.

> Note to paragraph (b)(1): These voltages include transient overvoltages, such as switching surges, as well as nominal line voltage. See appendix B to subpart V of this part for a discussion of transient overvoltages on electric power transmission and distribution systems. See IEEE Std 516-2009, *IEEE Guide for Maintenance Methods on Energized Power Lines,* for methods of determining the magnitude of transient overvoltages on an electrical system and for a discussion comparing the ability of insulation equipment to withstand a transient overvoltage based on its ability to withstand ac voltage testing.

    Equipment current.

        (i) Protective equipment used for the primary insulation of employees from energized circuit parts shall be capable of passing a current test when subjected to the highest nominal voltage on which the equipment is to be used.

        (ii) When insulating equipment is tested in accordance with paragraph (b)(2)(i) of this section, the equipment current may not exceed 1 microampere per kilovolt of phase-to-phase applied voltage.

> Note 1 to paragraph (b)(2): This paragraph applies to equipment that provides primary insulation of employees from energized parts. It does not apply to equipment used for secondary insulation or equipment used for brush contact only.

> Note 2 to paragraph (b)(2): For ac excitation, this current consists of three components: Capacitive current because of the dielectric properties of the insulating material itself, conduction current through the volume of the insulating equipment, and leakage current along the surface of the tool or equipment. The conduction current is normally negligible. For clean, dry insulating equipment, the leakage current is small, and the capacitive current predominates.

> Note to paragraph (b): Plastic guard equipment is deemed to conform to the performance requirements of paragraph (b) of this section if it meets, and is used in accordance with, ASTM F712-06 (2011), *Standard Test Methods and Specifications for Electrically Insulating Plastic Guard Equipment for Protection of Workers.*

(c) ***In-service care and use of electrical protective equipment -***

    (1) ***General.*** Electrical protective equipment shall be maintained in a safe, reliable condition.

(2) **Specific requirements.** The following specific requirements apply to rubber insulating blankets, rubber insulating covers, rubber insulating line hose, rubber insulating gloves, and rubber insulating sleeves:

    (i) Maximum use voltages shall conform to those listed in Table E-4.

    (ii) Insulating equipment shall be inspected for damage before each day's use and immediately following any incident that can reasonably be suspected of causing damage. Insulating glovesshall be given an air test, along with the inspection.

    Note to paragraph (c)(2)(ii): ASTM F1236-96 (2012), *Standard Guide for Visual Inspection of Electrical Protective Rubber Products,* presents methods and techniques for the visual inspection of electrical protective equipment made of rubber. This guide also contains descriptions and photographs of irregularities that can be found in this equipment.

    (iii) Insulating equipment with any of the following defects may not be used:

        (A) A hole, tear, puncture, or cut;

        (B) Ozone cutting or ozone checking (that is, a series of interlacing cracks produced by ozoneon rubber under mechanical stress);

        (C) An embedded foreign object;

        (D) Any of the following texture changes: Swelling, softening, hardening, or becoming sticky orinelastic.

        (E) Any other defect that damages the insulating properties.

    (iv) Insulating equipment found to have other defects that might affect its insulating propertiesshall be removed from service and returned for testing under paragraphs (c)(2)(viii) and (c)(2)(ix) of this section.

    (v) Insulating equipment shall be cleaned as needed to remove foreign substances.

    (vi) Insulating equipment shall be stored in such a location and in such a manner as to protect itfrom light, temperature extremes, excessive humidity, ozone, and other damaging substancesand conditions.

    (vii) Protector gloves shall be worn over insulating gloves, except as follows:

        (A) Protector gloves need not be used with Class 0 gloves, under limited-use conditions, whensmall equipment and parts manipulation necessitate unusually high finger dexterity.

        Note to paragraph (c)(2)(vii)(A): Persons inspecting rubber insulating gloves used under these conditions need to take extra care in visually examining them. Employees using rubber insulating gloves under these conditions need to take extra care to avoid handling sharp objects.

        (B) If the voltage does not exceed 250 volts, ac, or 375 volts, dc, protector gloves need not be used with Class 00 gloves, under limited-use conditions, when small equipment and partsmanipulation necessitate unusually high finger dexterity.

Note to paragraph (c)(2)(vii)(B): Persons inspecting rubber insulating gloves used under these conditions need to take extra care in visually examining them. Employees using rubber insulating gloves under these conditions need to take extra care to avoid handling sharp objects.

(C) Any other class of glove may be used without protector gloves, under limited-use conditions, when small equipment and parts manipulation necessitate unusually high finger dexterity but only if the employer can demonstrate that the possibility of physicaldamage to the gloves is small and if the class of glove is one class higher than that required for the voltage involved.

(D) Insulating gloves that have been used without protector gloves may not be reused until they have been tested under the provisions of paragraphs (c)(2)(viii) and (c)(2)(ix) of thissection.

(viii) Electrical protective equipment shall be subjected to periodic electrical tests. Test voltages and the maximum intervals between tests shall be in accordance with Table E-4 and Table E-5.

(ix) The test method used under paragraphs (c)(2)(viii) and (c)(2)(xi) of this section shall reliablyindicate whether the insulating equipment can withstand the voltages involved.

Note to paragraph (c)(2)(ix): Standard electrical test methods considered as meetingthis paragraph are given in the following national consensus standards: ASTM D120-09, *Standard Specification for Rubber Insulating Gloves.*

ASTM D178-01 (2010), *Standard Specification for Rubber Insulating Matting.*

ASTM D1048-12, *Standard Specification for Rubber Insulating Blankets.*

ASTM D1049-98 (2010), *Standard Specification for Rubber Insulating*

Covers. ASTM D1050-05 (2011), *Standard Specification for Rubber*

Insulating Line Hose. ASTM D1051-08, *Standard Specification for Rubber*

Insulating Sleeves.

ASTM F478-09, *Standard Specification for In-Service Care of Insulating Line Hose and Covers.*

ASTM F479-06 (2011), *Standard Specification for In-Service Care of Insulating Blankets.*

ASTM F496-08, *Standard Specification for In-Service Care of Insulating Gloves and Sleeves.*

(x) Insulating equipment failing to pass inspections or electrical tests may not be used by employees, except as follows:

    (A) Rubber insulating line hose may be used in shorter lengths with the defective portion cutoff.

    (B) Rubber insulating blankets may be salvaged by severing the defective area from the undamaged portion of the blanket. The resulting undamaged area may not be smaller than 560 millimeters by 560 millimeters (22 inches by 22 inches) for Class 1, 2, 3, and 4 blankets.

    (C) Rubber insulating blankets may be repaired using a compatible patch that results in physical and electrical properties equal to those of the blanket.

    (D) Rubber insulating gloves and sleeves with minor physical defects, such as small cuts, tears, or punctures, may be repaired by the application of a compatible patch. Also, rubber insulating gloves and sleeves with minor surface blemishes may be repaired with a

compatible liquid compound. The repaired area shall have electrical and physical properties equal to those of the surrounding material. Repairs to gloves are permitted only in the area between the wrist and the reinforced edge of the opening.

(xi) Repaired insulating equipment shall be retested before it may be used by employees.

(xii) The employer shall certify that equipment has been tested in accordance with the requirements of paragraphs (c)(2)(iv), (c)(2)(vii)(D), (c)(2)(viii), (c)(2)(ix), and (c)(2)(xi) of this section. The certification shall identify the equipment that passed the test and the date it was tested and shall be made available upon request to the Assistant Secretary for Occupational Safety and Health and to employees or their authorized representatives.

Note to paragraph (c)(2)(xii): Marking equipment with, and entering onto logs, the results of the tests and the dates of testing are two acceptable means of meeting the certification requirement.

Table E-1 - AC Proof-Test Requirements

Class of equipment	Proof-test voltage rms V	Maximum proof-test current, mA (gloves only)			
		280-mm (11-in) glove	360-mm (14-in) glove	410-mm (16-in) glove	460-mm (18-in) glove
00	2,500	8	12		
0	5,000	8	12	14	16
1	10,000		14	16	18
2	20,000		16	18	20
3	30,000		18	20	22
4	40,000			22	24

Table E-2 - DC Proof-Test Requirements

Class of equipment	Proof-test voltage
00	10,000
0	20,000
1	40,000
2	50,000
3	60,000
4	70,000

Note: The dc voltages listed in this table are not appropriate for proof testing rubber insulating linehose or covers. For this equipment, dc proof tests shall use a voltage high enough to indicate that the equipment can be safely used at the voltages listed in Table E-4. See ASTM D1050-05 (2011) and ASTM D1049-98 (2010) for further information on proof tests for rubber insulating line hose and covers, respectively.

Table E-3 - Glove Tests - Water Level[1,2]

Class of glove	AC proof test		DC proof test	
	mm	in	mm	in
00	38	1.5	38	1.5
0	38	1.5	38	1.5
1	38	1.5	51	2.0
2	64	2.5	76	3.0
3	89	3.5	102	4.0
4	127	5.0	153	6.0

[1] The water level is given as the clearance from the reinforced edge of the glove to the water line, with a tolerance of ±13 mm. (±0.5 in.).

[2] If atmospheric conditions make the specified clearances impractical, the clearances may be increased by a maximum of 25 mm. (1 in.).

Table E-4 - Rubber Insulating Equipment, Voltage Requirements

Class of equipment	Maximum use voltage[1] AC rms	Retest voltage[2] AC rms	Retest voltage[2] DC avg
00	500	2,500	10,000
0	1,000	5,000	20,000
1	7,500	10,000	40,000
2	17,000	20,000	50,000

		26,500	30,000	60,000
3				
4		36,000	40,000	70,000

[1] The maximum use voltage is the ac voltage (rms) classification of the protective equipment that designates the maximum nominal design voltage of the energized system that may be safely worked. The nominal design voltage is equal to the phase-to-phase voltage on multiphase circuits. However, the phase-to-ground potential is considered to be the nominal design voltage if:

(1) There is no multiphase exposure in a system area and the voltage exposure is limited to thephase-to-ground potential, or

(2) The electric equipment and devices are insulated or isolated or both so that the multiphaseexposure on a grounded wye circuit is removed.

[2] The proof-test voltage shall be applied continuously for at least 1 minute, but no more than 3minutes.

Table E-5 - Rubber Insulating Equipment, Test Intervals

Type of equipment	When to test
Rubber insulating line hose	Upon indication that insulating value is suspect and after repair.
Rubber insulating covers	Upon indication that insulating value is suspect and after repair.
Rubber insulating blankets	Before first issue and every 12 months thereafter;[1] upon indication that insulating value is suspect; and after repair.
Rubber insulating gloves	Before first issue and every 6 months thereafter;[1] upon indication that insulating value is suspect; after repair; and after use without protectors.
Rubber insulating sleeves	Before first issue and every 12 months thereafter;[1] upon indication that insulating value is suspect; and after repair.

[1] If the insulating equipment has been electrically tested but not issued for service, the insulatingequipment may not be placed into service unless it has been electrically tested within the previous12 months.

[79 FR 20693, Apr. 11, 2014]

§ 1926.98 [Reserved]

§ 1926.100 Head protection.

(a) Employees working in areas where there is a possible danger of head injury from impact, or from falling or flying objects, or from electrical shock and burns, shall be protected by protective helmets.

(b) **Criteria for head protection.**

(1) The employer must provide each employee with head protection that meets the specifications contained in any of the following consensus standards:

(i) American National Standards Institute (ANSI) Z89.1-2009, "American National Standard for Industrial Head Protection," incorporated by reference in § 1926.6;

(ii) American National Standards Institute (ANSI) Z89.1-2003, "American National Standard for Industrial Head Protection," incorporated by reference in § 1926.6; or

(iii) American National Standards Institute (ANSI) Z89.1-1997, "American National Standard for Personnel Protection - Protective Headwear for Industrial Workers - Requirements," incorporated by reference in § 1926.6.

(2) The employer must ensure that the head protection provided for each employee exposed to high-voltage electric shock and burns also meets the specifications contained in Section 9.7 ("Electrical Insulation") of any of the consensus standards identified in paragraph (b)(1) of this section.

(3) OSHA will deem any head protection device that the employer demonstrates is at least as effective as a head protection device constructed in accordance with one of the consensus standards identified in paragraph (b)(1) of this section to be in compliance with the requirements of this section.

[44 FR 8577, Feb. 9, 1979, as amended at 77 FR 37600, June 22, 2012; 77 FR 42988, July 23, 2012]

§ 1926.101 Hearing protection.

(a) Wherever it is not feasible to reduce the noise levels or duration of exposures to those specified in Table D-2, Permissible Noise Exposures, in § 1926.52, ear protective devices shall be provided and used.

(b) Ear protective devices inserted in the ear shall be fitted or determined individually by competent

(c) persons. Plain cotton is not an acceptable protective device.

§ 1926.102 Eye and face protection.

(a) **General.**

(1) The employer shall ensure that each affected employee uses appropriate eye or face protection when exposed to eye or face hazards from flying particles, molten metal, liquid chemicals, acids or caustic liquids, chemical gases or vapors, or potentially injurious light radiation.

(2) The employer shall ensure that each affected employee uses eye protection that provides side protection when there is a hazard from flying objects. Detachable side protectors (e.g. clip-on or slide-on side shields) meeting the pertinent requirements of this section are acceptable.

(3) The employer shall ensure that each affected employee who wears prescription lenses while engaged in operations that involve eye hazards wears eye protection that incorporates the prescription in its design, or wears eye protection that can be worn over the prescription lenses

without disturbing the proper position of the prescription lenses or the protective lenses.

(4) Eye and face PPE shall be distinctly marked to facilitate identification of the manufacturer.

(5) Protectors shall meet the following minimum requirements:

(i) They shall provide adequate protection against the particular hazards for which they are designed.

(ii) They shall be reasonably comfortable when worn under the designated

(iii) conditions. They shall fit snugly and shall not unduly interfere with the

(iv) movements of the wearer. They shall be durable.

(v) They shall be capable of being

(vi) disinfected. They shall be easily

(b) cleanable.

Criteria for protective eye and face protection.

(1) Protective eye and face protection devices must comply with any of the following consensus standards:

(i) ANSI/ISEA Z87.1-2010, Occupational and Educational Personal Eye and Face Protection Devices, incorporated by reference in § 1926.6;

(ii) ANSI Z87.1-2003, Occupational and Educational Personal Eye and Face Protection Devices, incorporated by reference in § 1926.6; or

(iii) ANSI Z87.1-1989 (R-1998), Practice for Occupational and Educational Eye and Face Protection, incorporated by reference in § 1926.6;

(2) Protective eye and face protection devices that the employer demonstrates are at least as effective as protective eye and face protection devices that are constructed in accordance with one of the above consensus standards will be deemed to be in compliance with the requirements of this section.

(c) *Protection against radiant energy* -

(1) *Selection of shade numbers for welding filter.* Table E-1 shall be used as a guide for the selection of the proper shade numbers of filter lenses or plates used in welding. Shades more dense than those listed may be used to suit the individual's needs.

Table E-1 - Filter Lens Shade Numbers for Protection Against Radiant Energy

Welding operation	Shade number
Shielded metal-arc welding 1/16-, 3/32-, 1/8-, 5/32-inch diameter electrodes	10
Gas-shielded arc welding (nonferrous) 1/16-, 3/32-, 1/8-, 5/32-inch diameter electrodes	11
Gas-shielded arc welding (ferrous) 1/16-, 3/32-, 1/8-, 5/32-inch diameter electrodes	12
Shielded metal-arc welding 3/16-, 7/32-, 1/4-inch diameter electrodes	12
5/16-, 3/8-inch diameter electrodes	14
Atomic hydrogen welding	10-14

Carbon-arc welding	14
Soldering	2
Torch-brazing	3 or 4
Light cutting, up to 1 inch	3 or 4
Medium cutting, 1 inch to 6 inches	4 or 5
Heavy cutting, over 6 inches	5 or 6
Gas welding (light), up to 1/8-inch	4 or 5
Gas welding (medium), 1/8-inch to 1/2-inch	5 or 6
Gas welding (heavy), over 1/2-inch	6 or 8

(2) **Laser protection.**

    (i) Employees whose occupation or assignment requires exposure to laser beams shall be furnished suitable laser safety goggles which will protect for the specific wavelength of the laser and be of optical density (O.D.) adequate for the energy involved. Table E-2 lists the

maximum power or energy density for which adequate protection is afforded by glasses of optical densities from 5 through 8. Output levels falling between lines in this table shall require the higher optical density.

Table E-2 - Selecting Laser Safety Glass

Intensity, CW maximum power density (watts/cm^2)	Attenuation	
	Optical density (O.D.)	Attenuation factor
10^{-2}	5	10^5
10^{-1}	6	10^6
1.0	7	10^7
10.0	8	10^8

    (ii) All protective goggles shall bear a label identifying the following

        (A) data: The laser wavelengths for which use is intended;

        (B) The optical density of those

        (C) wavelengths; The visible light transmission.

[44 FR 8577, Feb. 9, 1979; 44 FR 20940, Apr. 6, 1979, as amended at 58 FR 35160, June 30, 1993; 81 FR 16092, Mar. 25, 2016]

§ 1926.103 Respiratory protection.

Note: The requirements applicable to construction work under this section are identical to those set forth at 29 CFR 1910.134 of this chapter.

[63 FR 1297, Jan. 8, 1998]

§ 1926.104 Safety belts, lifelines, and lanyards.

(a) Lifelines, safety belts, and lanyards shall be used only for employee safeguarding. Any lifeline, safety belt, or lanyard actually subjected to in-service loading, as distinguished from static load testing, shall be immediately removed from service and shall not be used again for employee safeguarding.

(b) Lifelines shall be secured above the point of operation to an anchorage or structural member capable of supporting a minimum dead weight of 5,400 pounds.

(c) Lifelines used on rock-scaling operations, or in areas where the lifeline may be subjected to cutting or abrasion, shall be a minimum of $^7/_8$-inch wire core manila rope. For all other lifeline applications, a minimum of $^3/_4$-inch manila or equivalent, with a minimum breaking strength of 5,000 pounds, shall be used.

(d) Safety belt lanyard shall be a minimum of $^1/_2$-inch nylon, or equivalent, with a maximum length to provide for a fall of no greater than 6 feet. The rope shall have a nominal breaking strength of 5,400 pounds.

(e) All safety belt and lanyard hardware shall be drop forged or pressed steel, cadmium plated in accordance with type 1, Class B plating specified in Federal Specification QQ-P-416. Surface shall be smooth and free of sharp edges.

(f) All safety belt and lanyard hardware, except rivets, shall be capable of withstanding a tensile loading of 4,000 pounds without cracking, breaking, or taking a permanent deformation.

[44 FR 8577, Feb. 9, 1979; 44 FR 20940, Apr. 6, 1979, as amended at 84 FR 21577, May 14, 2019]

§ 1926.105 Safety nets.

(a) Safety nets shall be provided when workplaces are more than 25 feet above the ground or water surface, or other surfaces where the use of ladders, scaffolds, catch platforms, temporary floors, safety lines, or safety belts is impractical.

(b) Where safety net protection is required by this part, operations shall not be undertaken until the net is in place and has been tested.

(c)

    (1) Nets shall extend 8 feet beyond the edge of the work surface where employees are exposed and shall be installed as close under the work surface as practical but in no case more than 25 feet below such work surface. Nets shall be hung with sufficient clearance to prevent user's contact with the surfaces or structures below. Such clearances shall be determined by impact load testing.

    (2) It is intended that only one level of nets be required for bridge construction.

(d) The mesh size of nets shall not exceed 6 inches by 6 inches. All new nets shall meet accepted performance standards of 17,500 foot-pounds minimum impact resistance as determined and certified by the manufacturers, and shall bear a label of proof test. Edge ropes shall provide a minimum breaking strength of 5,000 pounds.

(e) Forged steel safety hooks or shackles shall be used to fasten the net to its

(f)

supports.Connections between net panels shall develop the full strength of the net.

§ 1926.106 Working over or near water.

(a) Employees working over or near water, where the danger of drowning exists, shall be provided with U.S.Coast Guard-approved life jacket or buoyant work vests.

(b) Prior to and after each use, the buoyant work vests or life preservers shall be inspected for defects whichwould alter their strength or buoyancy. Defective units shall not be used.

(c) Ring buoys with at least 90 feet of line shall be provided and readily available for emergency rescue operations. Distance between ring buoys shall not exceed 200 feet.

(d) At least one lifesaving skiff shall be immediately available at locations where employees are working overor adjacent to water.

§ 1926.107 Definitions applicable to this subpart.

(a) *Contaminant* means any material which by reason of its action upon, within, or to a person is likely tocause physical harm.

(b) *Lanyard* means a rope, suitable for supporting one person. One end is fastened to a safety belt or harness and the other end is secured to a substantial object or a safety line.

(c) *Lifeline* means a rope, suitable for supporting one person, to which a lanyard or safety belt (or harness) is attached.

(d) *O.D.* means optical density and refers to the light refractive characteristics of a lens.

(e) *Radiant energy* means energy that travels outward in all directions from its sources.

(f) *Safety belt* means a device, usually worn around the waist which, by reason of its attachment to a lanyardand lifeline or a structure, will prevent a worker from falling.

[44 FR 8577, Feb. 9, 1979]

Subpart F - Fire Protection and Prevention

Authority: Section 107 of the Contract Work Hours and Safety Standards Act (40 U.S.C. 3704); Sections 4, 6, and 8 of the Occupational Safety and Health Act of 1970 (29 U.S.C. 653, 655, 657); Secretary of Labor's Order No. 12-71 (36 FR 8754), 8-76 (41 FR 25059), 9-83 (48 FR 35736),1-90 (55 FR 9033), 6-96 (62 FR 111), 3-2000 (62 FR 50017), 5-2002 (67 FR 650008), 5-2007 (72 FR 31159), 4-2010 (75 FR 55355), or 1-2012 (77 FR 3912), as applicable; and

29 CFR part 1911.

§ 1926.150 Fire protection.

(a) *General requirements.*

(1) The employer shall be responsible for the development of a fire protection program to be followed throughout all phases of the construction and demolition work, and he shall provide for the firefighting equipment as specified in this subpart. As fire hazards occur, there shall be no

delay in providing the necessary equipment.

(2) Access to all available firefighting equipment shall be maintained at all times.

(3) All firefighting equipment, provided by the employer, shall be conspicuously located.

(4) All firefighting equipment shall be periodically inspected and maintained in operating condition. Defective equipment shall be immediately replaced.

(5) As warranted by the project, the employer shall provide a trained and equipped firefighting organization (Fire Brigade) to assure adequate protection to life.

(b) *Water supply.*

(1) A temporary or permanent water supply, of sufficient volume, duration, and pressure, required to properly operate the firefighting equipment shall be made available as soon as combustible materials accumulate.

(2) Where underground water mains are to be provided, they shall be installed, completed, and made available for use as soon as practicable.

(c) *Portable firefighting equipment* -

(1) *Fire extinguishers and small hose lines.*

(i) A fire extinguisher, rated not less than 2A, shall be provided for each 3,000 square feet of the protected building area, or major fraction thereof. Travel distance from any point of the protected area to the nearest fire extinguisher shall not exceed 100 feet.

(ii) One 55-gallon open drum of water with two fire pails may be substituted for a fire extinguisher having a 2A rating.

(iii) A $1/2$-inch diameter garden-type hose line, not to exceed 100 feet in length and equipped with a nozzle, may be substituted for a 2A-rated fire extinguisher, providing it is capable of discharging a minimum of 5 gallons per minute with a minimum hose stream range of 30 feet horizontally. The garden-type hose lines shall be mounted on conventional racks or reels. The number and location of hose racks or reels shall be such that at least one hose stream can be applied to all points in the area.

(iv) One or more fire extinguishers, rated not less than 2A, shall be provided on each floor. In multistory buildings, at least one fire extinguisher shall be located adjacent to stairway.

(v) Extinguishers and water drums, subject to freezing, shall be protected from freezing.

(vi) A fire extinguisher, rated not less than 10B, shall be provided within 50 feet of wherever more than 5 gallons of flammable or combustible liquids or 5 pounds of flammable gas are being used on the jobsite. This requirement does not apply to the integral fuel tanks of motor vehicles.

(vii) Carbon tetrachloride and other toxic vaporizing liquid fire extinguishers are prohibited.

(viii) Portable fire extinguishers shall be inspected periodically and maintained in accordance with Maintenance and Use of Portable Fire Extinguishers, NFPA No. 10A-1970.

(ix) Fire extinguishers which have been listed or approved by a nationally recognized testing laboratory, shall be used to meet the requirements of this subpart.

(x) Table F-1 may be used as a guide for selecting the appropriate portable fire extinguishers.

Table F-1 FIRE EXTINGUISHERS DATA

	WATER TYPE				FOAM	CARBON DIOXIDE	DRY CHEMICAL			
							SODIUM OR POTASSIUM BICARBONATE		MULTI-PURPOSE ABC	
	STORED PRESSURE	CARTRIDGE OPERATED	WATER PUMP TANK	SODA ACID	FOAM	CO2	CARTRIDGE OPERATED	STORED PRESSURE	STORED PRESSURE	CARTRIDGE OPERATED
CLASS A FIRES (WOOD, PAPER, TRASH HAVING GLOWING EMBERS)	YES	YES	YES	YES	YES	NO (BUT WILL CONTROL SMALL SURFACE FIRES)	NO (BUT WILL CONTROL SMALL SURFACE FIRES)	NO (BUT WILL CONTROL SMALL SURFACE FIRES)	YES	YES
CLASS B FIRES (FLAMMABLE LIQUIDS, GASOLINE, OIL, PAINTS, GREASE, ETC.)	NO	NO	NO	NO	YES	YES	YES	YES	YES	YES
CLASS C FIRES (ELECTRICAL EQUIPMENT)	NO	NO	NO	NO	NO	YES	YES	YES	YES	YES
CLASS D FIRES (COMBUSTIBLE METALS)	SPECIAL EXTINGUISHING AGENTS APPROVED BY RECOGNIZED TESTING LABORATORIES									
METHOD OF OPERATION	PULL PIN — SQUEEZE HANDLE	TURN UPSIDE DOWN AND BUMP	PUMP HANDLE	TURN UPSIDE DOWN	TURN UPSIDE DOWN	PULL PIN — SQUEEZE LEVER	RUPTURE CARTRIDGE — SQUEEZE LEVER	PULL PIN — SQUEEZE HANDLE	PULL PIN — SQUEEZE HANDLE	RUPTURE CARTRIDGE — SQUEEZE LEVER
RANGE	30'-40'	30'-40'	30'-40'	30'-40'	30'-40'	3'-8'	5'-20'	5'-20'	5'-20'	5'-20'
MAINTENANCE	CHECK AIR PRESSURE GAUGE MONTHLY	WEIGH GAS CARTRIDGE — ADD WATER IF REQUIRED ANNUALLY	DISCHARGE AND FILL WITH WATER ANNUALLY	DISCHARGE ANNUALLY RECHARGE	DISCHARGE ANNUALLY RECHARGE	WEIGH SEMI-ANNUALLY	WEIGH GAS CARTRIDGE — CHECK CONDITION OF DRY CHEMICAL ANNUALLY	CHECK PRESSURE GAUGE AND CONDITION OF DRY CHEMICAL ANNUALLY	CHECK PRESSURE GAUGE AND CONDITION OF DRY CHEMICAL ANNUALLY	WEIGH GAS CARTRIDGE — CHECK CONDITION OF DRY CHEMICAL ANNUALLY

(2) *Fire hose and*

(i) One hundred feet, or less, of 1½-inch hose, with a nozzle capable of discharging water at 25 gallons or more per minute, may be substituted for a fire extinguisher rated not more than 2A in the designated area provided that the hose line can reach all points in the area.

(ii) If fire hose connections are not compatible with local firefighting equipment, the contractor shall provide adapters, or equivalent, to permit connections.

(iii) During demolition involving combustible materials, charged hose lines, supplied by hydrants, water tank trucks with pumps, or equivalent, shall be made available.

(d) *Fixed firefighting equipment -*

(1) *Sprinkler protection.*

(i) If the facility being constructed includes the installation of automatic sprinkler protection, the installation shall closely follow the construction and be placed in service as soon as applicable laws permit following completion of each story.

(ii) During demolition or alterations, existing automatic sprinkler installations shall be retained in service as long as reasonable. The operation of sprinkler control valves shall be permitted only by properly authorized persons. Modification of sprinkler systems to permit alterations or additional demolition should be expedited so that the automatic protection may be returned to service as quickly as possible. Sprinkler control valves shall be checked daily at close of work to ascertain that the protection is in service.

(2) **Standpipes.** In all structures in which standpipes are required, or where standpipes exist in structures being altered, they shall be brought up as soon as applicable laws permit, and shall be maintained as construction progresses in such a manner that they are always ready for fire protection use. The standpipes shall be provided with Siamese fire department connections on the outside of the structure, at the street level, which shall be conspicuously marked. There shall be at least one standard hose outlet at each floor.

(e) **Fire alarm devices.**

(1) An alarm system, e.g., telephone system, siren, etc., shall be established by the employer whereby employees on the site and the local fire department can be alerted for an emergency.

(2) The alarm code and reporting instructions shall be conspicuously posted at phones and at employee entrances.

(f) **Fire cutoffs.**

(1) Fire walls and exit stairways, required for the completed buildings, shall be given construction priority. Fire doors, with automatic closing devices, shall be hung on openings as soon as practicable.

(2) Fire cutoffs shall be retained in buildings undergoing alterations or demolition until operations necessitate their removal.

[44 FR 8577, Feb. 9, 1979; 44 FR 20940, Apr. 6, 1979, as amended at 58 FR 35162, June 30, 1993; 61 FR 31432, June 20, 1996]

§ 1926.151 Fire prevention.

(a) **Ignition hazards.**

(1) Electrical wiring and equipment for light, heat, or power purposes shall be installed in compliance with the requirements of subpart K of this part.

(2) Internal combustion engine powered equipment shall be so located that the exhausts are well away from combustible materials. When the exhausts are piped to outside the building under construction, a clearance of at least 6 inches shall be maintained between such piping and combustible material.

(3) Smoking shall be prohibited at or in the vicinity of operations which constitute a fire hazard, and shall be conspicuously posted: "No Smoking or Open Flame."

(4) Portable battery powered lighting equipment, used in connection with the storage, handling, or use of flammable gases or liquids, shall be of the type approved for the hazardous locations.

(5) The nozzle of air, inert gas, and steam lines or hoses, when used in the cleaning or ventilation of tanks and vessels that contain hazardous concentrations of flammable gases or vapors, shall be bonded to the tank or vessel shell. Bonding devices shall not be attached or detached in hazardous concentrations of flammable gases or vapors.

(b) **Temporary buildings.**

(1) No temporary building shall be erected where it will adversely affect any means of exit.

(2) Temporary buildings, when located within another building or structure, shall be of either

noncombustible construction or of combustible construction having a fire resistance of not less than 1 hour.

(3) Temporary buildings, located other than inside another building and not used for the storage, handling, or use of flammable or combustible liquids, flammable gases, explosives, or blasting agents, or similar hazardous occupancies, shall be located at a distance of not less than 10 feet from another building or structure. Groups of temporary buildings, not exceeding 2,000 square feet in aggregate, shall, for the purposes of this part, be considered a single temporary building.

(c) **Open yard storage.**

(1) Combustible materials shall be piled with due regard to the stability of piles and in no case higher than 20 feet.

(2) Driveways between and around combustible storage piles shall be at least 15 feet wide and maintained free from accumulation of rubbish, equipment, or other articles or materials. Driveways shall be so spaced that a maximum grid system unit of 50 feet by 150 feet is produced.

(3) The entire storage site shall be kept free from accumulation of unnecessary combustible materials. Weeds and grass shall be kept down and a regular procedure provided for the periodic cleanup of the entire area.

(4) When there is a danger of an underground fire, that land shall not be used for combustible or flammable storage.

(5) Method of piling shall be solid wherever possible and in orderly and regular piles. No combustible material shall be stored outdoors within 10 feet of a building or structure.

(6) Portable fire extinguishing equipment, suitable for the fire hazard involved, shall be provided at convenient, conspicuously accessible locations in the yard area. Portable fire extinguishers, rated not less than 2A, shall be placed so that maximum travel distance to the nearest unit shall not exceed 100 feet.

(d) **Indoor storage.**

(1) Storage shall not obstruct, or adversely affect, means of exit.

(2) All materials shall be stored, handled, and piled with due regard to their fire characteristics.

(3) Noncompatible materials, which may create a fire hazard, shall be segregated by a barrier having a fire resistance of at least 1 hour.

(4) Material shall be piled to minimize the spread of fire internally and to permit convenient access for firefighting. Stable piling shall be maintained at all times. Aisle space shall be maintained to safely accommodate the widest vehicle that may be used within the building for firefighting purposes.

(5) Clearance of at least 36 inches shall be maintained between the top level of the stored material and the sprinkler deflectors.

(6) Clearance shall be maintained around lights and heating units to prevent ignition of combustible materials.

(7) A clearance of 24 inches shall be maintained around the path of travel of fire doors unless a

barricade is provided, in which case no clearance is needed. Material shall not be stored within 36inches of a fire door opening.

[44 FR 8577, Feb. 9, 1979; 44 FR 20940, Apr. 6, 1979, as amended at 51 FR 25318, July 11, 1986]

§ 1926.152 Flammable liquids.

(a) *General requirements.*

   (1) Only approved containers and portable tanks shall be used for storage and handling of flammable liquids. Approved safety cans or Department of Transportation approved containers shall be used forthe handling and use of flammable liquids in quantities of 5 gallons or less, except that this shall not apply to those flammable liquid materials which are highly viscid (extremely hard to pour), which may be used and handled in original shipping containers. For quantities of one gallon or less, the original container may be used, for storage, use and handling of flammable liquids.

   (2) Flammable liquids shall not be stored in areas used for exits, stairways, or normally used for the safe passage of people.

(b) *Indoor storage of flammable liquids.*

   (1) No more than 25 gallons of flammable liquids shall be stored in a room outside of an approved storage cabinet. For storage of liquefied petroleum gas, see § 1926.153.

   (2) Quantities of flammable liquid in excess of 25 gallons shall be stored in an acceptable or approvedcabinet meeting the following requirements:

      (i) Acceptable wooden storage cabinets shall be constructed in the following manner, or equivalent: The bottom, sides, and top shall be constructed of an exterior grade of plywood atleast 1 inch in thickness, which shall not break down or delaminate under standard fire test conditions. All joints shall be rabbeted and shall be fastened in two directions with flathead wood screws. When more than one door is used, there shall be a rabbeted overlap of not less than 1 inch. Steel hinges shall be mounted in such a manner as to not lose their holding capacity due to loosening or burning out of the screws when subjected to fire. Such cabinets shall be painted inside and out with fire retardant paint.

      (ii) Approved metal storage cabinets will be acceptable.

      (iii) Cabinets shall be labeled in conspicuous lettering, "Flammable-Keep Away from Open Flames."

   (3) Not more than 60 gallons of Category 1, 2 and/or 3 flammable liquids or 120 gallons of Category 4 flammable liquids shall be stored in any one storage cabinet. Not more than three such cabinets maybe located in a single storage area. Quantities in excess of this shall be stored in an inside storage room.

   (4)

      (i) Inside storage rooms shall be constructed to meet the required fire-resistive rating for their use.Such construction shall comply with the test specifications set forth in Standard Methods of Fire Test of Building Construction and Material, NFPA 251-1969.

      (ii) Where an automatic extinguishing system is provided, the system shall be designed and installed in an approved manner. Openings to other rooms or buildings shall be provided with noncombustible liquid-tight raised sills or ramps at least 4 inches in height, or the floor in the storage area shall be at least 4 inches below the surrounding floor. Openings shall be providedwith approved self-closing fire doors. The room shall be liquid-tight where the walls join the floor. A permissible alternate to the sill or ramp is an open-grated trench, inside of the room, which drains to a safe location. Where other portions of the building or other buildings are

exposed, windows shall be protected as set forth in the Standard for Fire Doors and Windows, NFPA No. 80-1970, for Class E or F openings. Wood of at least 1-inch nominal thickness may beused for shelving, racks, dunnage, scuffboards, floor overlay, and similar installations.

(iii) Materials which will react with water and create a fire hazard shall not be stored in the sameroom with flammable liquids.

(iv) Storage in inside storage rooms shall comply with Table F-2 following:

Table F-2

Fire protection provided	Fire resistance	Maximum size	Total allowable quantities gals./sq. ft./floor area
Yes	2 hrs	500 sq. ft	10
No	2 hrs	500 sq. ft	4
Yes	1 hr	150 sq. ft	5
No	1 hr	150 sq. ft	2

NOTE: Fire protection system shall be sprinkler, water spray, carbon dioxide or other system approved by a nationally recognized testing laboratory for this purpose.

(v) Electrical wiring and equipment located in inside storage rooms shall be approved for Class I,Division 1, Hazardous Locations. For definition of Class I, Division 1, Hazardous Locations, see
§ 1926.449.

(vi) Every inside storage room shall be provided with either a gravity or a mechanical exhausting system. Such system shall commence not more than 12 inches above the floor and be designed to provide for a complete change of air within the room at least 6 times per hour. If a mechanical exhausting system is used, it shall be controlled by a switch located outside of the door. The ventilating equipment and any lighting fixtures shall be operated by the same switch. An electric pilot light shall be installed adjacent to the switch if Category 1, 2, or 3 flammable liquids are dispensed within the room. Where gravity ventilation is provided, the fresh air intake,as well as the exhausting outlet from the room, shall be on the exterior of the building in which the room is located.

(vii) In every inside storage room there shall be maintained one clear aisle at least 3 feet wide. Containers over 30 gallons capacity shall not be stacked one upon the other.

(viii) Flammable liquids in excess of that permitted in inside storage rooms shall be stored outside ofbuildings in accordance with paragraph (c) of this section.

(5) **Quantity.** The quantity of flammable liquids kept in the vicinity of spraying operations shall be the minimum required for operations and should ordinarily not exceed a supply for 1 day or one shift. Bulk storage of portable containers of flammable liquids shall be in a separate, constructed

buildingdetached from other important buildings or cut off in a standard manner.

(c) **Storage outside buildings.**

    (1) Storage of containers (not more than 60 gallons each) shall not exceed 1,100 gallons in any one pileor area. Piles or groups of containers shall be separated by a 5-foot clearance. Piles or groups of containers shall not be nearer than 20 feet to a building.

    (2) Within 200 feet of each pile of containers, there shall be a 12-foot-wide access way to permit approach of fire control apparatus.

    (3) The storage area shall be graded in a manner to divert possible spills away from buildings or otherexposures, or shall be surrounded by a curb or earth dike at least 12 inches high. When curbs or dikes are used, provisions shall be made for draining off accumulations of ground or rain water, or spills of flammable liquids. Drains shall terminate at a safe location and shall be accessible to operation under fire conditions.

    (4) Outdoor portable tank storage:

        (i) Portable tanks shall not be nearer than 20 feet from any building. Two or more portable tanks,grouped together, having a combined capacity in excess of 2,200 gallons, shall be separated bya 5-foot-clear area. Individual portable tanks exceeding 1,100 gallons shall be separated by a
5-foot-clear area.

        (ii) Within 200 feet of each portable tank, there shall be a 12-foot-wide access way to permit approach of fire control apparatus.

    (5) Storage areas shall be kept free of weeds, debris, and other combustible material not necessary tothe storage.

    (6) Portable tanks, not exceeding 660 gallons, shall be provided with emergency venting and otherdevices, as required by chapters III and IV of NFPA 30-1969, The Flammable and Combustible Liquids Code.

    (7) Portable tanks, in excess of 660 gallons, shall have emergency venting and other devices, as required by chapters II and III of The Flammable and Combustible Liquids Code, NFPA 30-1969.

(d) **Fire control for flammable liquid storage.**

    (1) At least one portable fire extinguisher, having a rating of not less than 20-B units, shall be located outside of, but not more than 10 feet from, the door opening into any room used for storage of more than 60 gallons of flammable liquids.

    (2) At least one portable fire extinguisher having a rating of not less than 20-B units shall be located notless than 25 feet, nor more than 75 feet, from any flammable liquid storage area located outside.

    (3) When sprinklers are provided, they shall be installed in accordance with the Standard for the Installation of Sprinkler Systems, NFPA 13-1969.

    (4) At least one portable fire extinguisher having a rating of not less than 20-B:C units shall be providedon all tank trucks or other vehicles used for transporting and/or dispensing flammable liquids.

(e) **Dispensing liquids.**

(1) Areas in which flammable liquids are transferred at one time, in quantities greater than 5 gallons from one tank or container to another tank or container, shall be separated from other operations by 25-feet distance or by construction having a fire resistance of at least 1 hour. Drainage or other means shall be provided to control spills. Adequate natural or mechanical ventilation shall be provided to maintain the concentration of flammable vapor at or below 10 percent of the lower flammable limit.

(2) Transfer of Category 1, 2, or 3 flammable liquids from one container to another shall be done only when containers are electrically interconnected (bonded).

(3) Flammable liquids shall be drawn from or transferred into vessels, containers, or tanks within a building or outside only through a closed piping system, from safety cans, by means of a device drawing through the top, or from a container, or portable tanks, by gravity or pump, through an approved self-closing valve. Transferring by means of air pressure on the container or portable tanks is prohibited.

(4) The dispensing units shall be protected against collision damage.

(5) Dispensing devices and nozzles for Category 1, 2, or 3 flammable liquids shall be of an approved type.

(f) *Handling liquids at point of final use.*

(1) Category 1, 2, or 3 flammable liquids shall be kept in closed containers when not actually in use.

(2) Leakage or spillage of flammable liquids shall be disposed of promptly and safely.

(3) Category 1, 2, or 3 flammable liquids may be used only where there are no open flames or other sources of ignition within 50 feet of the operation, unless conditions warrant greater clearance.

(g) *Service and refueling areas.*

(1) Flammable liquids shall be stored in approved closed containers, in tanks located underground, or in aboveground portable tanks.

(2) The tank trucks shall comply with the requirements covered in the Standard for Tank Vehicles for Flammable and Combustible Liquids, NFPA No. 385-1966.

(3) The dispensing hose shall be an approved type.

(4) The dispensing nozzle shall be an approved automatic-closing type without a latch-open device.

(5) Underground tanks shall not be abandoned.

(6) Clearly identified and easily accessible switch(es) shall be provided at a location remote from dispensing devices to shut off the power to all dispensing devices in the event of an emergency.

(7)
  (i) Heating equipment of an approved type may be installed in the lubrication or service area where there is no dispensing or transferring of Category 1, 2, or 3 flammable liquids, provided the bottom of the heating unit is at least 18 inches above the floor and is protected from physical damage.

- (ii) Heating equipment installed in lubrication or service areas, where Category 1, 2, or 3 flammableliquids are dispensed, shall be of an approved type for garages, and shall be installed at least 8 feet above the floor.
- (8) There shall be no smoking or open flames in the areas used for fueling, servicing fuel systems for internal combustion engines, receiving or dispensing of flammable liquids.
- (9) Conspicuous and legible signs prohibiting smoking shall be posted.
- (10) The motors of all equipment being fueled shall be shut off during the fueling operation.
- (11) Each service or fueling area shall be provided with at least one fire extinguisher having a rating of not less than 20-B:C located so that an extinguisher will be within 75 feet of each pump, dispenser, underground fill pipe opening, and lubrication or service area.

(h) **Scope.** This section applies to the handling, storage, and use of flammable liquids with a flashpoint at orbelow 199.4 °F (93 °C). This section does not apply to:

(1) Bulk transportation of flammable liquids; and

(2) Storage, handling, and use of fuel oil tanks and containers connected with oil burning equipment.

(i) **Tank storage -**

(1) **Design and construction of tanks -**

- (i) **Materials.**
    - (A) Tanks shall be built of steel except as provided in paragraphs (i)(1)(i) (B) through (E) ofthis section.
    - (B) Tanks may be built of materials other than steel for installation underground or if required by the properties of the liquid stored. Tanks located above ground or inside buildings shallbe of noncombustible construction.
    - (C) Tanks built of materials other than steel shall be designed to specifications embodyingprinciples recognized as good engineering design for the material used.
    - (D) Unlined concrete tanks may be used for storing flammable liquids having a gravity of 40° API or heavier. Concrete tanks with special lining may be used for other services providedthe design is in accordance with sound engineering practice.
    - (E) [Reserved]
    - (F) Special engineering consideration shall be required if the specific gravity of the liquid to bestored exceeds that of water or if the tanks are designed to contain flammable liquids at a liquid temperature below 0 °F.

- (ii) **Fabrication.**
    - (A) [Reserved]
    - (B) Metal tanks shall be welded, riveted, and caulked, brazed, or bolted, or constructed by use of a combination of these methods. Filler metal used in brazing shall be nonferrous metalor an alloy having a melting point above 1000 °F. and below that of the metal joined.

- (iii) **Atmospheric tanks.**

(A) Atmospheric tanks shall be built in accordance with acceptable good standards of design. Atmospheric tanks may be built in accordance with:

    (*1*) Underwriters' Laboratories, Inc., Subjects No. 142, Standard for Steel Aboveground Tanks for Flammable and Combustible Liquids, 1968; No. 58, Standard for Steel Underground Tanks for Flammable and Combustible Liquids, Fifth Edition, December 1961; or No. 80, Standard for Steel Inside Tanks for Oil-Burner Fuel, September 1963.

    (*2*) American Petroleum Institute Standards No. 12A, Specification for Oil Storage Tanks with Riveted Shells, Seventh Edition, September 1951, or No. 650, Welded Steel Tanks for Oil Storage, Third Edition, 1966.

    (*3*) American Petroleum Institute Standards No. 12B, Specification for Bolted Production Tanks, Eleventh Edition, May 1958, and Supplement 1, March 1962; No. 12D, Specification for Large Welded Production Tanks, Seventh Edition, August 1957; or No. 12F, Specification for Small Welded Production Tanks, Fifth Edition, March 1961. Tanks built in accordance with these standards shall be used only as production tanks for storage of crude petroleum in oil-producing areas.

(B) Tanks designed for underground service not exceeding 2,500 gallons (9,462.5 L) capacity may be used aboveground.

(C) Low-pressure tanks and pressure vessels may be used as atmospheric tanks.

(D) Atmospheric tanks shall not be used for the storage of a flammable liquid at a temperature at or above its boiling point.

(iv) **Low pressure tanks.**

(A) The normal operating pressure of the tank shall not exceed the design pressure of the tank.

(B) Low-pressure tanks shall be built in accordance with acceptable standards of design. Low-pressure tanks may be built in accordance with:

    (*1*) American Petroleum Institute Standard No. 620. Recommended Rules for the Design and Construction of Large, Welded, Low-Pressure Storage Tanks, Third Edition, 1966.

    (*2*) The principles of the Code for Unfired Pressure Vessels, Section VIII of the ASME Boiler and Pressure Vessels Code, 1968.

(C) Atmospheric tanks built according to Underwriters' Laboratories, Inc., requirements in paragraph (i)(1)(iii)(A) of this section and shall be limited to 2.5 p.s.i.g. under emergency venting conditions.

This paragraph may be used for operating pressures not exceeding 1 p.s.i.g.

(D) Pressure vessels may be used as low-pressure tanks.

(v) **Pressure vessels.**

(A) The normal operating pressure of the vessel shall not exceed the design pressure of the vessel.

(B) Pressure vessels shall be built in accordance with the Code for Unfired Pressure Vessels, Section VIII of the ASME Boiler and Pressure Vessel Code 1968.

(vi) **Provisions for internal corrosion.** When tanks are not designed in accordance with the American Petroleum Institute, American Society of Mechanical Engineers, or the Underwriters' Laboratories, Inc.'s, standards, or if corrosion is anticipated beyond that provided for in the design formulas used, additional metal thickness or suitable protective coatings or linings shall be provided to compensate for the corrosion loss expected during the design life of the tank.

(2) **Installation of outside aboveground tanks.**

(i) [Reserved]

(ii) **Spacing (shell-to-shell) between aboveground tanks.**

(A) The distance between any two flammable liquid storage tanks shall not be less than 3 feet (0.912 m).

(B) Except as provided in paragraph (i)(2)(ii)(C) of this section, the distance between any two adjacent tanks shall not be less than one-sixth the sum of their diameters. When the diameter of one tank is less than one-half the diameter of the adjacent tank, the distance between the two tanks shall not be less than one-half the diameter of the smaller tank.

(C) Where crude petroleum in conjunction with production facilities are located in noncongested areas and have capacities not exceeding 126,000 gallons (3,000 barrels), the distance between such tanks shall not be less than 3 feet (0.912 m).

(D) Where unstable flammable liquids are stored, the distance between such tanks shall not be less than one-half the sum of their diameters.

(E) When tanks are compacted in three or more rows or in an irregular pattern, greater spacing or other means shall be provided so that inside tanks are accessible for firefighting purposes.

(F) The minimum separation between a liquefied petroleum gas container and a flammable liquid storage tank shall be 20 feet (6.08 m), except in the case of flammable liquid tanks operating at pressures exceeding 2.5 p.s.i.g. or equipped with emergency venting which will permit pressures to exceed 2.5 p.s.i.g. in which case the provisions of paragraphs (i)(2)(ii) (A) and (B) of this section shall apply. Suitable means shall be taken to prevent the accumulation of flammable liquids under adjacent liquefied petroleum gas containers such as by diversion curbs or grading. When flammable liquid storage tanks are within a diked area, the liquefied petroleum gas containers shall be outside the diked area and at least 10 feet (3.04 m) away from the centerline of the wall of the diked area. The foregoing provisions shall not apply when liquefied petroleum gas containers of 125 gallons (473.125 L) or less capacity are installed adjacent to fuel oil supply tanks of 550 gallons (2,081.75 L) or less capacity.

(iii) [Reserved]

(iv) **Normal venting for aboveground tanks.**

(A) Atmospheric storage tanks shall be adequately vented to prevent the development of vacuum or pressure sufficient to distort the roof of a cone roof tank or exceeding the design pressure in the case of other atmospheric tanks, as a result of filling or emptying, and atmospheric temperature changes.

(B) Normal vents shall be sized either in accordance with:

(1) The American Petroleum Institute Standard 2000 (1968), Venting Atmospheric and Low-Pressure Storage Tanks; or

(2) other accepted standard; or

(3) shall be at least as large as the filling or withdrawal connection, whichever is larger but in no case less than $1^{1}/_{4}$ inch (3.175 cm) nominal inside diameter.

(C) Low-pressure tanks and pressure vessels shall be adequately vented to prevent development of pressure or vacuum, as a result of filling or emptying and atmospheric temperature changes, from exceeding the design pressure of the tank or vessel.
Protection shall also be provided to prevent overpressure from any pump discharging into the tank or vessel when the pump discharge pressure can exceed the design pressure of the tank or vessel.

(D) If any tank or pressure vessel has more than one fill or withdrawal connection and simultaneous filling or withdrawal can be made, the vent size shall be based on the maximum anticipated simultaneous flow.

(E) Unless the vent is designed to limit the internal pressure 2.5 p.s.i. or less, the outlet of vents and vent drains shall be arranged to discharge in such a manner as to prevent localized overheating of any part of the tank in the event vapors from such vents are ignited.

(F) Tanks and pressure vessels storing Category 1 flammable liquids shall be equipped with venting devices that shall be normally closed except when venting to pressure or vacuum conditions. Tanks and pressure vessels storing Category 2 flammable liquids, or Category 3 flammable liquids with a flashpoint below 100 °F (37.8 °C), shall be equipped with venting devices that shall be normally closed except when venting under pressure or vacuum conditions, or with approved flame arresters.

Exemption: Tanks of 3,000 bbls (barrels) (84 m(3)) capacity or less containing crude petroleum in crude-producing areas; and, outside aboveground atmospheric tanks under 1,000 gallons (3,785 L) capacity containing other than Category 1 flammable liquids may have open vents. (*See* paragraph (i)(2)(vi)(B) of this section.)

(G) Flame arresters or venting devices required in paragraph (i)(2)(iv)(F) of this section may be omitted for Category 2 flammable liquids or Category 3 flammable liquids with a flashpoint below 100 °F (37.8 °C) where conditions are such that their use may, in case of obstruction, result in tank damage.

(v) *Emergency relief venting for fire exposure for aboveground tanks.*

(A) Every aboveground storage tank shall have some form of construction or device that will relieve excessive internal pressure caused by exposure fires.

(B) In a vertical tank the construction referred to in paragraph (i)(2)(v)(A) of this section maytake the form of a floating roof, lifter roof, a weak roof-to-shell seam, or other approvedpressure relieving construction. The weak roof-to-shell seam shall be constructed to fail preferential to any other seam.

(C) Where entire dependence for emergency relief is placed upon pressure relieving devices, the total venting capacity of both normal and emergency vents shall be enough to prevent rupture of the shell or bottom of the tank if vertical, or of the shell or heads if horizontal. If unstable liquids are stored, the effects of heat or gas resulting from polymerization, decomposition, condensation, or self-reactivity shall be taken into account. The total capacity of both normal and emergency venting devices shall be not less than that derivedfrom Table F-10 except as provided in paragraph (i)(2)(v) (E) or (F) of this section. Such device may be a self-closing manhole cover, or one using long bolts that permit the cover to lift under internal pressure, or an additional or larger relief valve or valves. The wetted area of the tank shall be calculated on the basis of 55 percent of the total exposed area of a sphere or spheroid, 75 percent of the total exposed area of a horizontal tank and the first30 feet (9.12 m) above grade of the exposed shell area of a vertical tank.

Table F-10 - Wetted Area Versus Cubic Feet (Meters) Free Air Per Hour

Square feet (m^2)	CFH (m^{3}H)	Square feet (m^2)	CFH (m^{3}H)	Square feet (m^2)	CFH (m^{3}H)
20 (1.84)	21,100 (590.8)	200 (18.4)	211,000 (5,908)	1,000 (90.2)	524,000 (14,672)
30 (2.76)	31,600 (884.8)	250 (23)	239,000 (6,692)	1,200 (110.4)	557,000 (15,596)
40 (3.68)	42,100 (1,178.8)	300 (27.6)	265,000 (7,420)	1,400 (128.8)	587,000 (16,436)
50 (4.6)	52,700 (1,475.6)	350 (32.2)	288,000 (8,064)	1,600 (147.2)	614,000 (17,192)
60 (5.52)	63,200 (1,769.6)	400 (36.8)	312,000 (8,736)	1,800 (165.6)	639,000 (17,892)
70 (6.44)	73,700 (2,063.6)	500 (46)	354,000 (9,912)	2,000 (180.4)	662,000 (18,536)
80 (7.36)	84,200 (2,357.6)	600 (55.2)	392,000 (10,976)	2,400 (220.8)	704,000 (19,712)
90 (8.28)	94,800 (2,654.4)	700 (64.4)	428,000 (11,984)	2,800 (257.6)	742,000 (20,776)
100 (9.2)	105,000 (2,940)	800 (73.6)	462,000 (12,936)	and	
120 (11.04)	126,000 (3,528)	900 (82.8)	493,000 (13,804)	over	
140 (12.88)	147,000 (4,116)	1,000 (90.2)	524,000 (14,672)		
160 (14.72)	168,000 (4,704)				
180 (16.56)	190,000				

	(5,320)				

(D) For tanks and storage vessels designed for pressure over 1 p.s.i.g., the total rate of

Square feet (m²)	CFH (m³H)	Square feet (m²)	CFH (m³H)	Square feet (m²)	CFH (m³H)
200 (18.4)	211,000 (5,908)				

venting shall be determined in accordance with Table F-10, except that when the exposed wetted area of the surface is greater than 2,800 square feet (257.6 m²), the total rate of venting shall be calculated by the following formula:

$CFH = 1,107 A^{0.82}$

Where:

CFH = Venting requirement, in cubic feet (meters) of free air per hour.

A = Exposed wetted surface, in square feet (m²).

Note: The foregoing formula is based on $Q = 21,000 A^{0.82}$.

(E) The total emergency relief venting capacity for any specific stable liquid may be determined by the following formula:

$V = 1337 \div L^{\sqrt{}} M$

V = Cubic feet (meters) of free air per hour from Table F-10.

L = Latent heat of vaporization of specific liquid in B.t.u. per pound.

M = Molecular weight of specific liquids.

(F) The required airflow rate of paragraph (i)(2)(v) (C) or (E) of this section may be multiplied by the appropriate factor listed in the following schedule when protection is provided as indicated. Only one factor may be used for any one tank.

0.5 for drainage in accordance with paragraph (i)(2)(vii)(B) of this section for tanks over 200 square feet (18.4 m²) of wetted area.

0.3 for approved water spray.

0.3 for approved insulation.

0.15 for approved water spray with approved insulation.

(G) The outlet of all vents and vent drains on tanks equipped with emergency venting to permit pressures exceeding 2.5 p.s.i.g. shall be arranged to discharge in such a way as to prevent localized overheating of any part of the tank, in the event vapors from such vents are ignited.

(H) Each commercial tank venting device shall have stamped on it the opening pressure, the pressure at which the valve reaches the full open position, and the flow capacity at the latter pressure, expressed in cubic feet (meters) per hour of air at 60 °F. (15.55 °C) and at apressure of 14.7 p.s.i.a.

(I) The flow capacity of tank venting devices 12 inches (30.48 cm) and smaller in nominal pipe size shall be determined by actual test of each type and size of vent. These flow tests may be conducted by the manufacturer if certified by a qualified impartial observer, or maybe conducted by an outside agency. The flow capacity of tank venting devices larger than 12 inches (30.48 cm) nominal pipe size, including manhole covers with long bolts or equivalent, may be calculated provided that the opening pressure is actually measured, therating pressure and corresponding free orifice area are stated, the word "calculated" appears on the nameplate, and the computation is based on a flow coefficient of 0.5 applied to the rated orifice area.

(vi) *Vent piping for aboveground tanks.*

(A) Vent piping shall be constructed in accordance with paragraph (c) of this section.

(B) Where vent pipe outlets for tanks storing Category 1 or 2 flammable liquids, or Category 3flammable liquids with a flashpoint below 100 °F (37.8 °C), are adjacent to buildings or public ways, they shall be located so that the vapors are released at a safe point outside ofbuildings and not less than 12 feet (3.658 m) above the adjacent ground level. In order to aid their dispersion, vapors shall be discharged upward or horizontally away from closelyadjacent walls. Vent outlets shall be located so that flammable vapors will not be trappedby eaves or other obstructions and shall be at least 5 feet (1.52 m) from building openings.

(C) When tank vent piping is manifolded, pipe sizes shall be such as to discharge, within thepressure limitations of the system, the vapors they may be required to handle when manifolded tanks are subject to the same fire exposure.

(vii) *Drainage, dikes, and walls for aboveground tanks -*

(A) *Drainage and diked areas.* The area surrounding a tank or a group of tanks shall be provided with drainage as in paragraph (i)(2)(vii)(B) of this section, or shall be diked asprovided in (i)(2)(vii)(C) of this section, to prevent accidental discharge of liquid from endangering adjoining property or reaching waterways.

(B) *Drainage.* Where protection of adjoining property or waterways is by means of a natural ormanmade drainage system, such systems shall comply with the following:

(*1*) [Reserved]

(*2*) The drainage system shall terminate in vacant land or other area or in an impoundingbasin having a capacity not smaller than that of the largest tank served. This termination area and the route of the drainage system shall be so located that, if the flammable liquids in the drainage system are ignited, the fire will not seriously expose tanks or adjoining property.

(C) *Diked areas.* Where protection of adjoining property or waterways is accomplished by retaining the liquid around the tank by means of a dike, the volume of the diked area shallcomply with the following requirements:

(1) Except as provided in paragraph (i)(2)(vii)(C)(2) of this section, the volumetric capacity of the diked area shall not be less than the greatest amount of liquid that can be released from the largest tank within the diked area, assuming a full tank. Thecapacity of the diked area enclosing more than one tank shall be calculated by deducting the volume of the tanks other than the largest tank below the height of the dike.

(2) For a tank or group of tanks with fixed roofs containing crude petroleum with boilovercharacteristics, the volumetric capacity of the diked area shall be not less than the capacity of the largest tank served by the enclosure, assuming a full tank. The capacity of the diked enclosure shall be calculated by deducting the volume below the height of the dike of all tanks within the enclosure.

(3) Walls of the diked area shall be of earth, steel, concrete or solid masonry designed tobe liquidtight and to withstand a full hydrostatic head. Earthen walls 3 feet (0.912 m) or more in height shall have a flat section at the top not less than 2 feet (0.608 m) wide. The slope of an earthen wall shall be consistent with the angle of repose of thematerial of which the wall is constructed.

(4) The walls of the diked area shall be restricted to an average height of 6 feet (1.824 m)above interior grade.

(5) [Reserved]

(6) No loose combustible material, empty or full drum or barrel, shall be permitted withinthe diked area.

(viii) **Tank openings other than vents for aboveground tanks.**

(A)-(C) [Reserved]

(D) Openings for gaging shall be provided with a vaportight cap or cover.

(E) For Category 2 flammable liquids or Category 3 flammable liquids with a flashpoint below 100 °F (37.8 °C), other than crude oils, gasolines, and asphalts, the fill pipe shall be so designed and installed as to minimize the possibility of generating static electricity. A fillpipe entering the top of a tank shall terminate within 6 inches (15.24 cm) of the bottom ofthe tank and shall be installed to avoid excessive vibration.

(F) Filling and emptying connections which are made and broken shall be located outside of buildings at a location free from any source of ignition and not less than 5 feet (1.52 m) away from any building opening. Such connection shall be closed and liquidtight when notin use. The connection shall be properly identified.

(3) **Installation of underground tanks -**

(i) **Location.** Evacuation for underground storage tanks shall be made with due care to avoid undermining of foundations of existing structures. Underground tanks or tanks under buildingsshall be so located with respect to existing building foundations and supports that the loads carried by the latter cannot be transmitted to the tank. The distance from any part of a tank storing Category 1 or 2 flammable liquids, or Category 3 flammable liquids with a flashpoint below 100 °F (37.8 °C), to the nearest wall of any basement or pit shall be not less than 1 foot(0.304 m), and to any property line that may be built upon, not less than 3 feet (0.912 m). The

distance from any part of a tank storing Category 3 flammable liquids with a flashpoint at or above 100 °F (37.8 °C) or Category 4 flammable liquids to the nearest wall of any basement, pitor property line shall be not less than 1 foot (0.304 m).

(ii) **Depth and cover.** Underground tanks shall be set on firm foundations and surrounded with at least 6 inches (15.24 cm) of noncorrosive, inert materials such as clean sand, earth, or gravel well tamped in place. The tank shall be placed in the hole with care since dropping or rolling thetank into the hole can break a weld, puncture or damage the tank, or scrape off the protective coating of coated tanks. Tanks shall be covered with a minimum of 2 feet (0.608 m) of earth, or shall be covered with not less than 1 foot (0.304 m) of earth, on top of which shall be placed a slab of reinforced concrete not less than 4 inches (10.16 cm) thick. When underground tanks are, or are likely to be, subject to traffic, they shall be protected against damage from vehicles passing over them by at least 3 feet (0.912 m) of earth cover, or 18 inches (45.72 cm) of well- tamped earth, plus 6 inches (15.24 cm) of reinforced concrete or 8 inches (20.32 cm) of asphaltic concrete. When asphaltic or reinforced concrete paving is used as part of the protection, it shall extend at least 1 foot (0.304 m) horizontally beyond the outline of the tank in all directions.

(iii) **Corrosion protection.** Corrosion protection for the tank and its piping shall be provided by one ormore of the following methods:

  (A) Use of protective coatings or wrappings;

  (B) Cathodic protection; or,

  (C) Corrosion resistant materials of construction.

(iv) **Vents.**

  (A) Location and arrangement of vents for Category 1 or 2 flammable liquids, or Category 3 flammable liquids with a flashpoint below 100 °F (37.8 °C). Vent pipes from tanks storing Category 1 or 2 flammable liquids, or Category 3 flammable liquids with a flashpoint below 100 °F (37.8 °C), shall be so located that the discharge point is outside of buildings,higher than the fill pipe opening, and not less than 12 feet (3.658 m) above the adjacent ground level. Vent pipes shall discharge only upward in order to disperse vapors. Vent pipes 2 inches (5.08 cm) or less in nominal inside diameter shall not be obstructed by devices that will cause excessive back pressure. Vent pipe outlets shall be so located thatflammable vapors will not enter building openings, or be trapped under eaves or other obstructions. If the vent pipe is less than 10 feet (3.04 m) in length, or greater than 2 inches (5.08 cm) in nominal inside diameter, the outlet shall be provided with a vacuum and pressure relief device or there shall be an approved flame arrester located in the vent line at the outlet or within the approved distance from the outlet.

  (B) Size of vents. Each tank shall be vented through piping adequate in size to prevent blow- back of vapor or liquid at the fill opening while the tank is being filled. Vent pipes shall benot less than $1^1/_4$ inch (3.175 cm) nominal inside diameter.

Table F-11 - Vent Line Diameters

Maximum flow GPM (L)	Pipe length[1]		
	50 feet (15.2 m)	100 feet (30.4 m)	200 feet (60.8 m)
	Inches (cm)	Inches (cm)	Inches (cm)
100 (378.5)	1 1/4 (3.175)	1 1/4 (3.175)	1 1/4 (3.175)
200 (757)	1 1/4 (3.175)	1 1/4 (3.175)	1 1/4 (3.175)
300 (1,135.5)	1 1/4 (3.175)	1 1/4 (3.175)	1 1/2 (3.81)
400 (1,514)	1 1/4 (3.175)	1 1/2 (3.81)	2 (5.08)
500 (1,892.5)	1 1/2 (3.81)	1 1/2 (3.81)	2 (5.08)
600 (2,271)	1 1/2 (3.81)	2 (5.08)	2 (5.08)
700 (2,649.5)	2 (5.08)	2 (5.08)	2 (5.08)
800 (3,028)	2 (5.08)	2 (5.08)	3 (7.62)
900 (3,406.5)	2 (5.08)	2 (5.08)	3 (7.62)
1,000 (3,785)	2 (5.08)	2 (5.08)	3 (7.62)

[1] Vent lines of 50 ft. (15.2 m), 100 ft. (30.4 m), and 200 ft. (60.8 m) of pipe plus 7 ells.

(C) Location and arrangement of vents for Category 3 flammable liquids with a flashpoint at or above 100 °F (37.8 °C) or Category 4 flammable liquids. Vent pipes from tanks storing Category 3 flammable liquids with a flashpoint at or above 100 °F (37.8 °C) or Category 4 flammable liquids shall terminate outside of the building and higher than the fill pipe opening. Vent outlets shall be above normal snow level. They may be fitted with return bends, coarse screens or other devices to minimize ingress of foreign material.

(D) Vent piping shall be constructed in accordance with paragraph (3)(iv)(C) of this section. Vent pipes shall be so laid as to drain toward the tank without sags or traps in which liquid can collect. They shall be located so that they will not be subjected to physical damage.
The tank end of the vent pipe shall enter the tank through the top.

(E) When tank vent piping is manifolded, pipe sizes shall be such as to discharge, within the pressure limitations of the system, the vapors they may be required to handle when manifolded tanks are filled simultaneously.

(v) **Tank openings other than vents.**

(A) Connections for all tank openings shall be vapor or liquid tight.

(B) Openings for manual gaging, if independent of the fill pipe, shall be provided with a liquid-tight cap or cover. If inside a building, each such opening shall be protected against liquid overflow and possible vapor release by means of a spring loaded check valve or other approved device.

(C) Fill and discharge lines shall enter tanks only through the top. Fill lines shall be sloped toward the tank.

(D) For Category 2 flammable liquids, or Category 3 flammable liquids with a flashpoint below 100 °F (37.8 °C), other than crude oils, gasolines, and asphalts, the fill pipe shall be so designed and installed as to minimize the possibility of generating static electricity by terminating within 6 inches (15.24 cm) of the bottom of the tank.

(E) Filling and emptying connections which are made and broken shall be located outside of buildings at a location free from any source of ignition and not less than 5 feet (1.52 m) away from any building opening. Such connection shall be closed and liquidtight when not in use. The connection shall be properly identified.

(4) *Installation of tanks inside of buildings -*

(i) *Location.* Tanks shall not be permitted inside of buildings except as provided in paragraphs (e),(g), (h), or (i) of this section.

(ii) *Vents.* Vents for tanks inside of buildings shall be as provided in paragraphs (i)(2) (iv), (v), (vi)(B), and (3)(iv) of this section, except that emergency venting by the use of weak roof seams on tanks shall not be permitted. Vents shall discharge vapors outside the buildings.

(iii) *Vent piping.* Vent piping shall be constructed in accordance with paragraph (c) of this section.

(iv) *Tank openings other than vents.*

(A) Connections for all tank openings shall be vapor or liquidtight. Vents are covered in paragraph (i)(4)(ii) of this section.

(B) Each connection to a tank inside of buildings through which liquid can normally flow shall be provided with an internal or an external valve located as close as practical to the shell of the tank. Such valves, when external, and their connections to the tank shall be of steel except when the chemical characteristics of the liquid stored are incompatible with steel. When materials other than steel are necessary, they shall be suitable for the pressures, structural stresses, and temperatures involved, including fire exposures.

(C) Flammable liquid tanks located inside of buildings, except in one-story buildings designed and protected for flammable liquid storage, shall be provided with an automatic-closing heat-actuated valve on each withdrawal connection below the liquid level, except for connections used for emergency disposal, to prevent continued flow in the event of fire in the vicinity of the tank. This function may be incorporated in the valve required in paragraph (i)(4)(iv)(B) of this section, and if a separate valve, shall be located adjacent to the valve required in paragraph (i)(4)(iv)(B) of this section.

(D) Openings for manual gaging, if independent of the fill pipe (see paragraph (i)(4)(iv)(F) of this section), shall be provided with a vaportight cap or cover. Each such opening shall be protected against liquid overflow and possible vapor release by means of a spring loaded check valve or other approved device.

(E) For Category 2 flammable liquids, or Category 3 flammable liquids with a flashpoint below 100 °F (37.8 °C), other than crude oils, gasolines, and asphalts, the fill pipe shall be so designed and installed as to minimize the possibility of generating static electricity by terminating within 6 inches (15.24 cm) of the bottom of the tank.

(F) The fill pipe inside of the tank shall be installed to avoid excessive vibration of the pipe.

(G) The inlet of the fill pipe shall be located outside of buildings at a location free from any source of ignition and not less than 5 feet (1.52 m) away from any building opening. The inlet of the fill pipe shall be closed and liquidtight when not in use. The fill connection shallbe properly identified.

(H) Tanks inside buildings shall be equipped with a device, or other means shall be provided,to prevent overflow into the building.

(5) *Supports, foundations, and anchorage for all tank locations* -

(i) *General.* Tank supports shall be installed on firm foundations. Tank supports shall be of concrete, masonry, or protected steel. Single wood timber supports (not cribbing) laid horizontally may be used for outside aboveground tanks if not more than 12 inches (30.48 cm)high at their lowest point.

(ii) *Fire resistance.* Steel supports or exposed piling shall be protected by materials having a fireresistance rating of not less than 2 hours, except that steel saddles need not be protected if less than 12 inches (30.48 cm) high at their lowest point. Water spray protection or its equivalent may be used in lieu of fire-resistive materials to protect supports.

(iii) *Spheres.* The design of the supporting structure for tanks such as spheres shall receive specialengineering consideration.

(iv) *Load distribution.* Every tank shall be so supported as to prevent the excessive concentration ofloads on the supporting portion of the shell.

(v) *Foundations.* Tanks shall rest on the ground or on foundations made of concrete, masonry, piling, or steel. Tank foundations shall be designed to minimize the possibility of uneven settling of the tank and to minimize corrosion in any part of the tank resting on the foundation.

(vi) *Flood areas.* Where a tank is located in an area that may be subjected to flooding, the applicableprecautions outlined in this subdivision shall be observed.

(A) No aboveground vertical storage tank containing a flammable liquid shall be located sothat the allowable liquid level within the tank is below the established maximum floodstage, unless the tank is provided with a guiding structure such as described in paragraphs (i)(5)(vi) (M), (N), and (O) of this section.

(B) Independent water supply facilities shall be provided at locations where there is no ample and dependable public water supply available for loading partially empty tanks with water.

(C) In addition to the preceding requirements, each tank so located that more than 70 percent,but less than 100 percent, of its allowable liquid storage capacity will be submerged at theestablished maximum flood stage, shall be safeguarded by one of the following methods: Tank shall be raised, or its height shall be increased, until its top extends above the maximum flood stage a distance equivalent to 30 percent or more of its allowable liquid storage capacity: *Provided, however,* That the submerged part of the tank shall not exceed two and one-half times the diameter. Or, as an alternative to the foregoing, adequate noncombustible structural guides,

(D) designed to permit the tank to float vertically without loss of product, shall be provided. Each horizontal tank so located that more than 70 percent of its storage capacity will be submerged at the established flood stage, shall be anchored, attached to a foundation of concrete or of steel and concrete, of sufficient weight to provide adequate load for the tank when filled with flammable liquid and submerged by flood waters to the established flood stage, or adequately secured by other means.

(E) [Reserved]

(F) At locations where there is no ample and dependable water supply, or where filling of underground tanks with liquids is impracticable because of the character of their contents, their use, or for other reasons, each tank shall be safeguarded against movement when empty and submerged by high ground water or flood waters by anchoring, weighting with concrete or other approved solid loading material, or securing by other means. Each such tank shall be so constructed and installed that it will safely resist external pressures due to high ground water or flood waters.

(G) At locations where there is an ample and dependable water supply available, underground tanks containing flammable liquids, so installed that more than 70 percent of their storage capacity will be submerged at the maximum flood stage, shall be so anchored, weighted, or secured by other means, as to prevent movement of such tanks when filled with flammable liquids, and submerged by flood waters to the established flood stage.

(H) Pipe connections below the allowable liquid level in a tank shall be provided with valves or cocks located as closely as practicable to the tank shell. Such valves and their connections to tanks shall be of steel or other material suitable for use with the liquid being stored. Cast iron shall not be permitted.

(I) At locations where an independent water supply is required, it shall be entirely independent of public power and water supply. Independent source of water shall be available when flood waters reach a level not less than 10 feet (3.04 m) below the bottom of the lowest tank on a property.

(J) The self-contained power and pumping unit shall be so located or so designed that pumping into tanks may be carried on continuously throughout the rise in flood waters from a level 10 feet (3.04 m) below the lowest tank to the level of the potential flood stage.

(K) Capacity of the pumping unit shall be such that the rate of rise of water in all tanks shall be equivalent to the established potential average rate of rise of flood waters at any stage.

(L) Each independent pumping unit shall be tested periodically to insure that it is in satisfactory operating condition.

(M) Structural guides for holding floating tanks above their foundations shall be so designed that there will be no resistance to the free rise of a tank, and shall be constructed of noncombustible material.

(N) The strength of the structure shall be adequate to resist lateral movement of a tank subject to a horizontal force in any direction equivalent to not less than 25 pounds per square foot (1.05 kg m^2) acting on the projected vertical cross-sectional area of the tank.

- (O) Where tanks are situated on exposed points or bends in a shoreline where swift currents inflood waters will be present, the structures shall be designed to withstand a unit force of not less than 50 pounds per square foot (2.1 kg m^2).

- (P) The filling of a tank to be protected by water loading shall be started as soon as flood waters reach a dangerous flood stage. The rate of filling shall be at least equal to the rateof rise of the floodwaters (or the established average potential rate of rise).

- (Q) Sufficient fuel to operate the water pumps shall be available at all times to insure adequatepower to fill all tankage with water.

- (R) All valves on connecting pipelines shall be closed and locked in closed position whenwater loading has been completed.

- (S) Where structural guides are provided for the protection of floating tanks, all rigid connections between tanks and pipelines shall be disconnected and blanked off or blinded before the floodwaters reach the bottom of the tank, unless control valves and their connections to the tank are of a type designed to prevent breakage between the valveand the tank shell.

- (T) All valves attached to tanks other than those used in connection with water loading operations shall be closed and locked.

- (U) If a tank is equipped with a swing line, the swing pipe shall be raised to and secured at itshighest position.

- (V) Inspections. The Assistant Secretary or his designated representative shall make periodicinspections of all plants where the storage of flammable liquids is such as to require compliance with the foregoing requirements, in order to assure the following:

    - (*1*) That all flammable liquid storage tanks are in compliance with these requirementsand so maintained.
    - (*2*) That detailed printed instructions of what to do in flood emergencies are properly posted.
    - (*3*) That station operators and other employees depended upon to carry out such instructions are thoroughly informed as to the location and operation of such valvesand other equipment necessary to effect these requirements.

- (vii) *Earthquake areas.* In areas subject to earthquakes, the tank supports and connections shall bedesigned to resist damage as a result of such shocks.

(6) *Sources of ignition.* In locations where flammable vapors may be present, precautions shall be takento prevent ignition by eliminating or controlling sources of ignition. Sources of ignition may include open flames, lightning, smoking, cutting and welding, hot surfaces, frictional heat, sparks (static, electrical, and mechanical), spontaneous ignition, chemical and physical-chemical reactions, and radiant heat.

(7) *Testing* -

- (i) *General.* All tanks, whether shop built or field erected, shall be strength tested before they are placed in service in accordance with the applicable paragraphs of the code under which they were built. The American Society of Mechanical Engineers (ASME) code stamp, American Petroleum Institute (API) monogram, or the label of the Underwriters' Laboratories, Inc., on atank shall be evidence of compliance with this strength test. Tanks not marked in accordance

with the above codes shall be strength tested before they are placed in service in accordance with good engineering principles and reference shall be made to the sections on testing in thecodes listed in paragraphs (i)(1) (iii)(A), (iv)(B), or (v)(B) of this section.

- (ii) *Strength.* When the vertical length of the fill and vent pipes is such that when filled with liquid the static head imposed upon the bottom of the tank exceeds 10 pounds per square inch (68.94 kPa), the tank and related piping shall be tested hydrostatically to a pressure equal to thestatic head thus imposed.

- (iii) *Tightness.* In addition to the strength test called for in paragraphs (i)(7) (i) and (ii) of this section, all tanks and connections shall be tested for tightness. Except for underground tanks, this tightness test shall be made at operating pressure with air, inert gas, or water prior to placing the tank in service. In the case of field-erected tanks the strength test may be considered to be the test for tank tightness. Underground tanks and piping, before being covered, enclosed, or placed in use, shall be tested for tightness hydrostatically, or with air pressure at not less than 3 pounds per square inch (20.68 kPa) and not more than 5 pounds persquare inch (34.47 kPa).

- (iv) *Repairs.* All leaks or deformations shall be corrected in an acceptable manner before the tank isplaced in service. Mechanical caulking is not permitted for correcting leaks in welded tanks except pinhole leaks in the roof.

- (v) *Derated operations.* Tanks to be operated at pressures below their design pressure may be tested by the applicable provisions of paragraphs (i)(7) (i) or (ii) of this section, based upon thepressure developed under full emergency venting of the tank.

(j) **Piping, valves, and fittings** -

  (1) *General* -

   - (i) *Design.* The design (including selection of materials) fabrication, assembly, test, and inspection of piping systems containing flammable liquids shall be suitable for the expected working pressures and structural stresses. Conformity with the applicable provisions of Pressure Piping,ANSI B31 series and the provisions of this paragraph, shall be considered prima facie evidence of compliance with the foregoing provisions.

   - (ii) *Exceptions.* This paragraph does not apply to any of the following:
     - (A) Tubing or casing on any oil or gas wells and any piping connected directly
     - (B) thereto.Motor vehicle, aircraft, boat, or portable or stationary engines.
     - (C) Piping within the scope of any applicable boiler and pressures vessel code.

   - (iii) *Definitions.* As used in this paragraph, piping systems consist of pipe, tubing, flanges, bolting,gaskets, valves, fittings, the pressure containing parts of other components such as expansionjoints and strainers, and devices which serve such purposes as mixing, separating, snubbing,distributing, metering, or controlling flow.

  (2) *Materials for piping, valves, and fittings* -

   - (i) *Required materials.* Materials for piping, valves, or fittings shall be steel, nodular iron, or malleable iron, except as provided in paragraphs (j)(2) (ii), (iii) and (iv) of this section.

- (ii) **Exceptions.** Materials other than steel, nodular iron, or malleable iron may be used underground, or if required by the properties of the flammable liquid handled. Material other than steel, nodular iron, or malleable iron shall be designed to specifications embodying principles recognized as good engineering practices for the material used.

- (iii) **Linings.** Piping, valves, and fittings may have combustible or noncombustible linings.

- (iv) **Low-melting materials.** When low-melting point materials such as aluminum and brass or materials that soften on fire exposure such as plastics, or non-ductile materials such as cast iron, are necessary, special consideration shall be given to their behavior on fire exposure. If such materials are used in above ground piping systems or inside buildings, they shall be suitably protected against fire exposure or so located that any spill resulting from the failure of these materials could not unduly expose persons, important buildings or structures or can be readily controlled by remote valves.

(3) **Pipe joints.** Joints shall be made liquid tight. Welded or screwed joints or approved connectors shall be used. Threaded joints and connections shall be made up tight with a suitable lubricant or piping compound. Pipe joints dependent upon the friction characteristics of combustible materials for mechanical continuity of piping shall not be used inside buildings. They may be used outside of buildings above or below ground. If used above ground, the piping shall either be secured to prevent disengagement at the fitting or the piping system shall be so designed that any spill resulting from such disengagement could not unduly expose persons, important buildings or structures, and could be readily controlled by remote valves.

(4) **Supports.** Piping systems shall be substantially supported and protected against physical damage and excessive stresses arising from settlement, vibration, expansion, or contraction.

(5) **Protection against corrosion.** All piping for flammable liquids, both aboveground and underground, where subject to external corrosion, shall be painted or otherwise protected.

(6) **Valves.** Piping systems shall contain a sufficient number of valves to operate the system properly and to protect the plant. Piping systems in connection with pumps shall contain a sufficient number of valves to control properly the flow of liquid in normal operation and in the event of physical damage. Each connection to pipelines, by which equipments such as tankcars or tank vehicles discharge liquids by means of pumps into storage tanks, shall be provided with a check valve for automatic protection against backflow if the piping arrangement is such that backflow from the system is possible.

(7) **Testing.** All piping before being covered, enclosed, or placed in use shall be hydrostatically tested to 150 percent of the maximum anticipated pressure of the system, or pneumatically tested to 110 percent of the maximum anticipated pressure of the system, but not less than 5 pounds per square inch gage at the highest point of the system. This test shall be maintained for a sufficient time to complete visual inspection of all joints and connections, but for at least 10 minutes.

(k) *Marine service stations -*

(1) ***Dispensing.***

- (i) The dispensing area shall be located away from other structures so as to provide room for safe ingress and egress of craft to be fueled. Dispensing units shall in all cases be at least 20 feet (6.08 m) from any activity involving fixed sources of ignition.

- (ii) Dispensing shall be by approved dispensing units with or without integral pumps and may be located on open piers, wharves, or floating docks or on shore or on piers of the solid fill type.
- (iii) Dispensing nozzles shall be automatic-closing without a hold-open latch.

(2) **Tanks and pumps.**
- (i) Tanks, and pumps not integral with the dispensing unit, shall be on shore or on a pier of the solid fill type, except as provided in paragraphs (k)(2) (ii) and (iii) of this section.
- (ii) Where shore location would require excessively long supply lines to dispensers, tanks may be installed on a pier provided that applicable portions of paragraph (b) of this section relative to spacing, diking, and piping are complied with and the quantity so stored does not exceed 1,100 gallons (4,163.5 L) aggregate capacity.
- (iii) Shore tanks supplying marine service stations may be located above ground, where rock ledges or high water table make underground tanks impractical.
- (iv) Where tanks are at an elevation which would produce gravity head on the dispensing unit, the tank outlet shall be equipped with a pressure control valve positioned adjacent to and outside the tank block valve specified in § 1926.152(c)(8) of this section, so adjusted that liquid cannot flow by gravity from the tank in case of piping or hose failure.

(3) **Piping.**
- (i) Piping between shore tanks and dispensing units shall be as described in paragraph (k)(2)(iii) of this section, except that, where dispensing is from a floating structure, suitable lengths of oil-resistant flexible hose may be employed between the shore piping and the piping on the floating structure as made necessary by change in water level or shoreline.

TABLE F-19 - ELECTRICAL EQUIPMENT HAZARDOUS AREAS - SERVICE STATIONS

Location	Class I Group D division	Extent of classified area
Underground tank:		
Fill opening	1	Any pit, box or space below grade level, any part of which is within the Division 1 or 2 classified area.
	2	Up to 18 inches (45.72 cm) above grade level within a horizontal radius of 10 feet (3.04 m) from a loose fill connection and within a horizontal radius of 5 feet (1.52 m) from a tight fill connection.
Vent - Discharging upward...	1	Within 3 feet (0.912 m) of open end of vent, extending in all directions.
	2	Area between 3 feet (0.912 m) and 5 feet (1.52 m) of open end of vent, extending in all directions.
Dispenser:		
Pits.......................	1	Any pit, box or space below grade level, any part of which is within the Division 1 or 2 classified area.
Dispenser enclosure........	1	The area 4 feet (1.216 m) vertically above base within the enclosure and 18 inches (45.72 cm) horizontally in all directions.
Outdoor....................	2	Up to 18 inches (45.72 cm) above grade level within 20 feet (6.08 m) horizontally of any edge of enclosure.
Indoor:		
With mechanical ventilation.	2	Up to 18 inches (45.72 cm) above grade level within 20 feet (6.08 m) horizontally of any edge of enclosure.
With gravity ventilation....	2	Up to 18 inches (45.72 cm) above grade or floor level within 25 feet (7.6 m) horizontally of any edge of enclosure.
Remote pump - Outdoor.......	1	Any pit, box or space below grade level if any part is within a horizontal distance

		in all directions. Also up to 18 inches (45.72 cm) above grade level within 10 feet (3.04 m) horizontally from any edge of pump.
Remote pump - Indoor........	1	Entire area within any pit.
	2	Within 5 feet (1.52 m) of any edge of pump, extending in all directions. Also up to 3 feet (3.04 m) above floor or grade level within 25 feet (6.08 m) horizontally from any edge of pump.
Lubrication or service room.	1	Entire area within any pit.
	2	Area up to 18 inches (45.72 cm) above floor or grade level within entire lubrication room.
Dispenser for Category 1 or 2 flammable liquids, or Category 3 flammable liquids with a flashpoint below 100 ° F (37.8 ° C)	2	Within 3 feet (0.912 m) of any fill or dispensing point, extending in all directions.
Special enclosure inside building per 1910.106(f)(1)(ii).	1	Entire enclosure.
Sales, storage and rest rooms................	(1)	If there is any opening to these rooms within the extent of a Division 1 area, the entire room shall be classified as Division 1.

(1) Ordinary.

* * * * *

(ii) A readily accessible valve to shut off the supply from shore shall be provided in each pipeline ator near the approach to the pier and at the shore end of each pipeline adjacent to the point where flexible hose is attached.

(iii) Piping shall be located so as to be protected from physical damage.

(iv) Piping handling Category 1 or 2 flammable liquids, or Category 3 flammable liquids with a flashpoint below 100 °F (37.8 °C), shall be grounded to control stray currents.

(4) ***Definition; as used in this section:*** Marine service station shall mean that portion of a property whereflammable liquids used as fuels are stored and dispensed from fixed equipment on shore, piers, wharves, or floating docks into the fuel tanks of self-propelled craft, and shall include all facilities used in connection therewith.

[44 FR 8577, Feb. 9, 1979; 44 FR 20940, Apr. 6, 1979, as amended at 51 FR 25318, July 11, 1986; 58 FR 35162, June 30, 1993; 63 FR 33469, June 18, 1998; 77 FR 17891, Mar. 26, 2012]

§ 1926.153 Liquefied petroleum gas (LP-Gas).

(a) *Approval of equipment and systems.*

   (1) Each system shall have containers, valves, connectors, manifold valve assemblies, and regulators of an approved type.

   (2) All cylinders shall meet the Department of Transportation specification identification requirements published in 49 CFR part 178, Shipping Container Specifications.

   (3) *Definition.* As used in this section, *Containers* - All vessels, such as tanks, cylinders, or drums, used for transportation or storing liquefied petroleum gases.

(b) *Welding on LP-Gas containers.* Welding is prohibited on containers.

(c) *Container valves and container accessories.*

   (1) Valves, fittings, and accessories connected directly to the container, including primary shut off valves, shall have a rated working pressure of at least 250 p.s.i.g. and shall be of material and design suitable for LP-Gas service.

   (2) Connections to containers, except safety relief connections, liquid level gauging devices, and plugged openings, shall have shutoff valves located as close to the container as practicable.

(d) *Safety devices.*

   (1) Every container and every vaporizer shall be provided with one or more approved safety relief valves or devices. These valves shall be arranged to afford free vent to the outer air with discharge not less than 5 feet horizontally away from any opening into a building which is below such discharge.

   (2) Shutoff valves shall not be installed between the safety relief device and the container, or the equipment or piping to which the safety relief device is connected, except that a shutoff valve may be used where the arrangement of this valve is such that full required capacity flow through the safety relief device is always afforded.

   (3) Container safety relief devices and regulator relief vents shall be located not less than 5 feet in any direction from air openings into sealed combustion system appliances or mechanical ventilation air intakes.

(e) *Dispensing.*

   (1) Filling of fuel containers for trucks or motor vehicles from bulk storage containers shall be performed not less than 10 feet from the nearest masonry-walled building, or not less than 25 feet from the nearest building or other construction and, in any event, not less than 25 feet from any building opening.

   (2) Filling of portable containers or containers mounted on skids from storage containers shall be performed not less than 50 feet from the nearest building.

(f) **Requirements for appliances.**

(1) LP-Gas consuming appliances shall be approved types.

(2) Any appliance that was originally manufactured for operation with a gaseous fuel other than LP-Gas, and is in good condition, may be used with LP-Gas only after it is properly converted, adapted, andtested for performance with LP-Gas before the appliance is placed in use.

(g) **Containers and regulating equipment installed outside of buildings or structures.** Containers shall be upright upon firm foundations or otherwise firmly secured. The possible effect on the outlet piping of settling shall be guarded against by a flexible connection or special fitting.

(h) **Containers and equipment used inside of buildings or structures.**

(1) When operational requirements make portable use of containers necessary, and their location outside of buildings or structures is impracticable, containers and equipment shall be permitted to be used inside of buildings or structures in accordance with paragraphs (h)(2) through (11) of this section.

(2) **Containers in use** means connected for use.

(3) Systems utilizing containers having a water capacity greater than $2^1/_2$ pounds (nominal 1 pound LP-Gas capacity) shall be equipped with excess flow valves. Such excess flow valves shall be either integral with the container valves or in the connections to the container valve outlets.

(4) Regulators shall be either directly connected to the container valves or to manifolds connected to thecontainer valves. The regulator shall be suitable for use with LP-Gas. Manifolds and fittings connecting containers to pressure regulator inlets shall be designed for at least 250 p.s.i.g. service pressure.

(5) Valves on containers having water capacity greater than 50 pounds (nominal 20 pounds LP-Gas capacity) shall be protected from damage while in use or storage.

(6) Aluminum piping or tubing shall not be used.

(7) Hose shall be designed for a working pressure of at least 250 p.s.i.g. Design, construction, and performance of hose, and hose connections shall have their suitability determined by listing by a nationally recognized testing agency. The hose length shall be as short as practicable. Hoses shallbe long enough to permit compliance with spacing provisions of paragraphs (h)(1) through (13) of this section, without kinking or straining, or causing hose to be so close to a burner as to be damaged by heat.

(8) Portable heaters, including salamanders, shall be equipped with an approved automatic device to shut off the flow of gas to the main burner, and pilot if used, in the event of flame failure. Such heaters, having inputs above 50,000 B.t.u. per hour, shall be equipped with either a pilot, which mustbe lighted and proved before the main burner can be turned on, or an electrical ignition system.

Note: The provisions of this subparagraph do not apply to portable heaters under 7,500 B.t.u. per hour input when used with containers having a maximum water capacity of $2^1/_2$ pounds.

Container valves, connectors, regulators, manifolds, piping, and tubing shall not be used as structural supports for heaters.

(9)

(10) Containers, regulating equipment, manifolds, pipe, tubing, and hose shall be located to minimize exposure to high temperatures or physical damage.

(11) Containers having a water capacity greater than 2$\frac{1}{2}$ pounds (nominal 1 pound LP-Gas capacity) connected for use shall stand on a firm and substantially level surface and, when necessary, shall be secured in an upright position.

(12) The maximum water capacity of individual containers shall be 245 pounds (nominal 100 pounds LP-Gas capacity).

(13) For temporary heating, heaters (other than integral heater-container units) shall be located at least 6 feet from any LP-Gas container. This shall not prohibit the use of heaters specifically designed for attachment to the container or to a supporting standard, provided they are designed and installed so as to prevent direct or radiant heat application from the heater onto the containers. Blower and radiant type heaters shall not be directed toward any LP-Gas container within 20 feet.

(14) If two or more heater-container units, of either the integral or nonintegral type, are located in an unpartitioned area on the same floor, the container or containers of each unit shall be separated from the container or containers of any other unit by at least 20 feet.

(15) When heaters are connected to containers for use in an unpartitioned area on the same floor, the total water capacity of containers, manifolded together for connection to a heater or heaters, shall not be greater than 735 pounds (nominal 300 pounds LP-Gas capacity). Such manifolds shall be separated by at least 20 feet.

(16) Storage of containers awaiting use shall be in accordance with paragraphs (j) and (k) of this section.

(i) *Multiple container systems.*

  (1) Valves in the assembly of multiple container systems shall be arranged so that replacement of containers can be made without shutting off the flow of gas in the system. This provision is not to be construed as requiring an automatic changeover device.

  (2) Heaters shall be equipped with an approved regulator in the supply line between the fuel cylinder and the heater unit. Cylinder connectors shall be provided with an excess flow valve to minimize the flow of gas in the event the fuel line becomes ruptured.

  (3) Regulators and low-pressure relief devices shall be rigidly attached to the cylinder valves, clyinders, supporting standards, the building walls, or otherwise rigidly secured, and shall be so installed or protected from the elements.

(j) *Storage of LPG containers.* Storage of LPG within buildings is prohibited.

(k) *Storage outside of buildings.*

  (1) Storage outside of buildings, for containers awaiting use, shall be located from the nearest building or group of buildings, in accordance with the following:

Table F-3

Quantity of LP-Gas stored	Distance (feet)
500 lbs. or less	0
501 to 6,000 lbs	10
6,001 to 10,000 lbs	20
Over 10,000 lbs	25

(2) Containers shall be in a suitable ventilated enclosure or otherwise protected against tampering.

(l) **Fire protection.** Storage locations shall be provided with at least one approved portable fire extinguisher having a rating of not less than 20-B:C.

(m) **Systems utilizing containers other than DOT containers** -

(1) **Application.** This paragraph applies specifically to systems utilizing storage containers other than those constructed in accordance with DOT specifications. Paragraph (b) of this section applies to this paragraph unless otherwise noted in paragraph (b) of this section.

(2) **Design pressure and classification of storage containers.** Storage containers shall be designed and classified in accordance with Table F-31.

Table F-31

Container type	For gases with vapor press. Not to exceed lb. per sq. in. gage at 100 °F. (37.8 °C.)	Minimum design pressure of container, lb. per sq. in. gage	
		1949 and earlier editions of ASME Code (Par. U-68, U-69)	1949 edition of ASME Code (Par. U-200, U-201); 1950, 1952, 1956, 1959, 1962, 1965, and 1968 (Division 1) editions of ASME Code; All editions of API-ASME Code[3]
[1] 80	[1] 80	[1] 80	[1] 100
100	100	100	125
125	125	125	156
150	150	150	187
175	175	175	219
[2] 200	215	200	250

[1] New storage containers of the 80 type have not been authorized since Dec. 31, 1947.

[2] Container type may be increased by increments of 25. The minimum design pressure of containers shall be 100% of the container type designation when constructed under 1949 or earlier editions of the ASME Code (Par. U-68 and U-69). The minimum design pressure of containers shall

be 125% of the container type designation when constructed under: (1) the 1949 ASME Code (Par. U-200 and U-201), (2) 1950, 1952, 1956, 1959, 1962, 1965, and 1968 (Division 1) editions of the ASME Code, and (3) all editions of the API-ASME Code.

[3] Construction of containers under the API-ASME Code is not authorized after July 1, 1961.

(3) Containers with foundations attached (portable or semiportable b containers with suitable steel "runners" or "skids" and popularly known in the industry as "skid tanks") shall be designed, installed, and used in accordance with these rules subject to the following provisions:

  (i) If they are to be used at a given general location for a temporary period not to exceed 6 monthsthey need not have fire-resisting foundations or saddles but shall have adequate ferrous metal supports.

  (ii) They shall not be located with the outside bottom of the container shell more than 5 feet (1.52m) above the surface of the ground unless fire-resisting supports are provided.

  (iii) The bottom of the skids shall not be less than 2 inches (5.08 cm) or more than 12 inches (30.48cm) below the outside bottom of the container shell.

  (iv) Flanges, nozzles, valves, fittings, and the like, having communication with the interior of the container, shall be protected against physical damage.

  (v) When not permanently located on fire-resisting foundations, piping connections shall be sufficiently flexible to minimize the possibility of breakage or leakage of connections if the container settles, moves, or is otherwise displaced.

  (vi) Skids, or lugs for attachment of skids, shall be secured to the container in accordance with thecode or rules under which the container is designed and built (with a minimum factor of safetyof four) to withstand loading in any direction equal to four times the weight of the container andattachments when filled to the maximum permissible loaded weight.

(4) Field welding where necessary shall be made only on saddle plates or brackets which were appliedby the manufacturer of the tank.

(n) When LP-Gas and one or more other gases are stored or used in the same area, the containers shall bemarked to identify their content. Marking shall be in compliance with American National Standard Z48.1-1954, "Method of Marking Portable Compressed Gas Containers To Identify the Material Contained."

(o) *Damage from vehicles.* When damage to LP-Gas systems from vehicular traffic is a possibility, precautionsagainst such damage shall be taken.

[44 FR 8577, Feb. 9, 1979; 44 FR 20940, Apr. 6, 1979, as amended at 58 FR 35170, June 30, 1993]

§ 1926.154 Temporary heating devices.

(a) *Ventilation.*

  (1) Fresh air shall be supplied in sufficient quantities to maintain the health and safety of workmen. Where natural means of fresh air supply is inadequate, mechanical ventilation shall be provided.

  (2) When heaters are used in confined spaces, special care shall be taken to provide sufficient

ventilation in order to ensure proper combustion, maintain the health and safety of workmen, and limit temperature rise in the area.

(b) **Clearance and mounting.**

(1) Temporary heating devices shall be installed to provide clearance to combustible material not less than the amount shown in Table F-4.

(2) Temporary heating devices, which are listed for installation with lesser clearances than specified in Table F-4, may be installed in accordance with their approval.

Table F-4

Heating appliances	Minimum clearance, (inches)		
	Sides	Rear	Chimney connector
Room heater, circulating type	12	12	18
Room heater, radiant type	36	36	18

(3) Heaters not suitable for use on wood floors shall not be set directly upon them or other combustible materials. When such heaters are used, they shall rest on suitable heat insulating material or at least 1-inch concrete, or equivalent. The insulating material shall extend beyond the heater 2 feet or more in all directions.

(4) Heaters used in the vicinity of combustible tarpaulins, canvas, or similar coverings shall be located at least 10 feet from the coverings. The coverings shall be securely fastened to prevent ignition or upsetting of the heater due to wind action on the covering or other material.

(c) **Stability.** Heaters, when in use, shall be set horizontally level, unless otherwise permitted by the manufacturer's markings.

(d) **Solid fuel salamanders.** Solid fuel salamanders are prohibited in buildings and on scaffolds.

(e) **Oil-fired heaters.**

(1) Flammable liquid-fired heaters shall be equipped with a primary safety control to stop the flow of fuel in the event of flame failure. Barometric or gravity oil feed shall not be considered a primary safety control.

(2) Heaters designed for barometric or gravity oil feed shall be used only with the integral tanks.

(3) [Reserved]

(4) Heaters specifically designed and approved for use with separate supply tanks may be directly connected for gravity feed, or an automatic pump, from a supply tank.

§ 1926.155 Definitions applicable to this subpart.

(a) **Approved,** for the purpose of this subpart, means equipment that has been listed or approved by a nationally recognized testing laboratory such as Factory Mutual Engineering Corp., or Underwriters' Laboratories, Inc., or Federal agencies such as Bureau of Mines, or U.S. Coast Guard, which issue approvals for such equipment.

(b) **Closed container** means a container so sealed by means of a lid or other device that neither liquid nor vapor will escape from it at ordinary temperatures.

(c) [Reserved]

(d) **Combustion** means any chemical process that involves oxidation sufficient to produce light or heat.

(e) **Fire brigade** means an organized group of employees that are knowledgeable, trained, and skilled in the safe evacuation of employees during emergency situations and in assisting in fire fighting operations.

(f) **Fire resistance** means so resistant to fire that, for specified time and under conditions of a standard heat intensity, it will not fail structurally and will not permit the side away from the fire to become hotter than a specified temperature. For purposes of this part, fire resistance shall be determined by the Standard Methods of Fire Tests of Building Construction and Materials, NFPA 251-1969.

(g) **Flammable** means capable of being easily ignited, burning intensely, or having a rapid rate of flamespread.

(h) **Flammable liquid** means any liquid having a vapor pressure not exceeding 40 pounds per square inch (absolute) at 100 °F (37.8 °C) and having a flashpoint at or below 199.4 °F (93 °C). Flammable liquids are divided into four categories as follows:

  (1) Category 1 shall include liquids having flashpoints below 73.4 °F (23 °C) and having a boiling point at or below 95 °F (35 °C).

  (2) Category 2 shall include liquids having flashpoints below 73.4 °F (23 °C) and having a boiling point above 95 °F (35 °C).

  (3) Category 3 shall include liquids having flashpoints at or above 73.4 °F (23 °C) and at or below 140 °F (60 °C).

  (4) Category 4 shall include liquids having flashpoints above 140 °F (60 °C) and at or below 199.4 °F (93 °C).

(i) **Flash point** of the liquid means the temperature at which it gives off vapor sufficient to form an ignitable mixture with the air near the surface of the liquid or within the vessel used as determined by appropriate test procedure and apparatus as specified below.

  (1) The flashpoint of liquids having a viscosity less than 45 Saybolt Universal Second(s) at 100 °F (37.8 °C) and a flashpoint below 175 °F (79.4 °C) shall be determined in accordance with the Standard Method of Test for Flash Point by the Tag Closed Tester, ASTM D-56-69 (incorporated by reference; See § 1926.6), or an equivalent method as defined by § 1910.1200 appendix B.

  (2) The flashpoints of liquids having a viscosity of 45 Saybolt Universal Second(s) or more at 175 °F (79.4 °C) or higher shall be determined in accordance with the Standard Method of Test for Flash Point by the Pensky Martens Closed Tester, ASTM D-93-69 (incorporated by reference; See § 1926.6), or an equivalent method as defined by § 1910.1200 appendix B.

(j) **Liquefied petroleum gases, LPG** and *LP Gas* mean and include any material which is composed predominantly of any of the following hydrocarbons, or mixtures of them, such as propane, propylene, butane (normal butane or iso-butane), and butylenes.

(k) **Portable tank** means a closed container having a liquid capacity more than 60 U.S. gallons, and not intended for fixed installation.

(l) **Safety can** means an approved closed container, of not more than 5 gallons capacity, having a flash- arresting screen, spring-closing lid and spout cover and so designed that it will safely relieve internalpressure when subjected to fire exposure.

(m) **Vapor pressure** means the pressure, measured in pounds per square inch (absolute), exerted by a volatileliquid as determined by the "Standard Method of Test for Vapor Pressure of Petroleum Products (Reid Method)." (ASTM D-323-58).

[44 FR 8577, Feb. 9, 1979; 44 FR 20940, Apr. 6, 1979, as amended at 77 FR 17894, Mar. 26, 2012]

Subpart G - Signs, Signals, and Barricades

Authority: 40 U.S.C. 333; 29 U.S.C. 653, 655, 657; Secretary of Labor's Order No. 12-71 (36 FR 8754), 8-76 (41 FR 25059), 9-83 (48 FR 35736), 3-2000 (65 FR 50017), 5-2002 (67 FR 65008), 5-2007 (72 FR 31159), 4-2010 (75 FR 55355), or 1-2012 (77 FR 3912), as applicable; and 29 CFR part 1911.

§ 1926.200 Accident prevention signs and tags.

(a) *General.* Signs and symbols required by this subpart shall be visible at all times when work is being performed, and shall be removed or covered promptly when the hazards no longer exist.

(b) *Danger signs.*

(1) Danger signs shall be used only where an immediate hazard exists, and shall follow the specifications illustrated in Figure 1 of ANSI Z35.1-1968 or in Figures 1 to 13 of ANSI Z535.2-2011,incorporated by reference in § 1926.6.

(2) Danger signs shall have red as the predominating color for the upper panel; black outline on the borders; and a white lower panel for additional sign wording.

(c) *Caution signs.*

(1) Caution signs shall be used only to warn against potential hazards or to caution against unsafe practices, and shall follow the specifications illustrated in Figure 4 of ANSI Z35.1-1968 or in Figures1 to 13 of ANSI Z535.2-2011, incorporated by reference in § 1926.6.

(2) Caution signs shall have yellow as the predominating color; black upper panel and borders: yellow lettering of "caution" on the black panel; and the lower yellow panel for additional sign wording. Blacklettering shall be used for additional wording.

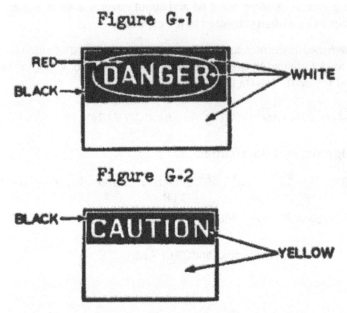

Figure G-1

Figure G-2

(3) The standard color of the background shall be yellow; and the panel, black with yellow letters. Any letters used against the yellow background shall be black. The colors shall be those of opaque glossy samples as specified in Table 1 of ANSI Z53.1-1967 or in Table 1 of ANSI Z535.1-2006(R2011), incorporated by reference in § 1926.6.

(d) **Exit signs.** Exit signs, when required, shall be lettered in legible red letters, not less than 6 inches high, on a white field and the principal stroke of the letters shall be at least three-fourths inch in width.

(e) **Safety instruction signs.** Safety instruction signs, when used, shall be white with green upper panel with white letters to convey the principal message. Any additional wording on the sign shall be black letters on the white background.

(f) **Directional signs.** Directional signs, other than automotive traffic signs specified in paragraph (g) of this section, shall be white with a black panel and a white directional symbol. Any additional wording on the sign shall be black letters on the white background.

(g) **Trafic control signs and devices.**

(1) At points of hazard, construction areas shall be posted with legible traffic control signs and protected by traffic control devices.

(2) The design and use of all traffic control devices, including signs, signals, markings, barricades, and other devices, for protection of construction workers shall conform to Part 6 of the MUTCD (incorporated by reference, see § 1926.6).

(h) **Accident prevention tags.**

(1) Accident prevention tags shall be used as a temporary means of warning employees of an existing hazard, such as defective tools, equipment, etc. They shall not be used in place of, or as a substitute for, accident prevention signs.

(2) For accident prevention tags, employers shall follow specifications that are similar to those in Figures 1 to 4 of ANSI Z35.2-1968 or Figures 1 to 8 of ANSI Z535.5-2011, incorporated by

referencein § 1926.6.

(i) **Additional rules.** ANSI Z35.1-1968, ANSI Z535.2-2011, ANSI Z35.2-1968, and ANSI Z535.5-2011, incorporated by reference in § 1926.6, contain rules in addition to those specifically prescribed in this subpart. The employer shall comply with ANSI Z35.1-1968 or ANSI Z535.2-2011, and ANSI Z35.2-1968 orZ535.5-2011, with respect to such additional rules.

[44 FR 8577, Feb. 9, 1979; 44 FR 20940, Apr. 6, 1979, as amended at 58 FR 35173, June 30, 1993; 67 FR 57736, Sept. 12, 2002; 69
FR 18803, Apr. 9, 2004; 78 FR 35567, June 13, 2013; 78 FR 66642, Nov. 6, 2013; 84 FR 21577, May 14, 2019]

§ 1926.201 Signaling.

(a) **Flaggers.** Signaling by flaggers and the use of flaggers, including warning garments worn by flaggers, shall conform to Part 6 of the MUTCD (incorporated by reference, see § 1926.6).

(b) **Crane and hoist signals.** Regulations for crane and hoist signaling will be found in applicable American National Standards Institute standards.

[44 FR 8577, Feb. 9, 1979; 44 FR 20940, Apr. 6, 1979, as amended at 67 FR 57736, Sept. 12, 2002; 78 FR 35567, June 13, 2013; 84
FR 21577, May 14, 2019]

Subpart H - Materials Handling, Storage, Use, and Disposal

Authority: 40 U.S.C. 3701; 29 U.S.C. 653, 655, 657; and Secretary of Labor's Order No. 12-71 (36 FR 8754), 8-76 (41 FR 25059), 9-83 (48 FR 35736), 1-90 (55 FR 9033), 4-2010 (75 FR 55355), or 1-2012 (77 FR 3912), as applicable. Section 1926.250 also issued under 29 CFR part 1911.

§ 1926.250 General requirements for storage.

(a) **General.**

(1) All materials stored in tiers shall be stacked, racked, blocked, interlocked, or otherwise secured toprevent sliding, falling or collapse.

(2)

   (i) The weight of stored materials on floors within buildings and structures shall not exceed maximum safe load limits.

   (ii) Employers shall conspicuously post maximum safe load limits of floors within buildings and structures, in pounds per square foot, in all storage areas, except when the storage area is on afloor or slab on grade. Posting is not required for storage areas in all single-family residentialstructures and wood-framed multi-family residential structures.

(3) Aisles and passageways shall be kept clear to provide for the free and safe movement of materialhandling equipment or employees. Such areas shall be kept in good repair.

(4) When a difference in road or working levels exist, means such as ramps, blocking, or grading shall beused to ensure the safe movement of vehicles between the two levels.

(b) **Material storage.**

(1) Material stored inside buildings under construction shall not be placed within 6 feet of any hoistway or inside floor openings, nor within 10 feet of an exterior wall which does not extend above the top of the material stored.

(2) Each employee required to work on stored material in silos, hoppers, tanks, and similar storage areas shall be equipped with personal fall arrest equipment meeting the requirements of subpart M of this part.

(3) Noncompatible materials shall be segregated in storage.

(4) Bagged materials shall be stacked by stepping back the layers and cross-keying the bags at least every 10 bags high.

(5) Materials shall not be stored on scaffolds or runways in excess of supplies needed for immediate operations.

(6) Brick stacks shall not be more than 7 feet in height. When a loose brick stack reaches a height of 4 feet, it shall be tapered back 2 inches in every foot of height above the 4-foot level.

(7) When masonry blocks are stacked higher than 6 feet, the stack shall be tapered back one-half block per tier above the 6-foot level.

(8) Lumber:

　(i) Used lumber shall have all nails withdrawn before stacking.

　(ii) Lumber shall be stacked on level and solidly supported sills.

　(iii) Lumber shall be so stacked as to be stable and self-

　(iv) supporting.

　Lumber piles shall not exceed 20 feet in height provided that lumber to be handled manually shall not be stacked more than 16 feet high.

(9) Structural steel, poles, pipe, bar stock, and other cylindrical materials, unless racked, shall be stacked and blocked so as to prevent spreading or tilting.

(c) *Housekeeping.* Storage areas shall be kept free from accumulation of materials that constitute hazards from tripping, fire, explosion, or pest harborage. Vegetation control will be exercised when necessary.

(d) *Dockboards (bridge plates).*

(1) Portable and powered dockboards shall be strong enough to carry the load imposed on them.

(2) Portable dockboards shall be secured in position, either by being anchored or equipped with devices which will prevent their slipping.

(3) Handholds, or other effective means, shall be provided on portable dockboards to permit safe handling.

(4) Positive protection shall be provided to prevent railroad cars from being moved while dockboards or bridge plates are in position.

[44 FR 8577, Feb. 9, 1979; 44 FR 20940, Apr. 6, 1979, as amended at 49 FR 18295, Apr. 30, 1984; 54 FR 24334, June 7, 1989;

58 FR 35173, June 30, 1993; 59 FR 40729, Aug. 9, 1994; 61 FR 5510, Feb. 13, 1996; 84 FR 21577, May 14, 2019]

§ 1926.251 Rigging equipment for material handling.

(a) *General.*

(1) Rigging equipment for material handling shall be inspected prior to use on each shift and as necessary during its use to ensure that it is safe. Defective rigging equipment shall be removed from service.

(2) Employers must ensure that rigging equipment:

(i) Has permanently affixed and legible identification markings as prescribed by the manufacturer that indicate the recommended safe working load;

(ii) Not be loaded in excess of its recommended safe working load as prescribed on the identification markings by the manufacturer; and

(iii) Not be used without affixed, legible identification markings, required by paragraph (a)(2)(i) of this section.

(3) Rigging equipment, when not in use, shall be removed from the immediate work area so as not to present a hazard to employees.

(4) Special custom design grabs, hooks, clamps, or other lifting accessories, for such units as modular panels, prefabricated structures and similar materials, shall be marked to indicate the safe working loads and shall be proof-tested prior to use to 125 percent of their rated load.

(5) *Scope.* This section applies to slings used in conjunction with other material handling equipment for the movement of material by hoisting, in employments covered by this part. The types of slings covered are those made from alloy steel chain, wire rope, metal mesh, natural or synthetic fiber rope (conventional three strand construction), and synthetic web (nylon, polyester, and polypropylene).

(6) *Inspections.* Each day before being used, the sling and all fastenings and attachments shall be inspected for damage or defects by a competent person designated by the employer. Additional inspections shall be performed during sling use, where service conditions warrant. Damaged or defective slings shall be immediately removed from service.

(b) *Alloy steel chains.*

(1) Welded alloy steel chain slings shall have permanently affixed durable identification stating size, grade, rated capacity, and sling manufacturer.

(2) Hooks, rings, oblong links, pear-shaped links, welded or mechanical coupling links, or other attachments, when used with alloy steel chains, shall have a rated capacity at least equal to that of the chain.

(3) Job or shop hooks and links, or makeshift fasteners, formed from bolts, rods, etc., or other such attachments, shall not be used.

(4) Employers must not use alloy steel-chain slings with loads in excess of the rated capacities (i.e., working load limits) indicated on the sling by permanently affixed and legible identification markings prescribed by the manufacturer.

(5) Whenever wear at any point of any chain link exceeds that shown in Table H-1, the assembly shall

beremoved from service.

(6) **Inspections.**

    (i) In addition to the inspection required by other paragraphs of this section, a thorough periodic inspection of alloy steel chain slings in use shall be made on a regular basis, to be determinedon the basis of

        (A) frequency of sling use;

        (B) severity of service

        (C) conditions;nature of

        (D) lifts being made; and

        experience gained on the service life of slings used in similar circumstances. Such inspections shall in no event be at intervals greater than once every 12 months.

    (ii) The employer shall make and maintain a record of the most recent month in which each alloysteel chain sling was thoroughly inspected, and shall make such record available for examination.

(c) **Wire rope.**

    (1) Employers must not use improved plow-steel wire rope and wire-rope slings with loads in excess ofthe rated capacities (i.e., working load limits) indicated on the sling by permanently affixed and legible identification markings prescribed by the manufacturer.

    (2) Protruding ends of strands in splices on slings and bridles shall be covered or

    (3) blunted.Wire rope shall not be secured by knots, except on haul back lines on

    (4) scrapers.

    The following limitations shall apply to the use of wire rope:

    (i) An eye splice made in any wire rope shall have not less than three full tucks. However, this requirement shall not operate to preclude the use of another form of splice or connection whichcan be shown to be as efficient and which is not otherwise prohibited.

    (ii) Except for eye splices in the ends of wires and for endless rope slings, each wire rope used inhoisting or lowering, or in pulling loads, shall consist of one continuous piece without knot or splice.

    (iii) Eyes in wire rope bridles, slings, or bull wires shall not be formed by wire rope clips or knots.

    (iv) Wire rope shall not be used if, in any length of eight diameters, the total number of visible broken wires exceeds 10 percent of the total number of wires, or if the rope shows other signsof excessive wear, corrosion, or defect.

    (5) When U-bolt wire rope clips are used to form eyes, Table H-2 shall be used to determine the numberand spacing of clips.

    (i) When used for eye splices, the U-bolt shall be applied so that the "U" section is in contact withthe dead end of the rope.

    (i1) [Reserved]

(6) Slings shall not be shortened with knots or bolts or other makeshift devices.

(7) Sling legs shall not be kinked.

(8) Slings used in a basket hitch shall have the loads balanced to prevent slippage.

(9) Slings shall be padded or protected from the sharp edges of their loads.

(10) Hands or fingers shall not be placed between the sling and its load while the sling is being tightened around the load.

(11) Shock loading is prohibited.

(12) A sling shall not be pulled from under a load when the load is resting on the sling.

(13) *Minimum sling lengths.*

   (i) Cable laid and 6 × 19 and 6 × 37 slings shall have a minimum clear length of wire rope 10 times the component rope diameter between splices, sleeves or end fittings.

   (ii) Braided slings shall have a minimum clear length of wire rope 40 times the component rope diameter between the loops or end fittings.

   (iii) Cable laid grommets, strand laid grommets and endless slings shall have a minimum circumferential length of 96 times their body diameter.

(14) *Safe operating temperatures.* Fiber core wire rope slings of all grades shall be permanently removed from service if they are exposed to temperatures in excess of 200 °F (93.33 °C). When nonfiber core wire rope slings of any grade are used at temperatures above 400 °F (204.44 °C) or below minus 60 °F (15.55 °C), recommendations of the sling manufacturer regarding use at that temperature shall be followed.

(15) *End attachments.*

   (i) Welding of end attachments, except covers to thimbles, shall be performed prior to the assembly of the sling.

   (ii) All welded end attachments shall not be used unless proof tested by the manufacturer or equivalent entity at twice their rated capacity prior to initial use. The employer shall retain a certificate of the proof test, and make it available for examination.

(16) Wire rope slings shall have permanently affixed, legible identification markings stating size, rated capacity for the type(s) of hitch(es) used and the angle upon which it is based, and the number of legs if more than one.

(d) *Natural rope, and synthetic fiber.*

   (1) Employers must not use natural- and synthetic-fiber rope slings with loads in excess of the rated capacities (i.e., working load limits) indicated on the sling by permanently affixed and legible identification markings prescribed by the manufacturer.

   (2) All splices in rope slings provided by the employer shall be made in accordance with fiber rope manufacturers recommendations.

(i) In manila rope, eye splices shall contain at least three full tucks, and short splices shall contain at least six full tucks (three on each side of the centerline of the splice).

(ii) In layed synthetic fiber rope, eye splices shall contain at least four full tucks, and short splices shall contain at least eight full tucks (four on each side of the centerline of the splice).

(iii) Strand end tails shall not be trimmed short (flush with the surface of the rope) immediately adjacent to the full tucks. This precaution applies to both eye and short splices and all types of fiber rope. For fiber ropes under 1-inch diameter, the tails shall project at least six rope diameters beyond the last full tuck. For fiber ropes 1-inch diameter and larger, the tails shall

project at least 6 inches beyond the last full tuck. In applications where the projecting tails may be objectionable, the tails shall be tapered and spliced into the body of the rope using at least two additional tucks (which will require a tail length of approximately six rope diameters beyond the last full tuck).

(iv) For all eye splices, the eye shall be sufficiently large to provide an included angle of not greater than 60° at the splice when the eye is placed over the load or support.

(v) Knots shall not be used in lieu of splices.

(3) **Safe operating temperatures.** Natural and synthetic fiber rope slings, except for wet frozen slings, may be used in a temperature range from minus 20 °F (−28.88 °C) to plus 180 °F (82.2 °C) without decreasing the working load limit. For operations outside this temperature range and for wet frozen slings, the sling manufacturer's recommendations shall be followed.

(4) **Splicing.** Spliced fiber rope slings shall not be used unless they have been spliced in accordance with the following minimum requirements and in accordance with any additional recommendations of the manufacturer:

(i) In manila rope, eye splices shall consist of at least three full tucks, and short splices shall consist of at least six full tucks, three on each side of the splice center line.

(ii) In synthetic fiber rope, eye splices shall consist of at least four full tucks, and short splices shall consist of at least eight full tucks, four on each side of the center line.

(iii) Strand end tails shall not be trimmed flush with the surface of the rope immediately adjacent to the full tucks. This applies to all types of fiber rope and both eye and short splices. For fiber rope under 1 inch (2.54 cm) in diameter, the tail shall project at least six rope diameters beyond the last full tuck. For fiber rope 1 inch (2.54 cm) in diameter and larger, the tail shall project at least 6 inches (15.24 cm) beyond the last full tuck. Where a projecting tail interferes with the use of the sling, the tail shall be tapered and spliced into the body of the rope using at least two additional tucks (which will require a tail length of approximately six rope diameters beyond the last full tuck).

(iv) Fiber rope slings shall have a minimum clear length of rope between eye splices equal to 10 times the rope diameter.

(v) Knots shall not be used in lieu of splices.

(vi) Clamps not designed specifically for fiber ropes shall not be used for splicing.

(vii) For all eye splices, the eye shall be of such size to provide an included angle of not greater than 60 degrees at the splice when the eye is placed over the load or support.

(5) **End attachments.** Fiber rope slings shall not be used if end attachments in contact with the rope have sharp edges or projections.

(6) **Removal from service.** Natural and synthetic fiber rope slings shall be immediately removed from service if any of the following conditions are present:

   (i) Abnormal wear.
   (ii) Powdered fiber between
   (iii) strands. Broken or cut

   fibers.

   (iv) Variations in the size or roundness of strands.
   (v) Discoloration or rotting.
   (vi) Distortion of hardware in the sling.

(7) Employers must use natural- and synthetic-fiber rope slings that have permanently affixed and legible identification markings that state the rated capacity for the type(s) of hitch(es) used and the angle upon which it is based, type of fiber material, and the number of legs if more than one.

(e) **Synthetic webbing** (*nylon, polyester, and polypropylene*).

   (1) The employer shall have each synthetic web sling marked or coded to
      (i) show: Name or trademark of manufacturer.
      (ii) Rated capacities for the type of
      (iii) hitch. Type of material.
   (2) Rated capacity shall not be exceeded.
   (3) **Webbing.** Synthetic webbing shall be of uniform thickness and width and selvage edges shall not be split from the webbing's width.
   (4) **Fittings.** Fittings shall be:
      (i) Of a minimum breaking strength equal to that of the sling;
      (ii) and Free of all sharp edges that could in any way damage the
   (5) webbing.

   Attachment of end fittings to webbing and formation of eyes. Stitching shall be the only method used to attach end fittings to webbing and to form eyes. The thread shall be in an even pattern and contain a sufficient number of stitches to develop the full breaking strength of the sling.

   (6) **Environmental conditions.** When synthetic web slings are used, the following precautions shall be taken:
      (i) Nylon web slings shall not be used where fumes, vapors, sprays, mists or liquids of acids or phenolics are present.

- (ii) Polyester and polypropylene web slings shall not be used where fumes, vapors, sprays, mists orliquids of caustics are present.
- (iii) Web slings with aluminum fittings shall not be used where fumes, vapors, sprays, mists orliquids of caustics are present.

(7) **Safe operating temperatures.** Synthetic web slings of polyester and nylon shall not be used at temperatures in excess of 180 °F (82.2 °C). Polypropylene web slings shall not be used at temperatures in excess of 200 °F (93.33 °C).

(8) **Removal from service.** Synthetic web slings shall be immediately removed from service if any of thefollowing conditions are present:

- (i) Acid or caustic burns;
- (ii) Melting or charring of any part of the sling
- (iii) surface;Snags, punctures, tears or cuts;
- (iv) Broken or worn
- (v) stitches; or

(f) Distortion of

fittings.

Shackles and hooks.

(1) Employers must not use shackles with loads in excess of the rated capacities (i.e., working load limits) indicated on the shackle by permanently affixed and legible identification markings prescribed by the manufacturer.

(2) The manufacturer's recommendations shall be followed in determining the safe working loads of thevarious sizes and types of specific and identifiable hooks. All hooks for which no applicable manufacturer's recommendations are available shall be tested to twice the intended safe working load before they are initially put into use. The employer shall maintain a record of the dates and results of such tests.

Table H-1 - Maximum Allowable Wear at any Point of Link

Chain size (inches)	Maximum allowable wear (inch)
$1/4$	$3/64$
$3/8$	$5/64$
$1/2$	$7/64$
$5/8$	$9/64$
$3/4$	$5/32$
$7/8$	$11/64$
1	$3/16$
$1\,1/8$	$7/32$
$1\,1/4$	$1/4$
$1\,3/8$	$9/32$
$1\,1/2$	$5/16$
$1\,3/4$	$11/32$

Table H-2 - Number and Spacing of U-Bolt Wire Rope Clips

Improved plow steel, rope diameter (inches)	Number of clips		Minimum spacing (inches)
	Drop forged	Other material	
$1/2$	3	4	3
$5/8$	3	4	$3\,3/4$
$3/4$	4	5	$4\,1/2$
$7/8$	4	5	$5\,1/4$
1	5	6	6
$1\,1/8$	6	6	$6\,3/4$
$1\,1/4$	6	7	$7\,1/2$
$1\,3/8$	7	7	$8\,1/4$
$1\,1/2$	7	8	9

[44 FR 8577, Feb. 9, 1979; 44 FR 20940, Apr. 6, 1979, as amended at 58 FR 35173, June 30, 1993; 76 FR 33611, June 8, 2011; 77 FR 23118, Apr. 18, 2012]

§ 1926.252 Disposal of waste materials.

(a) Whenever materials are dropped more than 20 feet to any point lying outside the exterior walls of thebuilding, an enclosed chute of wood, or equivalent material, shall be used. For the purpose of this paragraph, an enclosed chute is a slide, closed in on all sides, through which material is moved from ahigh place to a lower one.

(b) When debris is dropped through holes in the floor without the use of chutes, the area onto which the material is dropped shall be completely enclosed with barricades not less than 42 inches high and not less than 6 feet back from the projected edge of the opening above. Signs warning of the hazard of falling materials shall be posted at each level. Removal shall not be permitted in this lower area until debris handling ceases above.

(c) All scrap lumber, waste material, and rubbish shall be removed from the immediate work area as the work progresses.

(d) Disposal of waste material or debris by burning shall comply with local fire regulations.

(e) All solvent waste, oily rags, and flammable liquids shall be kept in fire resistant covered containers until removed from worksite.

Subpart I - Tools - Hand and Power

Authority: Sections 4, 6, and 8 of the Occupational Safety and Health Act of 1970 (29 U.S.C. 653, 655, 657); Secretary of Labor's Order No. 12-71 (36 FR 8754), 8-76 (41 FR 25059), 9-83 (48 FR 35736), 1-90 (55 FR 9033), or

5-2002 (67 FR 65008), as applicable; and 29 CFR part 1911. Section 1926.307 also issued under 5 U.S.C. 553.

§ 1926.300 General requirements.

(a) ***Condition of tools.*** All hand and power tools and similar equipment, whether furnished by the employer or the employee, shall be maintained in a safe condition.

(b) ***Guarding.***

    (1) When power operated tools are designed to accommodate guards, they shall be equipped with such guards when in use.

    (2) Belts, gears, shafts, pulleys, sprockets, spindles, drums, fly wheels, chains, or other reciprocating, rotating or moving parts of equipment shall be guarded if such parts are exposed to contact by employees or otherwise create a hazard. Guarding shall meet the requirements as set forth in American National Standards Institute, B15.1-1953 (R1958), Safety Code for Mechanical Power-Transmission Apparatus.

    (3) ***Types of guarding.*** One or more methods of machine guarding shall be provided to protect the operator and other employees in the machine area from hazards such as those created by point of operation, ingoing nip points, rotating parts, flying chips and sparks. Examples of guarding methods are - barrier guards, two-hand tripping devices, electronic safety devices, etc.

    (4) ***Point of operation guarding.***

        (i) Point of operation is the area on a machine where work is actually performed upon the material being processed.

        (ii) The point of operation of machines whose operation exposes an employee to injury, shall be guarded. The guarding device shall be in conformity with any appropriate standards therefor, or in the absence of applicable specific standards, shall be so designed and constructed as to prevent the operator from having any part of his body in the danger zone during the operating cycle.

(iii) Special handtools for placing and removing material shall be such as to permit easy handling of material without the operator placing a hand in the danger zone. Such tools shall not be in lieu of other guarding required by this section, but can only be used to supplement protection provided.

(iv) The following are some of the machines which usually require point of operation guarding:

    (a) Guillotine cutters.

    (b) Shears.

    (c) Alligator

    (d) shears.

    (e) Power

        presses.

        Milling

    (f) machines.

    (g) Power

        saws.

        Jointers.

    (h) Portable power tools.

    (i) Forming rolls and

(5)   calenders.

Exposure of blades. When the periphery of the blades of a fan is less than 7 feet (2.128 m) above the floor or working level, the blades shall be guarded. The guard shall have openings no larger than $1/2$ inch (1.27 cm).

(6) **Anchoring fixed machinery.** Machines designed for a fixed location shall be securely anchored to prevent walking or moving.

(7) **Guarding of abrasive wheel machinery - exposure adjustment.** Safety guards of the types

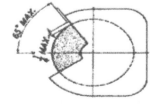

Figure I-1 Figure I-2

Correct

Showing adjustable tongue giving required angle protection for all sizes of wheel used.

described in paragraphs (b) (8) and (9) of this section, where the operator stands in front of the opening, shall be constructed so that the peripheral protecting member can be adjusted to the constantly decreasing diameter of the wheel. The maximum angular exposure above the horizontal plane of the wheel spindle as specified in paragraphs (b) (8) and (9) of this section shall never be exceeded, and the distance between the wheel periphery and the adjustable tongue or the end of the peripheral member at the top shall never exceed $1/4$ inch (0.635 cm). (See Figures I-1 through I-6.)

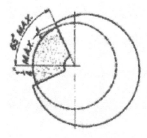 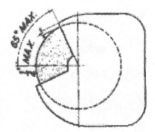

Figure I-3 Figure I-4

Correct

Showing movable guard with opening small enough to give required protection for the smallest size wheel used.

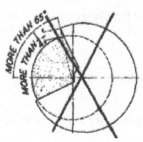 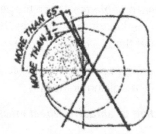

Figure I-5 Figure I-6

Incorrect

Showing movable guard with size of opening correct for full size wheel but too large for smaller wheel.

(8) **Bench and floor stands.** The angular exposure of the grinding wheel periphery and sides for safety guards used on machines known as bench and floor stands should not exceed 90° or one-fourth of the periphery. This exposure shall begin at a point not more than 65° above the horizontal plane of the wheel spindle. (See Figures I-7 and I-8 and paragraph (b)(7) of this section.)

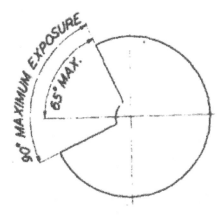 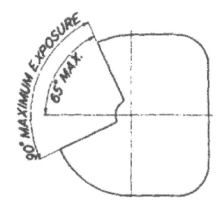

Figure I-7 Figure I-8

Wherever the nature of the work requires contact with the wheel below the horizontal plane of the spindle, the exposure shall not exceed 125° (See Figures I-9 and I-10.)

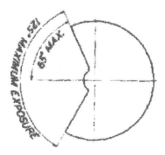 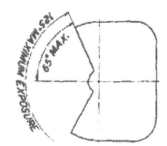

Figure I-9 Figure I-10

(9) **Cylindrical grinders.** The maximum angular exposure of the grinding wheel periphery and sides for safety guards used on cylindrical grinding machines shall not exceed 180°. This exposure shall begin at a point not more than 65° above the horizontal plane of the wheel spindle. (See Figures I-11 and
I-12 and paragraph (b)(7) of this section.)

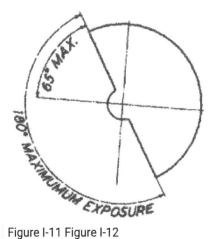 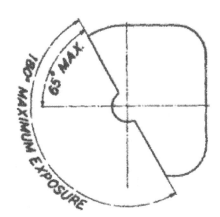

Figure I-11 Figure I-12

(c) ***Personal protective equipment.*** Employees using hand and power tools and exposed to the hazard of falling, flying, abrasive, and splashing objects, or exposed to harmful dusts, fumes, mists, vapors, or gases shall be provided with the particular personal protective equipment necessary to protect them from the hazard. All personal protective equipment shall meet the requirements and be maintained according to subparts D and E of this part.

(d) **Switches.**

    (1) All hand-held powered platen sanders, grinders with wheels 2-inch diameter or less, routers, planers, laminate trimmers, nibblers, shears, scroll saws, and jigsaws with blade shanks one-fourth of an inch wide or less may be equipped with only a positive "on-off" control.

    (2) All hand-held powered drills, tappers, fastener drivers, horizontal, vertical, and angle grinders with wheels greater than 2 inches in diameter, disc sanders, belt sanders, reciprocating saws, saber saws, and other similar operating powered tools shall be equipped with a momentary contact "on-off" control and may have a lock-on control provided that turnoff can be accomplished by a single motion of the same finger or fingers that turn it on.

    (3) All other hand-held powered tools, such as circular saws, chain saws, and percussion tools without positive accessory holding means, shall be equipped with a constant pressure switch that will shutoff the power when the pressure is released.

    (4) The requirements of this paragraph shall become effective on July 15, 1972.

    (5) Exception: This paragraph does not apply to concrete vibrators, concrete breakers, powered tampers, jack hammers, rock drills, and similar hand operated power tools.

[44 FR 8577, Feb. 9, 1979; 44 FR 20940, Apr. 6, 1979, as amended at 58 FR 35175, June 30, 1993; 61 FR 9250, Mar. 7, 1996]

§ 1926.301 Hand tools.

(a) Employers shall not issue or permit the use of unsafe hand tools.

(b) Wrenches, including adjustable, pipe, end, and socket wrenches shall not be used when jaws are sprung to the point that slippage occurs.

(c) Impact tools, such as drift pins, wedges, and chisels, shall be kept free of mushroomed heads.

(d) The wooden handles of tools shall be kept free of splinters or cracks and shall be kept tight in the tool.

§ 1926.302 Power-operated hand tools.

(a) ***Electric power-operated tools.***

    (1) Electric power operated tools shall either be of the approved double-insulated type or grounded in accordance with subpart K of this part.

    (2) The use of electric cords for hoisting or lowering tools shall not be permitted.

(b) ***Pneumatic power tools.***

    (1) Pneumatic power tools shall be secured to the hose or whip by some positive means to prevent the tool from becoming accidentally disconnected.

    (2) Safety clips or retainers shall be securely installed and maintained on pneumatic impact (percussion) tools to prevent attachments from being accidentally expelled.

(3) All pneumatically driven nailers, staplers, and other similar equipment provided with automatic fastener feed, which operate at more than 100 p.s.i. pressure at the tool shall have a safety device on the muzzle to prevent the tool from ejecting fasteners, unless the muzzle is in contact with the work surface.

(4) Compressed air shall not be used for cleaning purposes except where reduced to less than 30 p.s.i. and then only with effective chip guarding and personal protective equipment which meets the requirements of subpart E of this part. The 30 p.s.i. requirement does not apply for concrete form, mill scale and similar cleaning purposes.

(5) The manufacturer's safe operating pressure for hoses, pipes, valves, filters, and other fittings shall not be exceeded,

(6) The use of hoses for hoisting or lowering tools shall not be permitted.

(7) All hoses exceeding $1/2$-inch inside diameter shall have a safety device at the source of supply or branch line to reduce pressure in case of hose failure.

(8) Airless spray guns of the type which atomize paints and fluids at high pressures (1,000 pounds or more per square inch) shall be equipped with automatic or visible manual safety devices which will prevent pulling of the trigger to prevent release of the paint or fluid until the safety device is manually released.

(9) In lieu of the above, a diffuser nut which will prevent high pressure, high velocity release, while the nozzle tip is removed, plus a nozzle tip guard which will prevent the tip from coming into contact with the operator, or other equivalent protection, shall be provided.

(10) **Abrasive blast cleaning nozzles.** The blast cleaning nozzles shall be equipped with an operating valve which must be held open manually. A support shall be provided on which the nozzle may be mounted when it is not in use.

(c) **Fuel powered tools.**

(1) All fuel powered tools shall be stopped while being refueled, serviced, or maintained, and fuel shall be transported, handled, and stored in accordance with subpart F of this part.

(2) When fuel powered tools are used in enclosed spaces, the applicable requirements for concentrations of toxic gases and use of personal protective equipment, as outlined in subparts D and E of this part, shall apply.

(d) **Hydraulic power tools.**

(1) The fluid used in hydraulic powered tools shall be fire-resistant fluids approved under Schedule 30 of the U.S. Bureau of Mines, Department of the Interior, and shall retain its operating characteristics at the most extreme temperatures to which it will be exposed.

(2) The manufacturer's safe operating pressures for hoses, valves, pipes, filters, and other fittings shall not be exceeded.

(e) **Powder-actuated tools.**

(1) Only employees who have been trained in the operation of the particular tool in use shall be allowed to operate a powder-actuated tool.

(2) The tool shall be tested each day before loading to see that safety devices are in proper working condition. The method of testing shall be in accordance with the manufacturer's recommended procedure.

(3) Any tool found not in proper working order, or that develops a defect during use, shall be immediatelyremoved from service and not used until properly repaired.

(4) Personal protective equipment shall be in accordance with subpart E of this part.

(5) Tools shall not be loaded until just prior to the intended firing time. Neither loaded nor empty toolsare to be pointed at any employees. Hands shall be kept clear of the open barrel end.

(6) Loaded tools shall not be left unattended.

(7) Fasteners shall not be driven into very hard or brittle materials including, but not limited to, cast iron,glazed tile, surface-hardened steel, glass block, live rock, face brick, or hollow tile.

(8) Driving into materials easily penetrated shall be avoided unless such materials are backed by a substance that will prevent the pin or fastener from passing completely through and creating a flyingmissile hazard on the other side.

(9) No fastener shall be driven into a spalled area caused by an unsatisfactory

(10) fastening.Tools shall not be used in an explosive or flammable atmosphere.

(11) All tools shall be used with the correct shield, guard, or attachment recommended by the manufacturer.

(12) Powder-actuated tools used by employees shall meet all other applicable requirements of American National Standards Institute, A10.3-1970, Safety Requirements for Explosive-Actuated Fastening Tools.

[44 FR 8577, Feb. 9, 1979; 44 FR 20940, Apr. 6, 1979, as amended at 58 FR 35175, June 30, 1993]

§ 1926.303 Abrasive wheels and tools.

(a) **Power.** All grinding machines shall be supplied with sufficient power to maintain the spindle speed at safe levels under all conditions of normal operation.

(b) **Guarding.**

(1) Grinding machines shall be equipped with safety guards in conformance with the requirements of American National Standards Institute, B7.1-1970, Safety Code for the Use, Care and Protection ofAbrasive Wheels, and paragraph (d) of this section.

(2) **Guard design.** The safety guard shall cover the spindle end, nut, and flange projections. The safety guard shall be mounted so as to maintain proper alignment with the wheel, and the strength of thefastenings shall exceed the strength of the guard, except:

(i) Safety guards on all operations where the work provides a suitable measure of protection to theoperator, may be so constructed that the spindle end, nut, and outer flange are exposed; and where the nature of the work is such as to entirely cover the side of the wheel, the side covers of the guard may be omitted; and

(ii) The spindle end, nut, and outer flange may be exposed on machines designed as portablesaws.

(c) **Use of abrasive wheels.**

(1) Floor stand and bench mounted abrasive wheels, used for external grinding, shall be provided with safety guards (protection hoods). The maximum angular exposure of the grinding wheel periphery and sides shall be not more than 90°, except that when work requires contact with

the wheel below the horizontal plane of the spindle, the angular exposure shall not exceed 125°. In either case, the exposure shall begin not more than 65° above the horizontal plane of the spindle. Safety guards shall be strong enough to withstand the effect of a bursting wheel.

(2) Floor and bench-mounted grinders shall be provided with work rests which are rigidly supported and readily adjustable. Such work rests shall be kept at a distance not to exceed one-eighth inch from the surface of the wheel.

(3) Cup type wheels used for external grinding shall be protected by either a revolving cup guard or a band type guard in accordance with the provisions of the American National Standards Institute, B7.1-1970 Safety Code for the Use, Care, and Protection of Abrasive Wheels. All other portable abrasive wheels used for external grinding, shall be provided with safety guards (protection hoods) meeting the requirements of paragraph (c)(5) of this section, except as follows:

  (i) When the work location makes it impossible, a wheel equipped with safety flanges, as described in paragraph (c)(6) of this section, shall be used;

  (ii) When wheels 2 inches or less in diameter which are securely mounted on the end of a steel mandrel are used.

(4) Portable abrasive wheels used for internal grinding shall be provided with safety flanges (protection flanges) meeting the requirements of paragraph (c)(6) of this section, except as follows:

  (i) When wheels 2 inches or less in diameter which are securely mounted on the end of a steel mandrel are used;

  (ii) If the wheel is entirely within the work being ground while in use.

(5) When safety guards are required, they shall be so mounted as to maintain proper alignment with the wheel, and the guard and its fastenings shall be of sufficient strength to retain fragments of the wheel in case of accidental breakage. The maximum angular exposure of the grinding wheel periphery and sides shall not exceed 180°.

(6) When safety flanges are required, they shall be used only with wheels designed to fit the flanges. Only safety flanges, of a type and design and properly assembled so as to ensure that the pieces of the wheel will be retained in case of accidental breakage, shall be used.

(7) All abrasive wheels shall be closely inspected and ring-tested before mounting to ensure that they are free from cracks or defects.

(8) Grinding wheels shall fit freely on the spindle and shall not be forced on. The spindle nut shall be tightened only enough to hold the wheel in place.

(9) All employees using abrasive wheels shall be protected by eye protection equipment in accordance with the requirements of subpart E of this part, except when adequate eye protection is afforded by eye shields which are permanently attached to the bench or floor stand.

(d) **Other requirements.** All abrasive wheels and tools used by employees shall meet other applicable requirements of American National Standards Institute, B7.1-1970, Safety Code for the Use, Care and Protection of Abrasive Wheels.

(e) **Work rests.** On offhand grinding machines, work rests shall be used to support the work. They shall be of rigid construction and designed to be adjustable to compensate for wheel wear. Work rests shall be kept adjusted closely to the wheel with a maximum opening of $^1/_8$ inch (0.3175 cm) to prevent the work from being jammed between the wheel and the rest, which may cause wheel breakage. The work rest shall be securely clamped after each adjustment. The adjustment shall not be made with the wheel in motion.

[44 FR 8577, Feb. 9, 1979; 44 FR 20940, Apr. 6, 1979, as amended at 58 FR 35175, June 30, 1993]

§ 1926.304 Woodworking tools.

(a) **Disconnect switches.** All fixed power driven woodworking tools shall be provided with a disconnect switch that can either be locked or tagged in the off position.

(b) **Speeds.** The operating speed shall be etched or otherwise permanently marked on all circular saws over 20 inches in diameter or operating at over 10,000 peripheral feet per minute. Any saw so marked shall not be operated at a speed other than that marked on the blade. When a marked saw is retensioned for a different speed, the marking shall be corrected to show the new speed.

(c) **Self-feed.** Automatic feeding devices shall be installed on machines whenever the nature of the work will permit. Feeder attachments shall have the feed rolls or other moving parts covered or guarded so as to protect the operator from hazardous points.

(d) **Guarding.** All portable, power-driven circular saws shall be equipped with guards above and below the base plate or shoe. The upper guard shall cover the saw to the depth of the teeth, except for the minimum arc required to permit the base to be tilted for bevel cuts. The lower guard shall cover the saw to the depth of the teeth, except for the minimum arc required to allow proper retraction and contact with the work.
When the tool is withdrawn from the work, the lower guard shall automatically and instantly return to the covering position.

(e) **Personal protective equipment.** All personal protective equipment provided for use shall conform to subpart E of this part.

(f) **Other requirements.** All woodworking tools and machinery shall meet other applicable requirements of American National Standards Institute, 01.1-1961, Safety Code for Woodworking Machinery.

(g) **Radial saws.**

(1) The upper hood shall completely enclose the upper portion of the blade down to a point that will include the end of the saw arbor. The upper hood shall be constructed in such a manner and of such material that it will protect the operator from flying splinters, broken saw teeth, etc., and will deflect sawdust away from the operator. The sides of the lower exposed portion of the blade shall be guarded to the full diameter of the blade by a device that will automatically adjust itself to the thickness of the stock and remain in contact with stock being cut to give maximum protection possible for the operation being performed.

(h) **Hand-fed crosscut table saws.** (1) Each circular crosscut table saw shall be guarded by a hood which shall meet all the requirements of paragraph (i)(1) of this section for hoods for circular ripsaws.

(i) **Hand-fed ripsaws.**

(1) Each circular hand-fed ripsaw shall be guarded by a hood which shall completely enclose that portion of the saw above the table and that portion of the saw above the material being cut. The hood and mounting shall be arranged so that the hood will automatically adjust itself to the thickness of and remain in contact with the material being cut but it shall not offer any

considerable resistance to insertion of material to saw or to passage of the material being sawed. The hood shall be made of adequate strength to resist blows and strains incidental to reasonable operation, adjusting, and handling, and shall be so designed as to protect the operator from flying splinters and broken saw teeth. It shall be made of material that is soft enough so that it will be unlikely to cause tooth breakage. The hood shall be so mounted as to insure that its operation will be positive, reliable, and in true alignment with the saw; and the mounting shall be adequate in strength to resist any reasonable side thrust or other force tending to throw it out of line.

[44 FR 8577, Feb. 9, 1979; 44 FR 20940, Apr. 6, 1979, as amended at 58 FR 35175, June 30, 1993; 61 FR 9251, Mar. 7, 1996]

§ 1926.305 Jacks - lever and ratchet, screw, and hydraulic.

(a) *General requirements.*

 (1) The manufacturer's rated capacity shall be legibly marked on all jacks and shall not be

 (2) exceeded. All jacks shall have a positive stop to prevent overtravel.

(b) [Reserved]

(c) *Blocking.* When it is necessary to provide a firm foundation, the base of the jack shall be blocked or cribbed. Where there is a possibility of slippage of the metal cap of the jack, a wood block shall be placed between the cap and the load.

(d)

 (1) *Operation and maintenance.*

 (i) After the load has been raised, it shall be cribbed, blocked, or otherwise secured at once.

 (ii) Hydraulic jacks exposed to freezing temperatures shall be supplied with an adequate antifreeze liquid.

 (iii) All jacks shall be properly lubricated at regular intervals.

 (iv) Each jack shall be thoroughly inspected at times which depend upon the service conditions. Inspections shall be not less frequent than the following:

 (*a*) For constant or intermittent use at one locality, once every 6 months,

 (*b*) For jacks sent out of shop for special work, when sent out and when returned,

 (*c*) For a jack subjected to abnormal load or shock, immediately before and immediately thereafter.

 (v) Repair or replacement parts shall be examined for possible defects.

 (vi) Jacks which are out of order shall be tagged accordingly, and shall not be used until repairs are made.

[44 FR 8577, Feb. 9, 1979; 44 FR 20940, Apr. 6, 1979, as amended at 55 FR 42328, Oct. 18, 1990; 58 FR 35176, June 30, 1993]

§ 1926.306 Air receivers.

(a) *General requirements -*

 (1) *Application.* This section applies to compressed air receivers, and other equipment used in

providingand utilizing compressed air for performing operations such as cleaning, drilling, hoisting, and chipping. On the other hand, however, this section does not deal with the special problems createdby using compressed air to convey materials nor the problems created when men work in compressed air as in tunnels and caissons. This section is not intended to apply to compressed air machinery and equipment used on transportation vehicles such as steam railroad cars, electric railway cars, and automotive equipment.

(2) *New and existing equipment.*

(i) All new air receivers installed after the effective date of these regulations shall be constructedin accordance with the 1968 edition of the A.S.M.E. Boiler and Pressure Vessel Code SectionVIII.

(ii) All safety valves used shall be constructed, installed, and maintained in accordance with the A.S.M.E. Boiler and Pressure Vessel Code, Section VIII Edition 1968.

(b) *Installation and equipment requirements -*

(1) *Installation.* Air receivers shall be so installed that all drains, handholes, and manholes therein areeasily accessible. Under no circumstances shall an air receiver be buried underground or located inan inaccessible place.

(2) *Drains and traps.* A drain pipe and valve shall be installed at the lowest point of every air receiver to provide for the removal of accumulated oil and water. Adequate automatic traps may be installed inaddition to drain valves. The drain valve on the air receiver shall be opened and the receiver completely drained frequently and at such intervals as to prevent the accumulation of excessive amounts of liquid in the receiver.

(3) *Gages and valves.*

(i) Every air receiver shall be equipped with an indicating pressure gage (so located as to be readilyvisible) and with one or more spring-loaded safety valves. The total relieving capacity of such safety valves shall be such as to prevent pressure in the receiver from exceeding the maximum allowable working pressure of the receiver by more than 10 percent.

(ii) No valve of any type shall be placed between the air receiver and its safety valve or valves.

(iii) Safety appliances, such as safety valves, indicating devices and controlling devices, shall be constructed, located, and installed so that they cannot be readily rendered inoperative by anymeans, including the elements.

(iv) All safety valves shall be tested frequently and at regular intervals to determine whether theyare in good operating condition.

[58 FR 35176, June 30, 1993]

§ 1926.307 Mechanical power-transmission apparatus.

(a) *General requirements.*

(1) This section covers all types and shapes of power-transmission belts, except the following when operating at two hundred and fifty (250) feet per minute or less:

(i) Flat belts 1 inch (2.54 cm) or less in width,

- (ii) flat belts 2 inches (5.08 cm) or less in width which are free from metal lacings or
- (iii) fasteners, round belts $^1/_2$ inch (1.27 cm) or less in diameter; and
- (iv) single strand V-belts, the width of which is thirteen thirty-seconds ($^{13}/_{32}$) inch or less.

(2) Vertical and inclined belts (paragraphs (e) (3) and (4) of this section) if not more than $2^1/_2$ inches (6.35 cm) wide and running at a speed of less than one thousand (1,000) feet per minute, and if free from metal lacings or fastenings may be guarded with a nip-point belt and pulley guard.

(3) For the Textile Industry, because of the presence of excessive deposits of lint, which constitute a serious fire hazard, the sides and face sections only of nip-point belt and pulley guards are required, provided the guard shall extend at least 6 inches (15.24 cm) beyond the rim of the pulley on the in-running and off-running sides of the belt and at least 2 inches (5.08 cm) away from the rim and face of the pulley in all other directions.

(4) This section covers the principal features with which power transmission safeguards shall comply.

(b) *Prime-mover guards* -

(1) *Flywheels.* Flywheels located so that any part is 7 feet (2.128 m) or less above floor or platform shall be guarded in accordance with the requirements of this subparagraph:

- (i) With an enclosure of sheet, perforated, or expanded metal, or woven wire;
- (ii) With guard rails placed not less than 15 inches (38.1 cm) nor more than 20 inches (50.8 cm) from rim. When flywheel extends into pit or is within 12 inches (30.48 cm) of floor, a standard toeboard shall also be provided;
- (iii) When the upper rim of flywheel protrudes through a working floor, it shall be entirely enclosed or surrounded by a guardrail and toeboard.
- (iv) For flywheels with smooth rims 5 feet (1.52 m) or less in diameter, where the preceding methods cannot be applied, the following may be used: A disk attached to the flywheel in such manner as to cover the spokes of the wheel on the exposed side and present a smooth surface and edge, at the same time providing means for periodic inspection. An open space, not exceeding 4 inches (10.16 cm) in width, may be left between the outside edge of the disk and the rim of the wheel if desired, to facilitate turning the wheel over. Where a disk is used, the keys or other dangerous projections not covered by disk shall be cut off or covered. This subdivision does not apply to flywheels with solid web centers.
- (v) Adjustable guard to be used for starting engine or for running adjustment may be provided at the flywheel of gas or oil engines. A slot opening for jack bar will be permitted.
- (vi) Wherever flywheels are above working areas, guards shall be installed having sufficient strength to hold the weight of the flywheel in the event of a shaft or wheel mounting failure.

(2) *Cranks and connecting rods.* Cranks and connecting rods, when exposed to contact, shall be guarded in accordance with paragraphs (m) and (n) of this section, or by a guardrail as described in paragraph (o)(5) of this section.

(3) ***Tail rods or extension piston rods.*** Tail rods or extension piston rods shall be guarded in accordance with paragraphs (m) and (o) of this section, or by a guardrail on sides and end, with a clearance of not less than 15 (38.1 cm) nor more than 20 inches (50.8 cm) when rod is fully extended.

(c) ***Shafting*** -

   (1) ***Installation.***

      (i) Each continuous line of shafting shall be secured in position against excessive endwise movement.

      (ii) Inclined and vertical shafts, particularly inclined idler shafts, shall be securely held in position against endwise thrust.

   (2) ***Guarding horizontal shafting.***

      (i) All exposed parts of horizontal shafting 7 feet (2.128 m) or less from floor or working platform, excepting runways used exclusively for oiling, or running adjustments, shall be protected by a stationary casing enclosing shafting completely or by a trough enclosing sides and top or sides and bottom of shafting as location requires.

      (ii) Shafting under bench machines shall be enclosed by a stationary casing, or by a trough at sides and top or sides and bottom, as location requires. The sides of the trough shall come within at least 6 inches (15.24 cm) of the underside of table, or if shafting is located near floor within 6 inches (15.24 cm) of floor. In every case the sides of trough shall extend at least 2 inches (5.08 cm) beyond the shafting or protuberance.

   (3) ***Guarding vertical and inclined shafting.*** Vertical and inclined shafting 7 feet (2.128 m) or less from floor or working platform, excepting maintenance runways, shall be enclosed with a stationary casing in accordance with requirements of paragraphs (m) and (o) of this section.

   (4) ***Projecting shaft ends.***

      (i) Projecting shaft ends shall present a smooth edge and end and shall not project more than one-half the diameter of the shaft unless guarded by nonrotating caps or safety sleeves.

      (ii) Unused keyways shall be filled up or covered.

   (5) ***Power-transmission apparatus located in basements.*** All mechanical power transmission apparatus located in basements, towers, and rooms used exclusively for power transmission equipment shall be guarded in accordance with this section, except that the requirements for safeguarding belts, pulleys, and shafting need not be complied with when the following requirements are met:

      (i) The basement, tower, or room occupied by transmission equipment is locked against unauthorized entrance.

      (ii) The vertical clearance in passageways between the floor and power transmission beams, ceiling, or any other objects, is not less than 5 ft. 6 in. (1.672 m).

      (iii) The intensity of illumination conforms to the requirements of ANSI A11.1-1965 (R-1970).

      (iv) [Reserved]

      (v) The route followed by the oiler is protected in such manner as to prevent accident.

(d) *Pulleys* -

(1) *Guarding.* Pulleys, any parts of which are 7 feet (2.128 m) or less from the floor or working platform, shall be guarded in accordance with the standards specified in paragraphs (m) and (o) of this section. Pulleys serving as balance wheels (*e.g.,* punch presses) on which the point of contact between belt and pulley is more than 6 ft. 6 in. (1.976 m) from the floor or platform may be guarded with a disk covering the spokes.

(2) *Location of pulleys.*

(i) Unless the distance to the nearest fixed pulley, clutch, or hanger exceeds the width of the belt used, a guide shall be provided to prevent the belt from leaving the pulley on the side where insufficient clearance exists.

(ii) [Reserved]

(3) *Broken pulleys.* Pulleys with cracks, or pieces broken out of rims, shall not be used.

(4) *Pulley speeds.* Pulleys intended to operate at rim speed in excess of manufacturers normal recommendations shall be specially designed and carefully balanced for the speed at which they are to operate.

(e) *Belt, rope, and chain drives* -

(1) *Horizontal belts and ropes.*

(i) Where both runs of horizontal belts are 7 feet (2.128 m) or less from the floor level, the guard shall extend to at least 15 inches (38.1 cm) above the belt or to a standard height except that where both runs of a horizontal belt are 42 inches (106.68 cm) or less from the floor, the belt shall be fully enclosed.

(ii) In powerplants or power-development rooms, a guardrail may be used in lieu of the guard required by paragraph (e)(1)(i) of this section.

(2) *Overhead horizontal belts.*

(i) Overhead horizontal belts, with lower parts 7 feet (2.128 m) or less from the floor or platform, shall be guarded on sides and bottom in accordance with paragraph (o)(3) of this section.

(ii) Horizontal overhead belts more than 7 feet (2.128 m) above floor or platform shall be guarded for their entire length under the following conditions:

(*a*) If located over passageways or work places and traveling 1,800 feet or more per

(*b*) minute. If center to center distance between pulleys is 10 feet (3.04 m) or more.

(*c*) If belt is 8 inches (20.32 cm) or more in width.

(iii) Where the upper and lower runs of horizontal belts are so located that passage of persons between them would be possible, the passage shall be either:

(*a*) Completely barred by a guardrail or other barrier in accordance with paragraphs (m) and
(o) of this section; or

(*b*) Where passage is regarded as necessary, there shall be a platform over the lower run guarded on either side by a railing completely filled in with wire mesh or other filler, or by a solid barrier. The upper run shall be so guarded as to prevent contact therewith either by the worker or by objects carried by him. In powerplants only the lower run of the belt need be guarded.

  (iv) Overhead chain and link belt drives are governed by the same rules as overhead horizontal belts and shall be guarded in the same manner as belts.

(3) ***Vertical and inclined belts.***

  (i) Vertical and inclined belts shall be enclosed by a guard conforming to standards in paragraphs (m) and (o) of this section.

  (ii) All guards for inclined belts shall be arranged in such a manner that a minimum clearance of 7 feet (2.128 m) is maintained between belt and floor at any point outside of guard.

(4) ***Vertical belts.*** Vertical belts running over a lower pulley more than 7 feet (2.128 m) above floor or platform shall be guarded at the bottom in the same manner as horizontal overhead belts, if conditions are as stated in paragraphs (e)(2)(ii) (a) and (c) of this section.

(5) ***Cone-pulley belts.***

  (i) The cone belt and pulley shall be equipped with a belt shifter so constructed as to adequately guard the nip point of the belt and pulley. If the frame of the belt shifter does not adequately guard the nip point of the belt and pulley, the nip point shall be further protected by means of a vertical guard placed in front of the pulley and extending at least to the top of the largest step of the cone.

  (ii) If the belt is of the endless type or laced with rawhide laces, and a belt shifter is not desired, the belt will be considered guarded if the nip point of the belt and pulley is protected by a nip point guard located in front of the cone extending at least to the top of the largest step of the cone, and formed to show the contour of the cone in order to give the nip point of the belt and pulley the maximum protection.

  (iii) If the cone is located less than 3 feet (0.912 m) from the floor or working platform, the cone pulley and belt shall be guarded to a height of 3 feet (0.912 m) regardless of whether the belt is endless or laced with rawhide.

(6) ***Belt tighteners.***

  (i) Suspended counterbalanced tighteners and all parts thereof shall be of substantial construction and securely fastened; the bearings shall be securely capped. Means must be provided to prevent tightener from falling, in case the belt breaks.

  (ii) Where suspended counterweights are used and not guarded by location, they shall be so encased as to prevent accident.

(f) ***Gears, sprockets, and chains -***

 (1) ***Gears.*** Gears shall be guarded in accordance with one of the following

  (i) methods: By a complete enclosure; or

  (ii) By a standard guard as described in paragraph (o) of this section, at least 7 feet (2.128 m) high extending 6 inches (15.24 cm) above the mesh point of the gears; or

  (iii) By a band guard covering the face of gear and having flanges extended inward beyond the root of the teeth on the exposed side or sides. Where any portion of the train of gears guarded by a band guard is less than 6 feet (1.824 m) from the floor a disk guard or a complete enclosure tothe height of 6 feet (1.824 m) shall be required.

(2) **Hand-operated gears.** Paragraph (f)(1) of this section does not apply to hand-operated gears usedonly to adjust machine parts and which do not continue to move after hand power is removed.
However, the guarding of these gears is highly recommended.

(3) **Sprockets and chains.** All sprocket wheels and chains shall be enclosed unless they are more than 7feet (2.128 m) above the floor or platform. Where the drive extends over other machine or working areas, protection against falling shall be provided. This subparagraph does not apply to manually operated sprockets.

(4) **Openings for oiling.** When frequent oiling must be done, openings with hinged or sliding self-closing covers shall be provided. All points not readily accessible shall have oil feed tubes if lubricant is to be added while machinery is in motion.

(g) **Guarding friction drives.** The driving point of all friction drives when exposed to contact shall be guarded, all arm or spoke friction drives and all web friction drives with holes in the web shall be entirely enclosed, and all projecting belts on friction drives where exposed to contact shall be guarded.

(h) **Keys, setscrews, and other projections.**

(1) All projecting keys, setscrews, and other projections in revolving parts shall be removed or made flush or guarded by metal cover. This subparagraph does not apply to keys or setscrews within gearor sprocket casings or other enclosures, nor to keys, setscrews, or oilcups in hubs of pulleys less than 20 inches (50.8 cm) in diameter where they are within the plane of the rim of the pulley.

(2) It is recommended, however, that no projecting setscrews or oilcups be used in any revolving pulleyor part of machinery.

(i) **Collars and couplings -**

(1) **Collars.** All revolving collars, including split collars, shall be cylindrical, and screws or bolts used in collars shall not project beyond the largest periphery of the collar.

(2) **Couplings.** Shaft couplings shall be so constructed as to present no hazard from bolts, nuts, setscrews, or revolving surfaces. Bolts, nuts, and setscrews will, however, be permitted where they are covered with safety sleeves or where they are used parallel with the shafting and are countersunk or else do not extend beyond the flange of the coupling.

(j) **Bearings and facilities for oiling.** All drip cups and pans shall be securely fastened.

(k) *Guarding of clutches, cutoff couplings, and clutch pulleys -*

(1) **Guards.** Clutches, cutoff couplings, or clutch pulleys having projecting parts, where such clutches arelocated 7 feet (2.128 m) or less above the floor or working platform, shall be enclosed by a stationary guard constructed in accordance with this section. A "U" type guard is permissible.

(2) **Engine rooms.** In engine rooms a guardrail, preferably with toeboard, may be used instead of the guard required by paragraph (k)(1) of this section, provided such a room is occupied only by engineroom attendants.

(l) ***Belt shifters, clutches, shippers, poles, perches, and fasteners -***

   (1)   ***Belt shifters.***

      (i)   Tight and loose pulleys on all new installations made on or after August 31, 1971, shall be equipped with a permanent belt shifter provided with mechanical means to prevent belt fromcreeping from loose to tight pulley. It is recommended that old installations be changed to conform to this rule.

      (ii)   Belt shifter and clutch handles shall be rounded and be located as far as possible from danger of accidental contact, but within easy reach of the operator. Where belt shifters are not directlylocated over a machine or bench, the handles shall be cut off 6 ft. 6 in. (1.976 m) above floor level.

   (2)   ***Belt shippers and shipper poles.*** The use of belt poles as substitutes for mechanical shifters is notrecommended.

   (3)   ***Belt perches.*** Where loose pulleys or idlers are not practicable, belt perches in form of brackets,rollers, etc., shall be used to keep idle belts away from the shafts.

   (4)   ***Belt fasteners.*** Belts which of necessity must be shifted by hand and belts within 7 feet (2.128 m) ofthe floor or working platform which are not guarded in accordance with this section shall not be fastened with metal in any case, nor with any other fastening which by construction or wear will constitute an accident hazard.

(m)   ***Standard guards - general requirements -***

   (1)   ***Materials.***

      (i)   Standard conditions shall be secured by the use of the following materials. Expanded metal,perforated or solid sheet metal, wire mesh on a frame of angle iron, or iron pipe securely fastened to floor or to frame of machine.

      (ii)   All metal should be free from burrs and sharp edges.

   (2)   ***Methods of manufacture.***

      (i)   Expanded metal, sheet or perforated metal, and wire mesh shall be securely fastened to

(n)   frame.[Reserved]

(o)   ***Approved materials -***

   (1)   ***Minimum requirements.*** The materials and dimensions specified in this paragraph shall apply to allguards, except horizontal overhead belts, rope, cable, or chain guards more than 7 feet (2.128 m) above floor, or platform.

      (i)   [Reserved]

         (*a*)   All guards shall be rigidly braced every 3 feet (0.912 m) or fractional part of their height tosome fixed part of machinery or building structure. Where guard is exposed to contact with moving equipment additional strength may be necessary.

   (2)   ***Wood guards.***

      (i)   Wood guards may be used in the woodworking and chemical industries, in industries where thepresence of fumes or where manufacturing conditions would cause the rapid deterioration of metal guards; also in construction work and in locations outdoors where extreme cold or extreme heat make metal guards and railings undesirable. In all other industries, wood guards shall not be used.

(3) *Guards for horizontal overhead belts.*

(i) Guards for horizontal overhead belts shall run the entire length of the belt and follow the line ofthe pulley to the ceiling or be carried to the nearest wall, thus enclosing the belt effectively.
Where belts are so located as to make it impracticable to carry the guard to wall or ceiling, construction of guard shall be such as to enclose completely the top and bottom runs of beltand the face of pulleys.

(ii) [Reserved]

(iii) Suitable reinforcement shall be provided for the ceiling rafters or overhead floor beams, wheresuch is necessary, to sustain safely the weight and stress likely to be imposed by the guard.
The interior surface of all guards, by which is meant the surface of the guard with which a belt will come in contact, shall be smooth and free from all projections of any character, except where construction demands it; protruding shallow roundhead rivets may be used. Overhead belt guards shall be at least one-quarter wider than belt which they protect, except that this clearance need not in any case exceed 6 inches (15.24 cm) on each side. Overhead rope drive and block and roller-chain-drive guards shall be not less than 6 inches (15.24 cm) wider than the drive on each side. In overhead silent chain-drive guards where the chain is held from lateraldisplacement on the sprockets, the side clearances required on drives of 20 inch (50.8 cm) centers or under shall be not less than $1/4$ inch (0.635 cm) from the nearest moving chain part, and on drives of over 20 inch (50.8 cm) centers a minimum of $1/2$ inch (1.27 cm) from the nearest moving chain part.

(4) *Guards for horizontal overhead rope and chain drives.* Overhead-rope and chain-drive guard construction shall conform to the rules for overhead-belt guard.

(5) *Guardrails and toeboards.*

(i) Guardrail shall be 42 inches (106.68 cm) in height, with midrail between top rail and floor.

(ii) Posts shall be not more than 8 feet (2.432 m) apart; they are to be permanent and substantial, smooth, and free from protruding nails, bolts, and splinters. If made of pipe, the post shall be $1\frac{1}{4}$ inches (3.175 cm) inside diameter, or larger. If made of metal shapes or bars, their sectionshall be equal in strength to that of $1\frac{1}{2}$ (3.81 cm) by $1\frac{1}{2}$ (3.81 cm) by $3/16$ inch angle iron. If made of wood, the posts shall be two by four (2 × 4) inches or larger. The upper rail shall be twoby four (2 × 4) inches, or two one by four (1 × 4) strips, one at the top and one at the side of posts. The midrail may be one by four (1 × 4) inches or more. Where panels are fitted with expanded metal or wire mesh the middle rails may be omitted. Where guard is exposed to contact with moving equipment, additional strength may be necessary.

(iii) Toeboards shall be 4 inches (10.16 cm) or more in height, of wood, metal, or of metal grill notexceeding 1 inch (2.54 cm) mesh.

(p) *Care of equipment* -

(1) *General.* All power-transmission equipment shall be inspected at intervals not exceeding 60 days

andbe kept in good working condition at all times.

(2) **Shafting.**
- (i) Shafting shall be kept in alignment, free from rust and excess oil or grease.
- (ii) Where explosives, explosive dusts, flammable vapors or flammable liquids exist, the hazard ofstatic sparks from shafting shall be carefully considered.

(3) **Bearings.** Bearings shall be kept in alignment and properly adjusted.

(4) **Hangers.** Hangers shall be inspected to make certain that all supporting bolts and screws are tightand that supports of hanger boxes are adjusted properly.

(5) **Pulleys.**
- (i) Pulleys shall be kept in proper alignment to prevent belts from running off.

(6) **Care of belts.**
- (i) [Reserved]
- (ii) Inspection shall be made of belts, lacings, and fasteners and such equipment kept in goodrepair.

(7) **Lubrication.** The regular oilers shall wear tight-fitting clothing. Machinery shall be oiled when not in motion, wherever possible.

[58 FR 35176, June 30, 1993, as amended at 69 FR 31882, June 8, 2004]

Subpart J - Welding and Cutting

Authority: Sec. 107, Contract Work Hours and Safety Standards Act (Construction Safety Act) (40 U.S.C. 333); secs. 4, 6, 8, Occupational Safety and Health Act of 1970 (29 U.S.C. 653, 655, 657); Secretary of Labor's Order No.
12-71 (36 FR 8754), 8-76 (41 FR 25059), or 9-83 (48 FR 35736), as applicable.

§ 1926.350 Gas welding and cutting.

(a) *Transporting, moving, and storing compressed gas cylinders.*

- (1) Valve protection caps shall be in place and secured.
- (2) When cylinders are hoisted, they shall be secured on a cradle, slingboard, or pallet. They shall not behoisted or transported by means of magnets or choker slings.
- (3) Cylinders shall be moved by tilting and rolling them on their bottom edges. They shall not be intentionally dropped, struck, or permitted to strike each other violently.
- (4) When cylinders are transported by powered vehicles, they shall be secured in a vertical position.
- (5) Valve protection caps shall not be used for lifting cylinders from one vertical position to another.Bars shall not be used under valves or valve protection caps to pry cylinders loose when frozen. Warm, not boiling, water shall be used to thaw cylinders loose.
- (6) Unless cylinders are firmly secured on a special carrier intended for this purpose, regulators

shall beremoved and valve protection caps put in place before cylinders are moved.

(7) A suitable cylinder truck, chain, or other steadying device shall be used to keep cylinders from beingknocked over while in use.

(8) When work is finished, when cylinders are empty, or when cylinders are moved at any time, the cylinder valve shall be closed.

29 CFR Part 1926 (up to date as of 2/07/2023) Safety and Health Regulations for Construction

29 CFR 1926.350(a)(9)

(9) Compressed gas cylinders shall be secured in an upright position at all times except, if necessary, for short periods of time while cylinders are actually being hoisted or carried.

(10) Oxygen cylinders in storage shall be separated from fuel-gas cylinders or combustible materials (especially oil or grease), a minimum distance of 20 feet (6.1 m) or by a noncombustible barrier atleast 5 feet (1.5 m) high having a fire-resistance rating of at least one-half hour.

(11) Inside of buildings, cylinders shall be stored in a well-protected, well-ventilated, dry location, at least20 feet (6.1 m) from highly combustible materials such as oil or excelsior. Cylinders should be storedin definitely assigned places away from elevators, stairs, or gangways. Assigned storage places shall be located where cylinders will not be knocked over or damaged by passing or falling objects, or subject to tampering by unauthorized persons. Cylinders shall not be kept in unventilated enclosures such as lockers and cupboards.

(12) The in-plant handling, storage, and utilization of all compressed gases in cylinders, portable tanks, rail tankcars, or motor vehicle cargo tanks shall be in accordance with Compressed Gas AssociationPamphlet P-1-1965.

(b) *Placing cylinders.*

(1) Cylinders shall be kept far enough away from the actual welding or cutting operation so that sparks,hot slag, or flame will not reach them. When this is impractical, fire resistant shields shall be provided.

(2) Cylinders shall be placed where they cannot become part of an electrical circuit. Electrodes shall not be struck against a cylinder to strike an arc.

(3) Fuel gas cylinders shall be placed with valve end up whenever they are in use. They shall not beplaced in a location where they would be subject to open flame, hot metal, or other sources ofartificial heat.

(4) Cylinders containing oxygen or acetylene or other fuel gas shall not be taken into confined spaces.

(c) *Treatment of cylinders.*

(1) Cylinders, whether full or empty, shall not be used as rollers or supports.

(2) No person other than the gas supplier shall attempt to mix gases in a cylinder. No one except the owner of the cylinder or person authorized by him, shall refill a cylinder. No one shall use a cylinder's contents for purposes other than those intended by the supplier. All cylinders used shall meet the Department of Transportation requirements published in 49 CFR part 178, subpart C, Specificationfor Cylinders.

(3) No damaged or defective cylinder shall be used.

(d) **Use of fuel gas.** The employer shall thoroughly instruct employees in the safe use of fuel gas, as follows:

(1) Before a regulator to a cylinder valve is connected, the valve shall be opened slightly and closed immediately. (This action is generally termed "cracking" and is intended to clear the valve of dust or dirt that might otherwise enter the regulator.) The person cracking the valve shall stand to one side of the outlet, not in front of it. The valve of a fuel gas cylinder shall not be cracked where the gas would reach welding work, sparks, flame, or other possible sources of ignition.

(2) The cylinder valve shall always be opened slowly to prevent damage to the regulator. For quick closing, valves on fuel gas cylinders shall not be opened more than $1\frac{1}{2}$ turns. When a special wrench is required, it shall be left in position on the stem of the valve while the cylinder is in use so that the fuel gas flow can be shut off quickly in case of an emergency. In the case of manifolded or coupled cylinders, at least one such wrench shall always be available for immediate use. Nothing shall be placed on top of a fuel gas cylinder, when in use, which may damage the safety device or interfere with the quick closing of the valve.

(3) Fuel gas shall not be used from cylinders through torches or other devices which are equipped with shutoff valves without reducing the pressure through a suitable regulator attached to the cylinder valve or manifold.

(4) Before a regulator is removed from a cylinder valve, the cylinder valve shall always be closed and the gas released from the regulator.

(5) If, when the valve on a fuel gas cylinder is opened, there is found to be a leak around the valve stem, the valve shall be closed and the gland nut tightened. If this action does not stop the leak, the use of the cylinder shall be discontinued, and it shall be properly tagged and removed from the work area. In the event that fuel gas should leak from the cylinder valve, rather than from the valve stem, and the gas cannot be shut off, the cylinder shall be properly tagged and removed from the work area. If a regulator attached to a cylinder valve will effectively stop a leak through the valve seat, the cylinder need not be removed from the work area.

(6) If a leak should develop at a fuse plug or other safety device, the cylinder shall be removed from the work area.

(e) **Fuel gas and oxygen manifolds.**

(1) Fuel gas and oxygen manifolds shall bear the name of the substance they contain in letters at least 1-inch high which shall be either painted on the manifold or on a sign permanently attached to it.

(2) Fuel gas and oxygen manifolds shall be placed in safe, well ventilated, and accessible locations. They shall not be located within enclosed spaces.

(3) Manifold hose connections, including both ends of the supply hose that lead to the manifold, shall be such that the hose cannot be interchanged between fuel gas and oxygen manifolds and supply header connections. Adapters shall not be used to permit the interchange of hose. Hose connections shall be kept free of grease and oil.

(4) When not in use, manifold and header hose connections shall be capped.

(5) Nothing shall be placed on top of a manifold, when in use, which will damage the manifold or interfere with the quick closing of the valves.

(f) **Hose.**

(1) Fuel gas hose and oxygen hose shall be easily distinguishable from each other. The contrast may bemade by different colors or by surface characteristics readily distinguishable by the sense of touch. Oxygen and fuel gas hoses shall not be interchangeable. A single hose having more than one gas passage shall not be used.

(2) When parallel sections of oxygen and fuel gas hose are taped together, not more than 4 inches out of 12 inches shall be covered by tape.

(3) All hose in use, carrying acetylene, oxygen, natural or manufactured fuel gas, or any gas or substance which may ignite or enter into combustion, or be in any way harmful to employees, shall be inspectedat the beginning of each working shift. Defective hose shall be removed from service.

(4) Hose which has been subject to flashback, or which shows evidence of severe wear or damage, shallbe tested to twice the normal pressure to which it is subject, but in no case less than 300 p.s.i.
Defective hose, or hose in doubtful condition, shall not be used.

(5) Hose couplings shall be of the type that cannot be unlocked or disconnected by means of a straightpull without rotary motion.

(6) Boxes used for the storage of gas hose shall be ventilated.

(7) Hoses, cables, and other equipment shall be kept clear of passageways, ladders and stairs.

(g) *Torches.*

(1) Clogged torch tip openings shall be cleaned with suitable cleaning wires, drills, or other devices designed for such purpose.

(2) Torches in use shall be inspected at the beginning of each working shift for leaking shutoff valves, hose couplings, and tip connections. Defective torches shall not be used.

(3) Torches shall be lighted by friction lighters or other approved devices, and not by matches or fromhot work.

(h) *Regulators and gauges.* Oxygen and fuel gas pressure regulators, including their related gauges, shall be in proper working order while in use.

(i) *Oil and grease hazards.* Oxygen cylinders and fittings shall be kept away from oil or grease. Cylinders, cylinder caps and valves, couplings, regulators, hose, and apparatus shall be kept free from oil or greasysubstances and shall not be handled with oily hands or gloves. Oxygen shall not be directed at oily surfaces, greasy clothes, or within a fuel oil or other storage tank or vessel.

(j) *Additional rules.* For additional details not covered in this subpart, applicable technical portions of American National Standards Institute, Z49.1-1967, Safety in Welding and Cutting, shall apply.

[44 FR 8577, Feb. 9, 1979; 44 FR 20940, Apr. 6, 1979, as amended at 55 FR 42328, Oct. 18, 1990; 58 FR 35179, June 30, 1993]

§ 1926.351 Arc welding and cutting.

(a) *Manual electrode holders.*

(1) Only manual electrode holders which are specifically designed for arc welding and cutting, and are of a capacity capable of safely handling the maximum rated current required by the electrodes, shall be used.

(2) Any current-carrying parts passing through the portion of the holder which the arc welder or cuttergrips in his hand, and the outer surfaces of the jaws of the holder, shall be fully insulated against the maximum voltage encountered to ground.

(b) *Welding cables and connectors.*

(1) All arc welding and cutting cables shall be of the completely insulated, flexible type, capable of handling the maximum current requirements of the work in progress, taking into account the dutycycle under which the arc welder or cutter is working.

(2) Only cable free from repair or splices for a minimum distance of 10 feet from the cable end to which the electrode holder is connected shall be used, except that cables with standard insulated connectors or with splices whose insulating quality is equal to that of the cable are permitted.

(3) When it becomes necessary to connect or splice lengths of cable one to another, substantial insulated connectors of a capacity at least equivalent to that of the cable shall be used. If connections are effected by means of cable lugs, they shall be securely fastened together to givegood electrical contact, and the exposed metal parts of the lugs shall be completely insulated.

(4) Cables in need of repair shall not be used. When a cable, other than the cable lead referred to inparagraph (b)(2) of this section, becomes worn to the extent of exposing bare conductors, the portion thus exposed shall be protected by means of rubber and friction tape or other equivalentinsulation.

(c) *Ground returns and machine grounding.*

(1) A ground return cable shall have a safe current carrying capacity equal to or exceeding the specified maximum output capacity of the arc welding or cutting unit which it services. When a single ground return cable services more than one unit, its safe current-carrying capacity shall equal or exceed thetotal specified maximum output capacities of all the units which it services.

(2) Pipelines containing gases or flammable liquids, or conduits containing electrical circuits, shall not be used as a ground return. For welding on natural gas pipelines, the technical portions of regulations issued by the Department of Transportation, Office of Pipeline Safety, 49 CFR part 192, Minimum Federal Safety Standards for Gas Pipelines, shall apply.

(3) When a structure or pipeline is employed as a ground return circuit, it shall be determined that the required electrical contact exists at all joints. The generation of an arc, sparks, or heat at any pointshall cause rejection of the structures as a ground circuit.

(4) When a structure or pipeline is continuously employed as a ground return circuit, all joints shall be bonded, and periodic inspections shall be conducted to ensure that no condition of electrolysis or fire hazard exists by virtue of such use.

(5) The frames of all arc welding and cutting machines shall be grounded either through a third wire in the cable containing the circuit conductor or through a separate wire which is grounded at the source of the current. Grounding circuits, other than by means of the structure, shall be checked toensure that the circuit between the ground and the grounded power conductor has resistance low enough to permit sufficient current to flow to cause the fuse or circuit breaker to interrupt the current.

(6) All ground connections shall be inspected to ensure that they are mechanically strong and electrically adequate for the required current.

(d) **Operating instructions.** Employers shall instruct employees in the safe means of arc welding and cutting as follows:

   (1) When electrode holders are to be left unattended, the electrodes shall be removed and the holders shall be so placed or protected that they cannot make electrical contact with employees or conducting objects.

   (2) Hot electrode holders shall not be dipped in water; to do so may expose the arc welder or cutter to electric shock.

   (3) When the arc welder or cutter has occasion to leave his work or to stop work for any appreciable length of time, or when the arc welding or cutting machine is to be moved, the power supply switch to the equipment shall be opened.

   (4) Any faulty or defective equipment shall be reported to the

   (5) supervisor. See § 1926.406(c) for additional requirements.

(e) **Shielding.** Whenever practicable, all arc welding and cutting operations shall be shielded by noncombustible or flameproof screens which will protect employees and other persons working in the vicinity from the direct rays of the arc.

[44 FR 8577, Feb. 9, 1979; 44 FR 20940, Apr. 6, 1979, as amended at 51 FR 25318, July 11, 1986]

§ 1926.352 Fire prevention.

(a) When practical, objects to be welded, cut, or heated shall be moved to a designated safe location or, if the objects to be welded, cut, or heated cannot be readily moved, all movable fire hazards in the vicinity shall be taken to a safe place, or otherwise protected.

(b) If the object to be welded, cut, or heated cannot be moved and if all the fire hazards cannot be removed, positive means shall be taken to confine the heat, sparks, and slag, and to protect the immovable fire hazards from them.

(c) No welding, cutting, or heating shall be done where the application of flammable paints, or the presence of other flammable compounds, or heavy dust concentrations creates a hazard.

(d) Suitable fire extinguishing equipment shall be immediately available in the work area and shall be maintained in a state of readiness for instant use.

(e) When the welding, cutting, or heating operation is such that normal fire prevention precautions are not sufficient, additional personnel shall be assigned to guard against fire while the actual welding, cutting, or heating operation is being performed, and for a sufficient period of time after completion of the work to ensure that no possibility of fire exists. Such personnel shall be instructed as to the specific anticipated fire hazards and how the firefighting equipment provided is to be used.

(f) When welding, cutting, or heating is performed on walls, floors, and ceilings, since direct penetration of sparks or heat transfer may introduce a fire hazard to an adjacent area, the same precautions shall be taken on the opposite side as are taken on the side on which the welding is being performed.

(g) For the elimination of possible fire in enclosed spaces as a result of gas escaping through leaking or improperly closed torch valves, the gas supply to the torch shall be positively shut off at some point outside the enclosed space whenever the torch is not to be used or whenever the torch is left unattended for a substantial period of time, such as during the lunch period. Overnight and at the change of shifts, the

torch and hose shall be removed from the confined space. Open end fuel gas and oxygen hoses shall be immediately removed from enclosed spaces when they are disconnected from the torch or other gas-consuming device.

(h) Except when the contents are being removed or transferred, drums, pails, and other containers which contain or have contained flammable liquids shall be kept closed. Empty containers shall be removed to a safe area apart from hot work operations or open flames.

(i) Drums containers, or hollow structures which have contained toxic or flammable substances shall, before welding, cutting, or heating is undertaken on them, either be filled with water or thoroughly cleaned of such substances and ventilated and tested. For welding, cutting and heating on steel pipelines containing natural gas, the pertinent portions of regulations issued by the Department of Transportation, Office of Pipeline Safety, 49 CFR part 192, Minimum Federal Safety Standards for Gas Pipelines, shall apply.

(j) Before heat is applied to a drum, container, or hollow structure, a vent or opening shall be provided for the release of any built-up pressure during the application of heat.

§ 1926.353 Ventilation and protection in welding, cutting, and heating.

(a) *Mechanical ventilation.* For purposes of this section, mechanical ventilation shall meet the following requirements:

(1) Mechanical ventilation shall consist of either general mechanical ventilation systems or local exhaust systems.

(2) General mechanical ventilation shall be of sufficient capacity and so arranged as to produce the number of air changes necessary to maintain welding fumes and smoke within safe limits, as defined in subpart D of this part.

(3) Local exhaust ventilation shall consist of freely movable hoods intended to be placed by the welder or burner as close as practicable to the work. This system shall be of sufficient capacity and so arranged as to remove fumes and smoke at the source and keep the concentration of them in the breathing zone within safe limits as defined in subpart D of this part.

(4) Contaminated air exhausted from a working space shall be discharged into the open air or otherwise clear of the source of intake air.

(5) All air replacing that withdrawn shall be clean and respirable.

(6) Oxygen shall not be used for ventilation purposes, comfort cooling, blowing dust from clothing, or for cleaning the work area.

(b) *Welding, cutting, and heating in confined spaces.*

(1) Except as provided in paragraph (b)(2) of this section, and paragraph (c)(2) of this section, either general mechanical or local exhaust ventilation meeting the requirements of paragraph (a) of this section shall be provided whenever welding, cutting, or heating is performed in a confined space.

(2) When sufficient ventilation cannot be obtained without blocking the means of access, employees in the confined space shall be protected by air line respirators in accordance with the requirements of subpart E of this part, and an employee on the outside of such a confined

space shall be assigned tomaintain communication with those working within it and to aid them in an emergency.

(3) **Lifelines.** Where a welder must enter a confined space through a manhole or other small opening, means shall be provided for quickly removing him in case of emergency. When safety belts and lifelines are used for this purpose they shall be so attached to the welder's body that his body cannotbe jammed in a small exit opening. An attendant with a pre-planned rescue procedure shall be stationed outside to observe the welder at all times and be capable of putting rescue operations into effect.

(c) *Welding, cutting, or heating of metals of toxic significance.*

(1) Welding, cutting, or heating in any enclosed spaces involving the metals specified in this subparagraph shall be performed with either general mechanical or local exhaust ventilation meetingthe requirements of paragraph (a) of this section:

   (i) Zinc-bearing base or filler metals or metals coated with zinc-bearing
   (ii) materials; Lead base metals;
   (iii) Cadmium-bearing filler materials;
   (iv) Chromium-bearing metals or metals coated with chromium-bearing materials.

(2) Welding, cutting, or heating in any enclosed spaces involving the metals specified in this subparagraph shall be performed with local exhaust ventilation in accordance with the requirements of paragraph (a) of this section, or employees shall be protected by air line respirators in accordancewith the requirements of subpart E of this part:

   (i) Metals containing lead, other than as an impurity, or metals coated with lead-bearing
   (ii) materials; Cadmium-bearing or cadmium-coated base metals;
   (iii) Metals coated with mercury-bearing metals;
   (iv) Beryllium-containing base or filler metals. Because of its high toxicity, work involving beryllium shall be done with both local exhaust ventilation and air line respirators.

(3) Employees performing such operations in the open air shall be protected by filter-type respirators in accordance with the requirements of subpart E of this part, except that employees performing suchoperations on beryllium-containing base or filler metals shall be protected by air line respirators inaccordance with the requirements of subpart E of this part.

(4) Other employees exposed to the same atmosphere as the welders or burners shall be protected inthe same manner as the welder or burner.

(d) *Inert-gas metal-arc welding.*

(1) Since the inert-gas metal-arc welding process involves the production of ultra-violet radiation of intensities of 5 to 30 times that produced during shielded metal-arc welding, the decomposition ofchlorinated solvents by ultraviolet rays, and the liberation of toxic fumes and gases, employees shall not be permitted to engage in, or be exposed to the process until the following special precautions have been taken:

   (i) The use of chlorinated solvents shall be kept at least 200 feet, unless shielded, from the exposed arc, and surfaces prepared with chlorinated solvents shall be thoroughly dry beforewelding is permitted on such surfaces.

(ii) Employees in the area not protected from the arc by screening shall be protected by filter lenses meeting the requirements of subpart E of this part. When two or more welders are exposed to each other's arc, filter lens goggles of a suitable type, meeting the requirements of subpart E of this part, shall be worn under welding helmets. Hand shields to protect the welder against flashes and radiant energy shall be used when either the helmet is lifted or the shield is removed.

(iii) Welders and other employees who are exposed to radiation shall be suitably protected so that the skin is covered completely to prevent burns and other damage by ultraviolet rays. Welding helmets and hand shields shall be free of leaks and openings, and free of highly reflective surfaces.

(iv) When inert-gas metal-arc welding is being performed on stainless steel, the requirements of paragraph (c)(2) of this section shall be met to protect against dangerous concentrations of nitrogen dioxide.

(e) *General welding, cutting, and heating.*

(1) Welding, cutting, and heating, not involving conditions or materials described in paragraph (b), (c), or (d) of this section, may normally be done without mechanical ventilation or respiratory protective equipment, but where, because of unusual physical or atmospheric conditions, an unsafe accumulation of contaminants exists, suitable mechanical ventilation or respiratory protective equipment shall be provided.

(2) Employees performing any type of welding, cutting, or heating shall be protected by suitable eye protective equipment in accordance with the requirements of subpart E of this part.

[

§ 1926.354 Welding, cutting, and heating in way of preservative coatings.

(a) Before welding, cutting, or heating is commenced on any surface covered by a preservative coating whose flammability is not known, a test shall be made by a competent person to determine its flammability.
Preservative coatings shall be considered to be highly flammable when scrapings burn with extreme rapidity.

(b) Precautions shall be taken to prevent ignition of highly flammable hardened preservative coatings. When coatings are determined to be highly flammable, they shall be stripped from the area to be heated to prevent ignition.

(c) Protection against toxic preservative coatings:

(1) In enclosed spaces, all surfaces covered with toxic preservatives shall be stripped of all toxic coatings for a distance of at least 4 inches from the area of heat application, or the employees shall be protected by air line respirators, meeting the requirements of subpart E of this part.

(2) In the open air, employees shall be protected by a respirator, in accordance with requirements of subpart E of this part.

(d) The preservative coatings shall be removed a sufficient distance from the area to be heated to ensure that the temperature of the unstripped metal will not be appreciably raised. Artificial cooling of the metal surrounding the heating area may be used to limit the size of the area required to be cleaned.

Subpart K - Electrical

Authority: 29 U.S.C. 653, 655, 657; 40 U.S.C. 333; Secretary of Labor's Order No. 9-83 (48 FR 35736), 1-90 (55 FR

9033) or 1-2012 (77 FR 3912), as applicable; 29 CFR part 1911.

Source: 51 FR 25318, July 11, 1986, unless otherwise noted.

GENERAL

§ 1926.400 Introduction.

This subpart addresses electrical safety requirements that are necessary for the practical safeguarding of employees involved in construction work and is divided into four major divisions and applicable definitions as follows:

(a) **Installation safety requirements.** Installation safety requirements are contained in §§ 1926.402 through 1926.408. Included in this category are electric equipment and installations used to provide electric power and light on jobsites.

(b) **Safety-related work practices.** Safety-related work practices are contained in §§ 1926.416 and 1926.417. In addition to covering the hazards arising from the use of electricity at jobsites, these regulations also cover the hazards arising from the accidental contact, direct or indirect, by employees with all energized lines, above or below ground, passing through or near the jobsite.

(c) **Safety-related maintenance and environmental considerations.** Safety-related maintenance and environmental considerations are contained in §§ 1926.431 and 1926.432.

(d) **Safety requirements for special equipment.** Safety requirements for special equipment are contained in § 1926.441.

(e) **Definitions.** Definitions applicable to this subpart are contained in § 1926.449.

§ 1926.401 [Reserved]

INSTALLATION SAFETY REQUIREMENTS

§ 1926.402 Applicability.

(a) **Covered.** Sections 1926.402 through 1926.408 contain installation safety requirements for electrical equipment and installations used to provide electric power and light at the jobsite. These sections apply to installations, both temporary and permanent, used on the jobsite; but these sections do not apply to existing permanent installations that were in place before the construction activity commenced.

Note: If the electrical installation is made in accordance with the National Electrical Code ANSI/NFPA 70-1984, exclusive of Formal Interpretations and Tentative Interim Amendments, it will be deemed to be in compliance with §§ 1926.403 through 1926.408, except for §§ 1926.404(b)(1) and 1926.405(a)(2)(ii) (E), (F), (G), and (J).

(b) **Not covered.** Sections 1926.402 through 1926.408 do not cover installations used for the generation, transmission, and distribution of electric energy, including related communication, metering, control, and transformation installations. (However, these regulations do cover portable and vehicle-mounted generators used to provide power for equipment used at the jobsite.) See subpart V of this part for the construction of power distribution and transmission lines.

§ 1926.403 General requirements.

(a) *Approval.* All electrical conductors and equipment shall be approved.

(b) *Examination, installation, and use of equipment* -

    (1) *Examination.* The employer shall ensure that electrical equipment is free from recognized hazards that are likely to cause death or serious physical harm to employees. Safety of equipment shall be determined on the basis of the following considerations:

        (i) Suitability for installation and use in conformity with the provisions of this subpart. Suitability of equipment for an identified purpose may be evidenced by listing, labeling, or certification for that identified purpose.

        (ii) Mechanical strength and durability, including, for parts designed to enclose and protect other equipment, the adequacy of the protection thus provided.

        (iii) Electrical insulation.

        (iv) Heating effects under conditions of

        (v) use. Arcing effects.

        (vi) Classification by type, size, voltage, current capacity, specific use.

        (vii) Other factors which contribute to the practical safeguarding of employees using or likely to come in contact with the equipment.

    (2) *Installation and use.* Listed, labeled, or certified equipment shall be installed and used in accordance with instructions included in the listing, labeling, or certification.

(c) *Interrupting rating.* Equipment intended to break current shall have an interrupting rating at system voltage sufficient for the current that must be interrupted.

(d) *Mounting and cooling of equipment* -

    (1) *Mounting.* Electric equipment shall be firmly secured to the surface on which it is mounted. Wooden plugs driven into holes in masonry, concrete, plaster, or similar materials shall not be used.

    (2) *Cooling.* Electrical equipment which depends upon the natural circulation of air and convection principles for cooling of exposed surfaces shall be installed so that room air flow over such surfaces is not prevented by walls or by adjacent installed equipment. For equipment designed for floor mounting, clearance between top surfaces and adjacent surfaces shall be provided to dissipate rising warm air. Electrical equipment provided with ventilating openings shall be installed so that walls or other obstructions do not prevent the free circulation of air through the equipment.

(e) *Splices.* Conductors shall be spliced or joined with splicing devices designed for the use or by brazing, welding, or soldering with a fusible metal or alloy. Soldered splices shall first be so spliced or joined as to be mechanically and electrically secure without solder and then soldered. All splices and joints and the free ends of conductors shall be covered with an insulation equivalent to that of the conductors or with an insulating device designed for the purpose.

(f) **_Arcing parts._** Parts of electric equipment which in ordinary operation produce arcs, sparks, flames, or molten metal shall be enclosed or separated and isolated from all combustible material.

(g) **_Marking._** Electrical equipment shall not be used unless the manufacturer's name, trademark, or other descriptive marking by which the organization responsible for the product may be identified is placed on the equipment and unless other markings are provided giving voltage, current, wattage, or other ratings as necessary. The marking shall be of sufficient durability to withstand the environment involved.

(h) **_Identification of disconnecting means and circuits._** Each disconnecting means required by this subpart for motors and appliances shall be legibly marked to indicate its purpose, unless located and arranged so the purpose is evident. Each service, feeder, and branch circuit, at its disconnecting means or overcurrent device, shall be legibly marked to indicate its purpose, unless located and arranged so the purpose is evident. These markings shall be of sufficient durability to withstand the environment involved.

(i) **_600 Volts, nominal, or less._** This paragraph applies to equipment operating at 600 volts, nominal, or less.

   (1) **_Working space about electric equipment._** Sufficient access and working space shall be provided and maintained about all electric equipment to permit ready and safe operation and maintenance of such equipment.

      (i) **_Working clearances._** Except as required or permitted elsewhere in this subpart, the dimension of the working space in the direction of access to live parts operating at 600 volts or less and likely to require examination, adjustment, servicing, or maintenance while alive shall not be less than indicated in Table K-1. In addition to the dimensions shown in Table K-1, workspace shall not be less than 30 inches (762 mm) wide in front of the electric equipment. Distances shall be measured from the live parts if they are exposed, or from the enclosure front or opening if the live parts are enclosed. Walls constructed of concrete, brick, or tile are considered to be grounded. Working space is not required in back of assemblies such as dead-front switchboards or motor control centers where there are no renewable or adjustable parts such as fuses or switches on the back and where all connections are accessible from locations other than the back.

Table K-1 - Working Clearances

Nominal voltage to ground	Minimum clear distance for conditions[1]		
	(a)	(b)	(c)
	Feet[2]	Feet[2]	Feet[2]
0-150	3	3	3
151-600	3	3½	4

[1] Conditions (a), (b), and (c) are as follows: (a) Exposed live parts on one side and no live or grounded parts on the other side of the working space, or exposed live parts on both sides effectively guarded by insulating material. Insulated wire or insulated busbars operating at not

over 300 volts are not considered live parts. (b) Exposed live parts on one side and grounded parts on the other side. (c) Exposed live parts on both sides of the workspace [not guarded asprovided in Condition (a)] with the operator between.

[2] Note: For International System of Units (SI): one foot = 0.3048m.

- (ii) **Clear spaces.** Working space required by this subpart shall not be used for storage. When normally enclosed live parts are exposed for inspection or servicing, the working space, if in apassageway or general open space, shall be guarded.

- (iii) **Access and entrance to working space.** At least one entrance shall be provided to give access tothe working space about electric equipment.

- (iv) **Front working space.** Where there are live parts normally exposed on the front of switchboards or motor control centers, the working space in front of such equipment shall not be less than 3feet (914 mm).

- (v) **Headroom.** The minimum headroom of working spaces about service equipment, switchboards,panelboards, or motor control centers shall be 6 feet 3 inches (1.91 m).

(2) **Guarding of live parts.**

- (i) Except as required or permitted elsewhere in this subpart, live parts of electric equipment operating at 50 volts or more shall be guarded against accidental contact by cabinets or otherforms of enclosures, or by any of the following means:

    - (A) By location in a room, vault, or similar enclosure that is accessible only to qualified persons.

    - (B) By partitions or screens so arranged that only qualified persons will have access to the space within reach of the live parts. Any openings in such partitions or screens shall be sosized and located that persons are not likely to come into accidental contact with the liveparts or to bring conducting objects into contact with them.

    - (C) By location on a balcony, gallery, or platform so elevated and arranged as to exclude unqualified persons.

    - (D) By elevation of 8 feet (2.44 m) or more above the floor or other working surface and so installed as to exclude unqualified persons.

- (ii) In locations where electric equipment would be exposed to physical damage, enclosures or guards shall be so arranged and of such strength as to prevent such damage.

- (iii) Entrances to rooms and other guarded locations containing exposed live parts shall be markedwith conspicuous warning signs forbidding unqualified persons to enter.

(j) **Over 600 volts, nominal -**

(1) **General.** Conductors and equipment used on circuits exceeding 600 volts, nominal, shall comply withall applicable provisions of paragraphs (a) through (g) of this section and with the following provisions which supplement or modify those requirements. The provisions of paragraphs (j)(2), (j)(3), and (j)(4) of this section do not apply to equipment on the supply side of the service conductors.

(2) **Enclosure for electrical installations.** Electrical installations in a vault, room, closet or in an area surrounded by a wall, screen, or fence, access to which is controlled by lock and key or other equivalent means, are considered to be accessible to qualified persons only. A wall,

screen, or fence less than 8 feet (2.44 m) in height is not considered adequate to prevent access unless it has other features that provide a degree of isolation equivalent to an 8-foot (2.44-m) fence. The entrances to all buildings, rooms or enclosures containing exposed live parts or exposed conductors operating atover 600 volts, nominal, shall be kept locked or shall be under the observation of a qualified personat all times.

(i) **Installations accessible to qualified persons only.** Electrical installations having exposed live parts shall be accessible to qualified persons only and shall comply with the applicable provisions of paragraph (j)(3) of this section.

(ii) **Installations accessible to unqualified persons.** Electrical installations that are open to unqualified persons shall be made with metal-enclosed equipment or shall be enclosed in a vault or in an area, access to which is controlled by a lock. Metal-enclosed switchgear, unit substations, transformers, pull boxes, connection boxes, and other similar associated equipment shall be marked with appropriate caution signs. If equipment is exposed to physicaldamage from vehicular traffic, guards shall be provided to prevent such damage. Ventilating or similar openings in metal-enclosed equipment shall be designed so that foreign objects inserted through these openings will be deflected from energized parts.

(3) **Workspace about equipment.** Sufficient space shall be provided and maintained about electric equipment to permit ready and safe operation and maintenance of such equipment. Where energized parts are exposed, the minimum clear workspace shall not be less than 6 feet 6 inches (1.98 m) high (measured vertically from the floor or platform), or less than 3 feet (914 mm) wide(measured parallel to the equipment). The depth shall be as required in Table K-2. The workspaceshall be adequate to permit at least a 90-degree opening of doors or hinged panels.

(i) **Working space.** The minimum clear working space in front of electric equipment such as switchboards, control panels, switches, circuit breakers, motor controllers, relays, and similar equipment shall not be less than specified in Table K-2 unless otherwise specified in this subpart. Distances shall be measured from the live parts if they are exposed, or from the enclosure front or opening if the live parts are enclosed. However, working space is not requiredin back of equipment such as deadfront switchboards or control assemblies where there are no renewable or adjustable parts (such as fuses or switches) on the back and where all connections are accessible from locations other than the back. Where rear access is required to work on de-energized parts on the back of enclosed equipment, a minimum working space of 30 inches (762 mm) horizontally shall be provided.

Table K-2 - Minimum Depth of Clear Working Space in Front of ElectricEquipment

Nominal voltage to ground	Conditions[1]		
	(a)	(b)	(c)
	Feet[2]	Feet[2]	Feet[2]
601 to 2,500	3	4	5
2,501 to 9,000	4	5	6

Nominal voltage to ground	Conditions[1]		
	(a)	(b)	(c)
9,001 to 25,000	5	6	9
25,001 to 75 kV	6	8	10
Above 75kV	8	10	12

[1]Conditions (a), (b), and (c) are as follows: (a) Exposed live parts on one side and no live or grounded parts on the other side of the working space, or exposed live parts on both sides effectively guarded by insulating materials. Insulated wire or insulated busbars operating at notover 300 volts are not considered live parts. (b) Exposed live parts on one side and grounded parts on the other side. Walls constructed of concrete, brick, or tile are considered to be grounded surfaces. (c) Exposed live parts on both sides of the workspace [not guarded as provided in Condition (a)] with the operator between.

[2] NOTE: For SI units: one foot = 0.3048 m.

(ii) *Lighting outlets and points of control.* The lighting outlets shall be so arranged that persons changing lamps or making repairs on the lighting system will not be endangered by live parts or other equipment. The points of control shall be so located that persons are not likely to come incontact with any live part or moving part of the equipment while turning on the lights.

(iii) *Elevation of unguarded live parts.* Unguarded live parts above working space shall be maintained at elevations not less than specified in Table K-3.

Table K-3 - Elevation of Unguarded Energized Parts Above Working Space

Nominal voltage between phases	Minimum elevation
601-7,500	8 feet 6 inches.[1]
7,501-35,000	9 feet.
Over 35kV	9 feet + 0.37 inches per kV above 35kV.

[1] NOTE: For SI units: one inch = 25.4 mm; one foot = 0.3048 m.

(4) *Entrance and access to workspace.* At least one entrance not less than 24 inches (610 mm) wide and 6 feet 6 inches (1.98 m) high shall be provided to give access to the working space about electric equipment. On switchboard and control panels exceeding 48 inches (1.22 m) in width, there shall be one entrance at each end of such board where practicable. Where bare energized parts at any voltage or insulated energized parts above 600 volts are located adjacent to such entrance, they shall be guarded.

[51 FR 25318, July 11, 1986, as amended at 61 FR 5510, Feb. 13, 1996]

§ 1926.404 Wiring design and protection.

(a) *Use and identification of grounded and grounding conductors -*

(1) **Identification of conductors.** A conductor used as a grounded conductor shall be identifiable and distinguishable from all other conductors. A conductor used as an equipment grounding conductorshall be identifiable and distinguishable from all other conductors.

(2) **Polarity of connections.** No grounded conductor shall be attached to any terminal or lead so as toreverse designated polarity.

(3) **Use of grounding terminals and devices.** A grounding terminal or grounding-type device on a receptacle, cord connector, or attachment plug shall not be used for purposes other than grounding.

(b) *Branch circuits* -

(1) *Ground-fault protection* -

(i) **General.** The employer shall use either ground fault circuit interrupters as specified in paragraph(b)(1)(ii) of this section or an assured equipment grounding conductor program as specified inparagraph (b)(1)(iii) of this section to protect employees on construction sites. These requirements are in addition to any other requirements for equipment grounding conductors.

(ii) **Ground-fault circuit interrupters.** All 120-volt, single-phase, 15- and 20-ampere receptacle outlets on construction sites, which are not a part of the permanent wiring of the building or structure and which are in use by employees, shall have approved ground-fault circuit interrupters for personnel protection. Receptacles on a two-wire, single-phase portable or vehicle-mounted generator rated not more than 5kW, where the circuit conductors of the generator are insulated from the generator frame and all other grounded surfaces, need not beprotected with ground-fault circuit interrupters.

(iii) **Assured equipment grounding conductor program.** The employer shall establish and implement an assured equipment grounding conductor program on construction sites covering all cord sets, receptacles which are not a part of the building or structure, and equipment connected bycord and plug which are available for use or used by employees. This program shall comply with the following minimum requirements:

(A) A written description of the program, including the specific procedures adopted by theemployer, shall be available at the jobsite for inspection and copying by the Assistant Secretary and any affected employee.

(B) The employer shall designate one or more competent persons (as defined in § 1926.32(f))to implement the program.

(C) Each cord set, attachment cap, plug and receptacle of cord sets, and any equipment connected by cord and plug, except cord sets and receptacles which are fixed and not exposed to damage, shall be visually inspected before each day's use for external defects,such as deformed or missing pins or insulation damage, and for indications of possible internal damage. Equipment found damaged or defective shall not be used until repaired.

(D) The following tests shall be performed on all cord sets, receptacles which are not a part of the permanent wiring of the building or structure, and cord- and plug-connected equipment required to be grounded:

(*1*) All equipment grounding conductors shall be tested for continuity and shall be electrically continuous.

(2) Each receptacle and attachment cap or plug shall be tested for correct attachment of the equipment grounding conductor. The equipment grounding conductor shall be connected to its proper terminal.

(E) All required tests shall be

(1) performed: Before first

(2) use;

Before equipment is returned to service following any repairs;

(3) Before equipment is used after any incident which can be reasonably suspected to have caused damage (for example, when a cord set is run over); and

(4) At intervals not to exceed 3 months, except that cord sets and receptacles which are fixed and not exposed to damage shall be tested at intervals not exceeding 6 months.

(F) The employer shall not make available or permit the use by employees of any equipment which has not met the requirements of this paragraph (b)(1)(iii) of this section.

(G) Tests performed as required in this paragraph shall be recorded. This test record shall identify each receptacle, cord set, and cord- and plug-connected equipment that passed the test and shall indicate the last date it was tested or the interval for which it was tested. This record shall be kept by means of logs, color coding, or other effective means and shall be maintained until replaced by a more current record. The record shall be made available on the jobsite for inspection by the Assistant Secretary and any affected employee.

(2) **Outlet devices.** Outlet devices shall have an ampere rating not less than the load to be served and shall comply with the following:

(i) **Single receptacles.** A single receptacle installed on an individual branch circuit shall have an ampere rating of not less than that of the branch circuit.

(ii) **Two or more receptacles.** Where connected to a branch circuit supplying two or more receptacles or outlets, receptacle ratings shall conform to the values listed in Table K-4.

(iii) **Receptacles used for the connection of motors.** The rating of an attachment plug or receptacle used for cord- and plug-connection of a motor to a branch circuit shall not exceed 15 amperes at 125 volts or 10 amperes at 250 volts if individual overload protection is omitted.

Table K-4 - Receptacle Ratings for Various Size Circuits

Circuit rating amperes	Receptacle rating amperes
15	Not over 15.
20	15 or 20.
30	30.
40	40 or 50.
50	50.

(c) *Outside conductors and lamps* -

   (1) ***600 volts, nominal, or less.*** Paragraphs (c)(1)(i) through (c)(1)(iv) of this section apply to branch circuit, feeder, and service conductors rated 600 volts, nominal, or less and run outdoors as openconductors.

   (i) ***Conductors on poles.*** Conductors supported on poles shall provide a horizontal climbing spacenot less than the following:

   (A) Power conductors below communication conductors - 30 inches (762 mm).

   (B) Power conductors alone or above communication conductors: 300 volts or less - 24inches (610 mm); more than 300 volts - 30 inches (762 mm).

   (C) Communication conductors below power conductors: with power conductors 300 volts orless - 24 inches (610 mm); more than 300 volts - 30 inches (762 mm).

   (ii) ***Clearance from ground.*** Open conductors shall conform to the following minimum clearances:

   (A) 10 feet (3.05 m) - above finished grade, sidewalks, or from any platform or projection fromwhich they might be reached.

   (B) 12 feet (3.66 m) - over areas subject to vehicular traffic other than truck traffic.

   (C) 15 feet (4.57 m) - over areas other than those specified in paragraph (c)(1)(ii)(D) of thissection that are subject to truck traffic.

   (D) 18 feet (5.49 m) - over public streets, alleys, roads, and driveways.

   (iii) ***Clearance from building openings.*** Conductors shall have a clearance of at least 3 feet (914 mm) from windows, doors, fire escapes, or similar locations. Conductors run above the top level of a window are considered to be out of reach from that window and, therefore, do nothave to be 3 feet (914 mm) away.

   (iv) ***Clearance over roofs.*** Conductors above roof space accessible to employees on foot shall have a clearance from the highest point of the roof surface of not less than 8 feet (2.44 m) vertical clearance for insulated conductors, not less than 10 feet (3.05 m) vertical or diagonal clearancefor covered conductors, and not less than 15 feet (4.57 m) for bare conductors, except that:

   (A) Where the roof space is also accessible to vehicular traffic, the vertical clearance shall notbe less than 18 feet (5.49 m), or

   (B) Where the roof space is not normally accessible to employees on foot, fully insulated conductors shall have a vertical or diagonal clearance of not less than 3 feet (914 mm), or

   (C) Where the voltage between conductors is 300 volts or less and the roof has a slope of notless than 4 inches (102 mm) in 12 inches (305 mm), the clearance from roofs shall be atleast 3 feet (914 mm), or

   (D) Where the voltage between conductors is 300 volts or less and the conductors do not passover more than 4 feet (1.22 m) of the overhang portion of the roof and they are terminated at a through-the-roof raceway or support, the clearance from roofs shall be at least 18 inches (457 mm).

(2) **Location of outdoor lamps.** Lamps for outdoor lighting shall be located below all live conductors, transformers, or other electric equipment, unless such equipment is controlled by a disconnecting means that can be locked in the open position or unless adequate clearances or other safeguards are provided for relamping operations.

(d) **Services** -

(1) **connecting means** -

    (i) **General.** Means shall be provided to disconnect all conductors in a building or other structure from the service-entrance conductors. The disconnecting means shall plainly indicate whether it is in the open or closed position and shall be installed at a readily accessible location nearest the point of entrance of the service-entrance conductors.

    (ii) **Simultaneous opening of poles.** Each service disconnecting means shall simultaneously disconnect all ungrounded conductors.

(2) **Services over 600 volts, nominal.** The following additional requirements apply to services over 600 volts, nominal.

    (i) **Guarding.** Service-entrance conductors installed as open wires shall be guarded to make them accessible only to qualified persons.

    (ii) **Warning signs.** Signs warning of high voltage shall be posted where unauthorized employees might come in contact with live parts.

(e) **Overcurrent protection** -

(1) **600 volts, nominal, or less.** The following requirements apply to overcurrent protection of circuits rated 600 volts, nominal, or less.

    (i) **Protection of conductors and equipment.** Conductors and equipment shall be protected from overcurrent in accordance with their ability to safely conduct current. Conductors shall have sufficient ampacity to carry the load.

    (ii) **Grounded conductors.** Except for motor-running overload protection, overcurrent devices shall not interrupt the continuity of the grounded conductor unless all conductors of the circuit are opened simultaneously.

    (iii) **Disconnection of fuses and thermal cutouts.** Except for devices provided for current-limiting on the supply side of the service disconnecting means, all cartridge fuses which are accessible to other than qualified persons and all fuses and thermal cutouts on circuits over 150 volts to ground shall be provided with disconnecting means. This disconnecting means shall be installed so that the fuse or thermal cutout can be disconnected from its supply without disrupting service to equipment and circuits unrelated to those protected by the overcurrent device.

    (iv) **Location in or on premises.** Overcurrent devices shall be readily accessible. Overcurrent devices shall not be located where they could create an employee safety hazard by being exposed to physical damage or located in the vicinity of easily ignitible material.

    (v) **Arcing or suddenly moving parts.** Fuses and circuit breakers shall be so located or shielded that employees will not be burned or otherwise injured by their operation.

    (vi) *cuit breakers*

- (A) Circuit breakers shall clearly indicate whether they are in the open (off) or closed (on) position.
- (B) Where circuit breaker handles on switchboards are operated vertically rather than horizontally or rotationally, the up position of the handle shall be the closed (on) position.
- (C) If used as switches in 120-volt, fluorescent lighting circuits, circuit breakers shall be marked "SWD."

(2) **Over 600 volts, nominal.** Feeders and branch circuits over 600 volts, nominal, shall have short-circuit protection.

(f) **Grounding.** Paragraphs (f)(1) through (f)(11) of this section contain grounding requirements for systems, circuits, and equipment.

- (1) **Systems to be grounded.** The following systems which supply premises wiring shall be grounded:
  - (i) **Three-wire DC systems.** All 3-wire DC systems shall have their neutral conductor grounded.
  - (ii) **Two-wire DC systems.** Two-wire DC systems operating at over 50 volts through 300 volts between conductors shall be grounded unless they are rectifier-derived from an AC system complying with paragraphs (f)(1)(iii), (f)(1)(iv), and (f)(1)(v) of this section.
  - (iii) **AC circuits, less than 50 volts.** AC circuits of less than 50 volts shall be grounded if they are installed as overhead conductors outside of buildings or if they are supplied by transformersand the transformer primary supply system is ungrounded or exceeds 150 volts to ground.
  - (iv) **AC systems, 50 volts to 1000 volts.** AC systems of 50 volts to 1000 volts shall be grounded under any of the following conditions, unless exempted by paragraph (f)(1)(v) of this section:
    - (A) If the system can be so grounded that the maximum voltage to ground on the ungroundedconductors does not exceed 150 volts;
    - (B) If the system is nominally rated 480Y/277 volt, 3-phase, 4-wire in which the neutral is usedas a circuit conductor;
    - (C) If the system is nominally rated 240/120 volt, 3-phase, 4-wire in which the midpoint of onephase is used as a circuit conductor; or
    - (D) If a service conductor is uninsulated.
  - (v) **Exceptions.** AC systems of 50 volts to 1000 volts are not required to be grounded if the systemis separately derived and is supplied by a transformer that has a primary voltage rating less than 1000 volts, provided all of the following conditions are met:
    - (A) The system is used exclusively for control circuits,
    - (B) The conditions of maintenance and supervision assure that only qualified persons willservice the installation,
    - (C) Continuity of control power is required, and
    - (D) Ground detectors are installed on the control system.

(2) **Separately derived systems.** Where paragraph (f)(1) of this section requires grounding of wiring systems whose power is derived from generator, transformer, or converter windings and has no direct electrical connection, including a solidly connected grounded circuit conductor, to supply conductors originating in another system, paragraph (f)(5) of this section shall also apply.

(3) **Portable and vehicle-mounted generators** -

   (i) **Portable generators.** Under the following conditions, the frame of a portable generator need not be grounded and may serve as the grounding electrode for a system supplied by the generator:

   (A) The generator supplies only equipment mounted on the generator and/or cord- and plug- connected equipment through receptacles mounted on the generator, and

   (B) The noncurrent-carrying metal parts of equipment and the equipment grounding conductor terminals of the receptacles are bonded to the generator frame.

   (ii) **Vehicle-mounted generators.** Under the following conditions the frame of a vehicle may serve as the grounding electrode for a system supplied by a generator located on the vehicle:

   (A) The frame of the generator is bonded to the vehicle frame, and

   (B) The generator supplies only equipment located on the vehicle and/or cord- and plug- connected equipment through receptacles mounted on the vehicle or on the generator, and

   (C) The noncurrent-carrying metal parts of equipment and the equipment grounding conductor terminals of the receptacles are bonded to the generator frame, and

   (D) The system complies with all other provisions of this section.

   (iii) **Neutral conductor bonding.** A neutral conductor shall be bonded to the generator frame if the generator is a component of a separately derived system. No other conductor need be bonded to the generator frame.

(4) **Conductors to be grounded.** For AC premises wiring systems the identified conductor shall be grounded.

(5) **Grounding connections** -

   (i) **Grounded system.** For a grounded system, a grounding electrode conductor shall be used to connect both the equipment grounding conductor and the grounded circuit conductor to the grounding electrode. Both the equipment grounding conductor and the grounding electrode conductor shall be connected to the grounded circuit conductor on the supply side of the service disconnecting means, or on the supply side of the system disconnecting means or overcurrent devices if the system is separately derived.

   (ii) **Ungrounded systems.** For an ungrounded service-supplied system, the equipment grounding conductor shall be connected to the grounding electrode conductor at the service equipment. For an ungrounded separately derived system, the equipment grounding conductor shall be connected to the grounding electrode conductor at, or ahead of, the system disconnecting means or overcurrent devices.

(6) *Grounding path.* The path to ground from circuits, equipment, and enclosures shall be permanent and continuous.

(7) ***Supports, enclosures, and equipment to be grounded -***

    (i) *Supports and enclosures for conductors.* Metal cable trays, metal raceways, and metal enclosures for conductors shall be grounded, except that:

        (A) Metal enclosures such as sleeves that are used to protect cable assemblies from physical damage need not be grounded; and

        (B) Metal enclosures for conductors added to existing installations of open wire, knob-and-tube wiring, and nonmetallic-sheathed cable need not be grounded if all of the following conditions are met:

            (*1*) Runs are less than 25 feet (7.62 m);

            (*2*) Enclosures are free from probable contact with ground, grounded metal, metal laths, or other conductive materials; and

            (*3*) Enclosures are guarded against employee contact.

    (ii) *Service equipment enclosures.* Metal enclosures for service equipment shall be grounded.

    (iii) *Fixed equipment.* Exposed noncurrent-carrying metal parts of fixed equipment which may become energized shall be grounded under any of the following conditions:

        (A) If within 8 feet (2.44 m) vertically or 5 feet (1.52 m) horizontally of ground or grounded metal objects and subject to employee contact.

        (B) If located in a wet or damp location and subject to employee

        (C) contact. If in electrical contact with metal.

        (D) If in a hazardous (classified) location.

        (E) If supplied by a metal-clad, metal-sheathed, or grounded metal raceway wiring method.

        (F) If equipment operates with any terminal at over 150 volts to ground; however, the following need not be grounded:

            (*1*) Enclosures for switches or circuit breakers used for other than service equipment and accessible to qualified persons only;

            (*2*) Metal frames of electrically heated appliances which are permanently and effectively insulated from ground; and

            (*3*) The cases of distribution apparatus such as transformers and capacitors mounted on wooden poles at a height exceeding 8 feet (2.44 m) above ground or grade level.

    (iv) *Equipment connected by cord and plug.* Under any of the conditions described in paragraphs (f)(7)(iv)(A) through (f)(7)(iv)(C) of this section, exposed noncurrent-carrying metal parts of cord- and plug-connected equipment which may become energized shall be grounded:

        (A) If in a hazardous (classified) location (see § 1926.407).

(B) If operated at over 150 volts to ground, except for guarded motors and metal frames of electrically heated appliances if the appliance frames are permanently and effectively insulated from ground.

(C) If the equipment is one of the types listed in paragraphs (f)(7)(iv)(C)(1) through (f)(7)(iv)(C)(5) of this section. However, even though the equipment may be one of these types, it need not be grounded if it is exempted by paragraph (f)(7)(iv)(C)(6).

- (1) Hand held motor-operated tools;
- (2) Cord- and plug-connected equipment used in damp or wet locations or by employees standing on the ground or on metal floors or working inside of metal tanks or boilers;
- (3) Portable and mobile X-ray and associated equipment;
- (4) Tools likely to be used in wet and/or conductive locations; and
- (5) Portable hand lamps.
- (6) Tools likely to be used in wet and/or conductive locations need not be grounded if supplied through an isolating transformer with an ungrounded secondary of not over 50 volts. Listed or labeled portable tools and appliances protected by a system of double insulation, or its equivalent, need not be grounded. If such a system is employed, the equipment shall be distinctively marked to indicate that the tool or appliance utilizes a system of double insulation.

(v) **Nonelectrical equipment.** The metal parts of the following nonelectrical equipment shall be grounded: Frames and tracks of electrically operated cranes; frames of nonelectrically driven elevator cars to which electric conductors are attached; hand-operated metal shifting ropes or cables of electric elevators, and metal partitions, grill work, and similar metal enclosures around equipment of over 1kV between conductors.

(8) **Methods of grounding equipment -**

(i) **With circuit conductors.** Noncurrent-carrying metal parts of fixed equipment, if required to be grounded by this subpart, shall be grounded by an equipment grounding conductor which is contained within the same raceway, cable, or cord, or runs with or encloses the circuit conductors. For DC circuits only, the equipment grounding conductor may be run separately from the circuit conductors.

(ii) **Grounding conductor.** A conductor used for grounding fixed or movable equipment shall have capacity to conduct safely any fault current which may be imposed on it.

(iii) **Equipment considered effectively grounded.** Electric equipment is considered to be effectively grounded if it is secured to, and in electrical contact with, a metal rack or structure that is provided for its support and the metal rack or structure is grounded by the method specified for the noncurrent-carrying metal parts of fixed equipment in paragraph (f)(8)(i) of this section.
Metal car frames supported by metal hoisting cables attached to or running over metal sheaves or drums of grounded elevator machines are also considered to be effectively grounded.

(9) **Bonding.** If bonding conductors are used to assure electrical continuity, they shall have the

capacity to conduct any fault current which may be imposed.

(10) **Made electrodes.** If made electrodes are used, they shall be free from nonconductive coatings, such as paint or enamel; and, if practicable, they shall be embedded below permanent moisture level. A single electrode consisting of a rod, pipe or plate which has a resistance to ground greater than 25 ohms shall be augmented by one additional electrode installed no closer than 6 feet (1.83 m) to the first electrode.

(11) **Grounding of systems and circuits of 1000 volts and over (high voltage) -**

(i) **General.** If high voltage systems are grounded, they shall comply with all applicable provisions of paragraphs (f)(1) through (f)(10) of this section as supplemented and modified by this paragraph (f)(11).

(ii) **Grounding of systems supplying portable or mobile equipment.** Systems supplying portable or mobile high voltage equipment, other than substations installed on a temporary basis, shall comply with the following:

(A) Portable and mobile high voltage equipment shall be supplied from a system having its neutral grounded through an impedance. If a delta-connected high voltage system is used to supply the equipment, a system neutral shall be derived.

(B) Exposed noncurrent-carrying metal parts of portable and mobile equipment shall be connected by an equipment grounding conductor to the point at which the system neutral impedance is grounded.

(C) Ground-fault detection and relaying shall be provided to automatically de-energize any high voltage system component which has developed a ground fault. The continuity of the equipment grounding conductor shall be continuously monitored so as to de-energize automatically the high voltage feeder to the portable equipment upon loss of continuity of the equipment grounding conductor.

(D) The grounding electrode to which the portable or mobile equipment system neutral impedance is connected shall be isolated from and separated in the ground by at least 20 feet (6.1 m) from any other system or equipment grounding electrode, and there shall be no direct connection between the grounding electrodes, such as buried pipe, fence or like objects.

(iii) **Grounding of equipment.** All noncurrent-carrying metal parts of portable equipment and fixed equipment including their associated fences, housings, enclosures, and supporting structures shall be grounded. However, equipment which is guarded by location and isolated from ground need not be grounded. Additionally, pole-mounted distribution apparatus at a height exceeding 8 feet (2.44 m) above ground or grade level need not be grounded.

[51 FR 25318, July 11, 1986, as amended at 54 FR 24334, June 7, 1989; 61 FR 5510, Feb. 13, 1996]

§ 1926.405 Wiring methods, components, and equipment for general use.

(a) **Wiring methods.** The provisions of this paragraph do not apply to conductors which form an integral part of equipment such as motors, controllers, motor control centers and like equipment.

(1) **General requirements -**

(i) ***Electrical continuity of metal raceways and enclosures.*** Metal raceways, cable armor, and other metal enclosures for conductors shall be metallically joined together into a continuous electric conductor and shall be so connected to all boxes, fittings, and cabinets as to provide effective electrical continuity.

(ii) ***Wiring in ducts.*** No wiring systems of any type shall be installed in ducts used to transport dust, loose stock or flammable vapors. No wiring system of any type shall be installed in any duct used for vapor removal or in any shaft containing only such ducts.

(2) **Temporary wiring** -

    (i) ***Scope.*** The provisions of paragraph (a)(2) of this section apply to temporary electrical power and lighting wiring methods which may be of a class less than would be required for a permanent installation. Except as specifically modified in paragraph (a)(2) of this section, all other requirements of this subpart for permanent wiring shall apply to temporary wiring installations. Temporary wiring shall be removed immediately upon completion of construction or the purpose for which the wiring was installed.

    (ii) ***General requirements for temporary wiring.***

        (A) Feeders shall originate in a distribution center. The conductors shall be run as multiconductor cord or cable assemblies or within raceways; or, where not subject to physical damage, they may be run as open conductors on insulators not more than 10 feet (3.05 m) apart.

        (B) Branch circuits shall originate in a power outlet or panelboard. Conductors shall be run as multiconductor cord or cable assemblies or open conductors, or shall be run in raceways. All conductors shall be protected by overcurrent devices at their ampacity. Runs of open conductors shall be located where the conductors will not be subject to physical damage, and the conductors shall be fastened at intervals not exceeding 10 feet (3.05 m). No branch-circuit conductors shall be laid on the floor. Each branch circuit that supplies receptacles or fixed equipment shall contain a separate equipment grounding conductor if the branch circuit is run as open conductors.

        (C) Receptacles shall be of the grounding type. Unless installed in a complete metallic raceway, each branch circuit shall contain a separate equipment grounding conductor, and all receptacles shall be electrically connected to the grounding conductor. Receptacles for uses other than temporary lighting shall not be installed on branch circuits which supply temporary lighting. Receptacles shall not be connected to the same ungrounded conductor of multiwire circuits which supply temporary lighting.

        (D) Disconnecting switches or plug connectors shall be installed to permit the disconnection of all ungrounded conductors of each temporary circuit.

        (E) All lamps for general illumination shall be protected from accidental contact or breakage. Metal-case sockets shall be grounded.

        (F) Temporary lights shall not be suspended by their electric cords unless cords and lights are designed for this means of suspension.

        (G) Portable electric lighting used in wet and/or other conductive locations, as for example, drums, tanks, and vessels, shall be operated at 12 volts or less. However,

120-volt lights may be used if protected by a ground-fault circuit interrupter.

- (H) A box shall be used wherever a change is made to a raceway system or a cable system which is metal clad or metal sheathed.

- (I) Flexible cords and cables shall be protected from damage. Sharp corners and projections shall be avoided. Flexible cords and cables may pass through doorways or other pinch points, if protection is provided to avoid damage.

- (J) Extension cord sets used with portable electric tools and appliances shall be of three-wire type and shall be designed for hard or extra-hard usage. Flexible cords used with temporary and portable lights shall be designed for hard or extra-hard usage.

Note: The National Electrical Code, ANSI/NFPA 70, in Article 400, Table 400-4, lists various types of flexible cords, some of which are noted as being designed for hard or extra-hard usage. Examples of these types of flexible cords include hard service cord (types S, ST, SO, STO) and junior hard service cord (types SJ, SJO, SJT, SJTO).

(iii) *Guarding.* For temporary wiring over 600 volts, nominal, fencing, barriers, or other effective means shall be provided to prevent access of other than authorized and qualified personnel.

(b) *Cabinets, boxes, and fittings -*

(1) *Conductors entering boxes, cabinets, or fittings.* Conductors entering boxes, cabinets, or fittings shall be protected from abrasion, and openings through which conductors enter shall be effectively closed. Unused openings in cabinets, boxes, and fittings shall also be effectively closed.

(2) *Covers and canopies.* All pull boxes, junction boxes, and fittings shall be provided with covers. If metal covers are used, they shall be grounded. In energized installations each outlet box shall have a cover, faceplate, or fixture canopy. Covers of outlet boxes having holes through which flexible cord pendants pass shall be provided with bushings designed for the purpose or shall have smooth, well-rounded surfaces on which the cords may bear.

(3) *Pull and junction boxes for systems over 600 volts, nominal.* In addition to other requirements in this section for pull and junction boxes, the following shall apply to these boxes for systems over 600 volts, nominal:

(i) *Complete enclosure.* Boxes shall provide a complete enclosure for the contained conductors or cables.

(ii) *Covers.* Boxes shall be closed by covers securely fastened in place. Underground box covers that weigh over 100 pounds (43.6 kg) meet this requirement. Covers for boxes shall be permanently marked "HIGH VOLTAGE." The marking shall be on the outside of the box cover and shall be readily visible and legible.

(c) *Knife switches.* Single-throw knife switches shall be so connected that the blades are dead when the switch is in the open position. Single-throw knife switches shall be so placed that gravity will not tend to close them. Single-throw knife switches approved for use in the inverted position shall be provided with a locking device that will ensure that the blades remain in the open position when so set. Double-throw knife switches may be mounted so that the throw will be either vertical or horizontal. However, if the throw is vertical, a locking device shall be provided to ensure that the blades remain in the open position when so set.

(d) **Switchboards and panelboards.** Switchboards that have any exposed live parts shall be located in permanently dry locations and accessible only to qualified persons. Panelboards shall be mounted in cabinets, cutout boxes, or enclosures designed for the purpose and shall be dead front. However, panelboards other than the dead front externally-operable type are permitted where accessible only to qualified persons. Exposed blades of knife switches shall be dead when open.

(e) **Enclosures for damp or wet locations** -

 (1) **Cabinets, fittings, and boxes.** Cabinets, cutout boxes, fittings, boxes, and panelboard enclosures indamp or wet locations shall be installed so as to prevent moisture or water from entering and accumulating within the enclosures. In wet locations the enclosures shall be weatherproof.

 (2) **Switches and circuit breakers.** Switches, circuit breakers, and switchboards installed in wet locations shall be enclosed in weatherproof enclosures.

(f) **Conductors for general wiring.** All conductors used for general wiring shall be insulated unless otherwise permitted in this subpart. The conductor insulation shall be of a type that is suitable for the voltage, operating temperature, and location of use. Insulated conductors shall be distinguishable by appropriate color or other means as being grounded conductors, ungrounded conductors, or equipment grounding conductors.

(g) **Flexible cords and cables** -

 (1) **Use of flexible cords and cables** -

  (i) **Permitted uses.** Flexible cords and cables shall be suitable for conditions of use and location.Flexible cords and cables shall be used only for:

   (A) Pendants;

   (B) Wiring of fixtures;

   (C) Connection of portable lamps or

   (D) appliances;Elevator cables;

   (E) Wiring of cranes and hoists;

   (F) Connection of stationary equipment to facilitate their frequent

   (G) interchange;Prevention of the transmission of noise or vibration; or

   (H) Appliances where the fastening means and mechanical connections are designed topermit removal for maintenance and repair.

  (ii) **Attachment plugs for cords.** If used as permitted in paragraphs (g)(1)(i)(C), (g)(1)(i)(F), or (g)(1)(i)(H) of this section, the flexible cord shall be equipped with an attachment plug and shallbe energized from a receptacle outlet.

  (iii) **Prohibited uses.** Unless necessary for a use permitted in paragraph (g)(1)(i) of this section,flexible cords and cables shall not be used:

   (A) As a substitute for the fixed wiring of a

   (B) structure; Where run through holes in walls,

   (C)

ceilings, or floors;

Where run through doorways, windows, or similar openings, except as permitted in paragraph (a)(2)(ii)(I) of this section;

- (D) Where attached to building surfaces; or
- (E) Where concealed behind building walls, ceilings, or floors.

(2) *Identification, splices, and terminations* -

- (i) *Identification.* A conductor of a flexible cord or cable that is used as a grounded conductor or anequipment grounding conductor shall be distinguishable from other conductors.
- (ii) *Marking.* Type SJ, SJO, SJT, SJTO, S, SO, ST, and STO cords shall not be used unless durably marked on the surface with the type designation, size, and number of conductors.
- (iii) *Splices.* Flexible cords shall be used only in continuous lengths without splice or tap. Hard service flexible cords No. 12 or larger may be repaired if spliced so that the splice retains theinsulation, outer sheath properties, and usage characteristics of the cord being spliced.
- (iv) *Strain relief.* Flexible cords shall be connected to devices and fittings so that strain relief is provided which will prevent pull from being directly transmitted to joints or terminal screws.
- (v) *Cords passing through holes.* Flexible cords and cables shall be protected by bushings orfittings where passing through holes in covers, outlet boxes, or similar enclosures.

(h) *Portable cables over 600 volts, nominal.* Multiconductor portable cable for use in supplying power to portable or mobile equipment at over 600 volts, nominal, shall consist of No. 8 or larger conductors employing flexible stranding. Cables operated at over 2000 volts shall be shielded for the purpose of confining the voltage stresses to the insulation. Grounding conductors shall be provided. Connectors for these cables shall be of a locking type with provisions to prevent their opening or closing while energized.Strain relief shall be provided at connections and terminations. Portable cables shall not be operated withsplices unless the splices are of the permanent molded, vulcanized, or other equivalent type. Termination enclosures shall be marked with a high voltage hazard warning, and terminations shall be accessible only to authorized and qualified personnel.

(i) *Fixture wires* -

(1) *General.* Fixture wires shall be suitable for the voltage, temperature, and location of use. A fixturewire which is used as a grounded conductor shall be identified.

(2) *Uses permitted.* Fixture wires may be used:

- (i) For installation in lighting, fixtures and in similar equipment where enclosed or protected andnot subject to bending or twisting in use; or
- (ii) For connecting lighting fixtures to the branch-circuit conductors supplying the fixtures.

(3) *Uses not permitted.* Fixture wires shall not be used as branch-circuit conductors except as permittedfor Class 1 power-limited circuits.

(j) *Equipment for general use* -

(1) *Lighting fixtures, lampholders, lamps, and receptacles* -

    (i) **Live parts.** Fixtures, lampholders, lamps, rosettes, and receptacles shall have no live parts normally exposed to employee contact. However, rosettes and cleat-type lampholders and receptacles located at least 8 feet (2.44 m) above the floor may have exposed parts.

    (ii) **Support.** Fixtures, lampholders, rosettes, and receptacles shall be securely supported. A fixture that weighs more than 6 pounds (2.72 kg) or exceeds 16 inches (406 mm) in any dimension shall not be supported by the screw shell of a lampholder.

    (iii) **Portable lamps.** Portable lamps shall be wired with flexible cord and an attachment plug of the polarized or grounding type. If the portable lamp uses an Edison-based lampholder, the grounded conductor shall be identified and attached to the screw shell and the identified blade of the attachment plug. In addition, portable handlamps shall comply with the following:

        (A) Metal shell, paperlined lampholders shall not be used;

        (B) Handlamps shall be equipped with a handle of molded composition or other insulating material;

        (C) Handlamps shall be equipped with a substantial guard attached to the lampholder or handle;

        (D) Metallic guards shall be grounded by the means of an equipment grounding conductor run within the power supply cord.

    (iv) **Lampholders.** Lampholders of the screw-shell type shall be installed for use as lampholders only. Lampholders installed in wet or damp locations shall be of the weatherproof type.

    (v) **Fixtures.** Fixtures installed in wet or damp locations shall be identified for the purpose and shall be installed so that water cannot enter or accumulate in wireways, lampholders, or other electrical parts.

(2) *Receptacles, cord connectors, and attachment plugs (caps)* -

    (i) **Configuration.** Receptacles, cord connectors, and attachment plugs shall be constructed so that no receptacle or cord connector will accept an attachment plug with a different voltage or current rating than that for which the device is intended. However, a 20-ampere T-slot receptacle or cord connector may accept a 15-ampere attachment plug of the same voltage rating. Receptacles connected to circuits having different voltages, frequencies, or types of current (ac or dc) on the same premises shall be of such design that the attachment plugs used on these circuits are not interchangeable.

    (ii) **Damp and wet locations.** A receptacle installed in a wet or damp location shall be designed for the location.

(3) *Appliances* -

    (i) **Live parts.** Appliances, other than those in which the current-carrying parts at high temperatures are necessarily exposed, shall have no live parts normally exposed to employee contact.

    (ii) **Disconnecting means.** A means shall be provided to disconnect each appliance.

(iii) **Rating.** Each appliance shall be marked with its rating in volts and amperes or volts and watts.

(4) **Motors.** This paragraph applies to motors, motor circuits, and controllers.

    (i) **In sight from.** If specified that one piece of equipment shall be "in sight from" another piece of equipment, one shall be visible and not more than 50 feet (15.2 m) from the other.

    (ii) **Disconnecting means**

        (A) A disconnecting means shall be located in sight from the controller location. The controller disconnecting means for motor branch circuits over 600 volts, nominal, may be out of sight of the controller, if the controller is marked with a warning label giving the location and identification of the disconnecting means which is to be locked in the open position.

        (B) The disconnecting means shall disconnect the motor and the controller from all ungrounded supply conductors and shall be so designed that no pole can be operated independently.

        (C) If a motor and the driven machinery are not in sight from the controller location, the installation shall comply with one of the following conditions:

            (*1*) The controller disconnecting means shall be capable of being locked in the open position.

            (*2*) A manually operable switch that will disconnect the motor from its source of supply shall be placed in sight from the motor location.

        (D) The disconnecting means shall plainly indicate whether it is in the open (off) or closed (on) position.

        (E) The disconnecting means shall be readily accessible. If more than one disconnect is provided for the same equipment, only one need be readily accessible.

        (F) An individual disconnecting means shall be provided for each motor, but a single disconnecting means may be used for a group of motors under any one of the following conditions:

            (*1*) If a number of motors drive special parts of a single machine or piece of apparatus, such as a metal or woodworking machine, crane, or hoist;

            (*2*) If a group of motors is under the protection of one set of branch-circuit protective devices; or

            (*3*) If a group of motors is in a single room in sight from the location of the disconnecting means.

    (iii) **Motor overload, short-circuit, and ground-fault protection.** Motors, motor-control apparatus, and motor branch-circuit conductors shall be protected against overheating due to motor overloads or failure to start, and against short-circuits or ground faults. These provisions do not require overload protection that will stop a motor where a shutdown is likely to introduce additional or increased hazards, as in the case of fire pumps, or where continued operation of a motor is necessary for a safe shutdown of equipment or process and motor overload sensing devices are connected to a supervised alarm.

(iv) **Protection of live parts - all voltages.**

- (A) Stationary motors having commutators, collectors, and brush rigging located inside of motor end brackets and not conductively connected to supply circuits operating at more than 150 volts to ground need not have such parts guarded. Exposed live parts of motorsand controllers operating at 50 volts or more between terminals shall be guarded against accidental contact by any of the following:

    - (*1*) By installation in a room or enclosure that is accessible only to qualified persons;
    - (*2*) By installation on a balcony, gallery, or platform, so elevated and arranged as to exclude unqualified persons; or
    - (*3*) By elevation 8 feet (2.44 m) or more above the floor.

- (B) Where live parts of motors or controllers operating at over 150 volts to ground are guardedagainst accidental contact only by location, and where adjustment or other attendance may be necessary during the operation of the apparatus, insulating mats or platforms shall be provided so that the attendant cannot readily touch live parts unless standing onthe mats or platforms.

(5) **Transformers -**

- (i) **Application.** The following paragraphs cover the installation of all transformers,

    - (A) except:Current transformers;
    - (B) Dry-type transformers installed as a component part of other apparatus;
    - (C) Transformers which are an integral part of an X-ray, high frequency, or electrostatic-coating apparatus;
    - (D) Transformers used with Class 2 and Class 3 circuits, sign and outline lighting, electric discharge lighting, and power-limited fire-protective signaling circuits.

- (ii) **Operating voltage.** The operating voltage of exposed live parts of transformer installations shallbe indicated by warning signs or visible markings on the equipment or structure.

- (iii) **Transformers over 35 kV.** Dry-type, high fire point liquid-insulated, and askarel-insulated transformers installed indoors and rated over 35 kV shall be in a vault.

- (iv) **Oil-insulated transformers.** If they present a fire hazard to employees, oil-insulated transformers installed indoors shall be in a vault.

- (v) **Fire protection.** Combustible material, combustible buildings and parts of buildings, fire escapes, and door and window openings shall be safeguarded from fires which may originate inoil-insulated transformers attached to or adjacent to a building or combustible material.

- (vi) **Transformer vaults.** Transformer vaults shall be constructed so as to contain fire and combustible liquids within the vault and to prevent unauthorized access. Locks and latchesshall be so arranged that a vault door can be readily opened from the inside.

- (vii) **Pipes and ducts.** Any pipe or duct system foreign to the vault installation shall not enter or passthrough a transformer vault.

(viii) **Material storage.** Materials shall not be stored in transformer vaults.

(6) **Capacitors -**

    (i) **Drainage of stored charge.** All capacitors, except surge capacitors or capacitors included as a component part of other apparatus, shall be provided with an automatic means of draining the stored charge and maintaining the discharged state after the capacitor is disconnected from its source of supply.

    (ii) **Over 600 volts.** Capacitors rated over 600 volts, nominal, shall comply with the following additional requirements:

        (A) Isolating or disconnecting switches (with no interrupting rating) shall be interlocked with the load interrupting device or shall be provided with prominently displayed caution signs to prevent switching load current.

        (B) For series capacitors the proper switching shall be assured by use of at least one of the following:

            (*1*) Mechanically sequenced isolating and bypass switches,

            (*2*) Interlocks, or

            (*3*) Switching procedure prominently displayed at the switching location.

[51 FR 25318, July 11, 1986, as amended at 61 FR 5510, Feb. 13, 1996; 85 FR 8736, Feb. 18, 2020]

§ 1926.406 Specific purpose equipment and installations.

(a) **Cranes and hoists.** This paragraph applies to the installation of electric equipment and wiring used in connection with cranes, monorail hoists, hoists, and all runways.

    (1) **Disconnecting means -**

        (i) **Runway conductor disconnecting means.** A readily accessible disconnecting means shall be provided between the runway contact conductors and the power supply.

        (ii) **Disconnecting means for cranes and monorail hoists.** A disconnecting means, capable of being locked in the open position, shall be provided in the leads from the runway contact conductors or other power supply on any crane or monorail hoist.

           (A) If this additional disconnecting means is not readily accessible from the crane or monorail hoist operating station, means shall be provided at the operating station to open the power circuit to all motors of the crane or monorail hoist.

           (B) The additional disconnect may be omitted if a monorail hoist or hand-propelled crane bridge installation meets all of the following:

                (*1*) The unit is floor controlled;

                (*2*) The unit is within view of the power supply disconnecting means;

                (*3*) and No fixed work platform has been provided for servicing the unit.

    (2) **Control.** A limit switch or other device shall be provided to prevent the load block from passing the safe upper limit of travel of any hoisting mechanism.

(3) **Clearance.** The dimension of the working space in the direction of access to live parts which may require examination, adjustment, servicing, or maintenance while alive shall be a minimum of 2 feet 6 inches (762 mm). Where controls are enclosed in cabinets, the door(s) shall open at least 90 degrees or be removable, or the installation shall provide equivalent access.

(4) **Grounding.** All exposed metal parts of cranes, monorail hoists, hoists and accessories including pendant controls shall be metallically joined together into a continuous electrical conductor so that the entire crane or hoist will be grounded in accordance with § 1926.404(f). Moving parts, other than

removable accessories or attachments, having metal-to-metal bearing surfaces shall be considered to be electrically connected to each other through the bearing surfaces for grounding purposes. The trolley frame and bridge frame shall be considered as electrically grounded through the bridge and trolley wheels and its respective tracks unless conditions such as paint or other insulating materials prevent reliable metal-to-metal contact. In this case a separate bonding conductor shall be provided.

(b) **Elevators, escalators, and moving walks** -

(1) **Disconnecting means.** Elevators, escalators, and moving walks shall have a single means for disconnecting all ungrounded main power supply conductors for each unit.

(2) **Control panels.** If control panels are not located in the same space as the drive machine, they shall be located in cabinets with doors or panels capable of being locked closed.

(c) **Electric welders - disconnecting means** -

(1) **Motor-generator, AC transformer, and DC rectifier arc welders.** A disconnecting means shall be provided in the supply circuit for each motor-generator arc welder, and for each AC transformer and DC rectifier arc welder which is not equipped with a disconnect mounted as an integral part of the welder.

(2) **Resistance welders.** A switch or circuit breaker shall be provided by which each resistance welder and its control equipment can be isolated from the supply circuit. The ampere rating of this disconnecting means shall not be less than the supply conductor ampacity.

(d) **X-Ray equipment** -

(1) **Disconnecting means** -

(i) **General.** A disconnecting means shall be provided in the supply circuit. The disconnecting means shall be operable from a location readily accessible from the X-ray control. For equipment connected to a 120-volt branch circuit of 30 amperes or less, a grounding-type attachment plug cap and receptacle of proper rating may serve as a disconnecting means.

(ii) **More than one piece of equipment.** If more than one piece of equipment is operated from the same high-voltage circuit, each piece or each group of equipment as a unit shall be provided with a high-voltage switch or equivalent disconnecting means. This disconnecting means shall be constructed, enclosed, or located so as to avoid contact by employees with its live parts.

(2) **Control - Radiographic and fluoroscopic types.** Radiographic and fluoroscopic-type equipment shall be effectively enclosed or shall have interlocks that deenergize the

equipment automatically to prevent ready access to live current-carrying parts.

§ 1926.407 Hazardous (classified) locations.

(a) **Scope.** This section sets forth requirements for electric equipment and wiring in locations which are classified depending on the properties of the flammable vapors, liquids or gases, or combustible dusts orfibers which may be present therein and the likelihood that a flammable or combustible concentration or quantity is present. Each room, section or area shall be considered individually in determining its classification. These hazardous (classified) locations are assigned six designations as follows:

Class I, Division 1

Class I, Division 2

Class II, Division 1

Class II, Division 2

Class III, Division I

Class III, Division 2

For definitions of these locations see § 1926.449. All applicable requirements in this subpart apply to all hazardous (classified) locations, unless modified by provisions of this section.

(b) **Electrical installations.** Equipment, wiring methods, and installations of equipment in hazardous (classified) locations shall be approved as intrinsically safe or approved for the hazardous (classified) location or safe for the hazardous (classified) location. Requirements for each of these options are as follows:

(1) **Intrinsically safe.** Equipment and associated wiring approved as intrinsically safe is permitted in anyhazardous (classified) location included in its listing or labeling.

(2) *Approved for the hazardous (classified) location* -

(i) ***General.*** Equipment shall be approved not only for the class of location but also for the ignitible or combustible properties of the specific gas, vapor, dust, or fiber that will be present.

> Note: NFPA 70, the National Electrical Code, lists or defines hazardous gases, vapors, and dusts by "Groups" characterized by their ignitible or combustible properties.

(ii) ***Marking.*** Equipment shall not be used unless it is marked to show the class, group, and operating temperature or temperature range, based on operation in a 40-degree C ambient, for which it is approved. The temperature marking shall not exceed the ignition temperature of the specific gas, vapor, or dust to be encountered. However, the following provisions modify this marking requirement for specific equipment:

 (A) Equipment of the non-heat-producing type (such as junction boxes, conduit, and fitting) and equipment of the heat-producing type having a maximum temperature of not more than 100 degrees C (212 degrees F) need not have a marked operating temperature or temperature range.

 (B) Fixed lighting fixtures marked for use only in Class I, Division 2 locations need not be marked to indicate the group.

 (C) Fixed general-purpose equipment in Class I locations, other than lighting fixtures, which is acceptable for use in Class I, Division 2 locations need not be marked with the class, group, division, or operating temperature.

 (D) Fixed dust-tight equipment, other than lighting fixtures, which is acceptable for use in Class II, Division 2 and Class III locations need not be marked with the class, group, division, or operating temperature.

(3) ***Safe for the hazardous (classified) location.*** Equipment which is safe for the location shall be of a type and design which the employer demonstrates will provide protection from the hazards arising from the combustibility and flammability of vapors, liquids, gases, dusts, or fibers.

Note: The National Electrical Code, NFPA 70, contains guidelines for determining the type and design of equipment and installations which will meet this requirement. The guidelines of this document address electric wiring, equipment, and systems installed in hazardous (classified) locations and contain specific provisions for the following: wiring methods, wiring connections, conductor insulation, flexible cords, sealing and drainage, transformers, capacitors, switches, circuit breakers, fuses, motor controllers, receptacles, attachment plugs, meters, relays, instruments, resistors, generators, motors, lighting fixtures, storage battery charging equipment, electric cranes, electric hoists and similar equipment, utilization equipment, signaling systems, alarm systems, remote control systems, local loud speaker and communication systems, ventilation piping, live parts, lightning surge protection, and grounding. Compliance with these guidelines will constitute one means, but not the only means, of compliance with this

Conduits. All conduits shall be threaded and shall be made wrench-tight. Where it is impractical to make a threaded joint tight, a bonding jumper shall be utilized.

(c)

[51 FR 25318, July 11, 1986, as amended at 61 FR 5510, Feb. 13, 1996]

§ 1926.408 Special systems.

(a) **Systems over 600 volts, nominal.** Paragraphs (a)(1) through (a)(4) of this section contain general requirements for all circuits and equipment operated at over 600 volts.

   (1) *Wiring methods for fixed installations* -

      (i) *Above ground.* Above-ground conductors shall be installed in rigid metal conduit, in intermediate metal conduit, in cable trays, in cablebus, in other suitable raceways, or as open runs of metal-clad cable designed for the use and purpose. However, open runs of non-metallic- sheathed cable or of bare conductors or busbars may be installed in locations which are accessible only to qualified persons. Metallic shielding components, such as tapes, wires, or braids for conductors, shall be grounded. Open runs of insulated wires and cables having a barelead sheath or a braided outer covering shall be supported in a manner designed to prevent physical damage to the braid or sheath.

      (ii) *Installations emerging from the ground.* Conductors emerging from the ground shall be enclosed in raceways. Raceways installed on poles shall be of rigid metal conduit, intermediatemetal conduit, PVC schedule 80 or equivalent extending from the ground line up to a point 8 feet (2.44 m) above finished grade. Conductors entering a building shall be protected by an enclosure from the ground line to the point of entrance. Metallic enclosures shall be grounded.

   (2) *Interrupting and isolating devices* -

      (i) *Circuit breakers.* Circuit breakers located indoors shall consist of metal-enclosed or fire- resistant, cell-mounted units. In locations accessible only to qualified personnel, open mounting of circuit breakers is permitted. A means of indicating the open and closed position of circuit breakers shall be provided.

      (ii) *Fused cutouts.* Fused cutouts installed in buildings or transformer vaults shall be of a type identified for the purpose. They shall be readily accessible for fuse replacement.

      (iii) *Equipment isolating means.* A means shall be provided to completely isolate equipment for inspection and repairs. Isolating means which are not designed to interrupt the load current ofthe circuit shall be either interlocked with a circuit interrupter or provided with a sign warning against opening them under load.

   (3) *Mobile and portable equipment* -

      (i) *Power cable connections to mobile machines.* A metallic enclosure shall be provided on the mobile machine for enclosing the terminals of the power cable. The enclosure shall include provisions for a solid connection for the ground wire(s) terminal to ground effectively the machine frame. The method of cable termination used shall prevent any strain or pull on the cable from stressing the electrical connections. The enclosure shall have provision for lockingso only authorized qualified persons may open it and shall be marked with a sign warning of thepresence of energized parts.

      (ii) *Guarding live parts.* All energized switching and control parts shall be enclosed in effectively grounded metal cabinets or enclosures. Circuit breakers and protective equipment shall have the operating means projecting through the metal cabinet or enclosure so these units can bereset without locked doors being opened. Enclosures and metal cabinets shall be locked so that only authorized qualified persons have access and shall be marked with a sign warning ofthe presence of energized parts. Collector ring assemblies on revolving-type machines (shovels, draglines, etc.) shall be guarded.

(4) **Tunnel installations** -

    (i) **Application.** The provisions of this paragraph apply to installation and use of high-voltage powerdistribution and utilization equipment which is associated with tunnels and which is portable and/or mobile, such as substations, trailers, cars, mobile shovels, draglines, hoists, drills, dredges, compressors, pumps, conveyors, and underground excavators.

    (ii) **Conductors.** Conductors in tunnels shall be installed in one or more of the

        (A) following:Metal conduit or other metal raceway,

        (B) Type MC cable, or

        (C) Other suitable multiconductor cable. Conductors shall also be so located or guarded as toprotect them from physical damage. Multiconductor portable cable may supply mobile equipment. An equipment grounding conductor shall be run with circuit conductors insidethe metal raceway or inside the multiconductor cable jacket. The equipment grounding conductor may be insulated or bare.

    (iii) **Guarding live parts.** Bare terminals of transformers, switches, motor controllers, and other equipment shall be enclosed to prevent accidental contact with energized parts. Enclosures foruse in tunnels shall be drip-proof, weatherproof, or submersible as required by the environmental conditions.

    (iv) **Disconnecting means.** A disconnecting means that simultaneously opens all ungrounded conductors shall be installed at each transformer or motor location.

    (v) **Grounding and bonding.** All nonenergized metal parts of electric equipment and metal raceways and cable sheaths shall be grounded and bonded to all metal pipes and rails at the portal and atintervals not exceeding 1000 feet (305 m) throughout the tunnel.

(b) ***Class 1, Class 2, and Class 3 remote control, signaling, and power-limited circuits*** -

    (1) **Classification.** Class 1, Class 2, or Class 3 remote control, signaling, or power-limited circuits are characterized by their usage and electrical power limitation which differentiates them from light andpower circuits. These circuits are classified in accordance with their respective voltage and power limitations as summarized in paragraphs (b)(1)(i) through (b)(1)(iii) of this section.

        (i) **Class 1 circuits.**

            (A) A Class 1 power-limited circuit is supplied from a source having a rated output of not morethan 30 volts and 1000 volt-amperes.

            (B) A Class 1 remote control circuit or a Class 1 signaling circuit has a voltage which does notexceed 600 volts; however, the power output of the source need not be limited.

        (ii) **Class 2 and Class 3 circuits.**

            (A) Power for Class 2 and Class 3 circuits is limited either inherently (in which no overcurrent protection is required) or by a combination of a power source and overcurrent protection.

- (B) The maximum circuit voltage is 150 volts AC or DC for a Class 2 inherently limited powersource, and 100 volts AC or DC for a Class 3 inherently limited power source.
- (C) The maximum circuit voltage is 30 volts AC and 60 volts DC for a Class 2 power sourcelimited by overcurrent protection, and 150 volts AC or DC for a Class 3 power source limited by overcurrent protection.

(iii) *Application.* The maximum circuit voltages in paragraphs (b)(1)(i) and (b)(1)(ii) of this section apply to sinusoidal AC or continuous DC power sources, and where wet contact occurrence isnot likely.

(2) *Marking.* A Class 2 or Class 3 power supply unit shall not be used unless it is durably marked whereplainly visible to indicate the class of supply and its electrical rating.

(c) *Communications systems* -

(1) *Scope.* These provisions for communication systems apply to such systems as central-station-connected and non-central-station-connected telephone circuits, radio receiving and transmitting equipment, and outside wiring for fire and burglar alarm, and similar central station systems. Theseinstallations need not comply with the provisions of §§ 1926.403 through 1926.408(b), except § 1926.404(c)(1)(ii) and § 1926.407.

(2) *Protective devices* -

(i) *Circuits exposed to power conductors.* Communication circuits so located as to be exposed toaccidental contact with light or power conductors operating at over 300 volts shall have each circuit so exposed provided with an approved protector.

(ii) *Antenna lead-ins.* Each conductor of a lead-in from an outdoor antenna shall be provided withan antenna discharge unit or other means that will drain static charges from the antenna system.

(3) *Conductor location.*

(i) *Outside of buildings* -

- (A) Receiving distribution lead-in or aerial-drop cables attached to buildings and lead-in conductors to radio transmitters shall be so installed as to avoid the possibility ofaccidental contact with electric light or power conductors.
- (B) The clearance between lead-in conductors and any lightning protection conductors shall not be less than 6 feet (1.83 m).

(ii) *On poles.* Where practicable, communication conductors on poles shall be located below thelight or power conductors. Communications conductors shall not be attached to a crossarm that carries light or power conductors.

(iii) *Inside of buildings.* Indoor antennas, lead-ins, and other communication conductors attached as open conductors to the inside of buildings shall be located at least 2 inches (50.8 mm) from conductors of any light or power or Class 1 circuits unless a special and equally protective method of conductor separation is employed.

(4) *Equipment location.* Outdoor metal structures supporting antennas, as well as self-supporting antennas such as vertical rods or dipole structures, shall be located as far away

from overhead conductors of electric light and power circuits of over 150 volts to ground as necessary to avoid thepossibility of the antenna or structure falling into or making accidental contact with such circuits.

(5) *Grounding* -

(i) *Lead-in conductors.* If exposed to contact with electric light or power conductors, the metal sheath of aerial cables entering buildings shall be grounded or shall be interrupted close to the entrance to the building by an insulating joint or equivalent device. Where protective devices areused, they shall be grounded.

(ii) *Antenna structures.* Masts and metal structures supporting antennas shall be permanently and effectively grounded without splice or connection in the grounding conductor.

(iii) *Equipment enclosures.* Transmitters shall be enclosed in a metal frame or grill or separated from the operating space by a barrier, all metallic parts of which are effectively connected to ground. All external metal handles and controls accessible to the operating personnel shall beeffectively grounded. Unpowered equipment and enclosures shall be considered grounded where connected to an attached coaxial cable with an effectively grounded metallic shield.

[51 FR 25318, July 11, 1986, as amended at 61 FR 5510, Feb. 13, 1996]

§§ 1926.409-1926.415 [Reserved]

SAFETY-RELATED WORK PRACTICES

§ 1926.416 General requirements.

(a) *Protection of employees.*

(1) No employer shall permit an employee to work in such proximity to any part of an electric powercircuit that the employee could contact the electric power circuit in the course of work, unless the employee is protected against electric shock by deenergizing the circuit and grounding it or by guarding it effectively by insulation or other means.

(2) In work areas where the exact location of underground electric powerlines is unknown, employeesusing jack-hammers, bars, or other hand tools which may contact a line shall be provided with insulated protective gloves.

(3) Before work is begun the employer shall ascertain by inquiry or direct observation, or by instruments,whether any part of an energized electric power circuit, exposed or concealed, is so located that the performance of the work may bring any person, tool, or machine into physical or electrical contact with the electric power circuit. The employer shall post and maintain proper warning signs where such a circuit exists. The employer shall advise employees of the location of such lines, the hazards involved, and the protective measures to be taken.

(b) *Passageways and open spaces* --(1) Barriers or other means of guarding shall be provided to ensure thatworkspace for electrical equipment will not be used as a passageway during periods when energized parts of electrical equipment are exposed.

(2) Working spaces, walkways, and similar locations shall be kept clear of cords so as not to create a hazard to employees.

(c) **Load ratings.** In existing installations, no changes in circuit protection shall be made to increase the load in excess of the load rating of the circuit wiring.

(d) **Fuses.** When fuses are installed or removed with one or both terminals energized, special tools insulated for the voltage shall be used.

(e) **Cords and cables.**

   (1) Worn or frayed electric cords or cables shall not be used.

   (2) Extension cords shall not be fastened with staples, hung from nails, or suspended by wire.

[44 FR 8577, Feb. 9, 1979; 44 FR 20940, Apr. 6, 1979, as amended at 55 FR 42328, Oct. 18, 1990; 58 FR 35179, June 30, 1993; 61
FR 9251, Mar. 7, 1996; 61 FR 41738, Aug. 12, 1996]

§ 1926.417 Lockout and tagging of circuits.

(a) **Controls.** Controls that are to be deactivated during the course of work on energized or deenergized equipment or circuits shall be tagged.

(b) **Equipment and circuits.** Equipment or circuits that are deenergized shall be rendered inoperative and shall have tags attached at all points where such equipment or circuits can be energized.

(c) **Tags.** Tags shall be placed to identify plainly the equipment or circuits being worked on.

[44 FR 8577, Feb. 9, 1979; 44 FR 20940, Apr. 6, 1979, as amended at 55 FR 42328, Oct. 18, 1990; 58 FR 35181, June 30, 1993; 61
FR 9251, Mar. 7, 1996; 61 FR 41739, Aug. 12, 1996]]

§§ 1926.418-1926.430 [Reserved]

SAFETY-RELATED MAINTENANCE AND ENVIRONMENTAL CONSIDERATIONS

§ 1926.431 Maintenance of equipment.

The employer shall ensure that all wiring components and utilization equipment in hazardous locations are maintained in a dust-tight, dust-ignition-proof, or explosion-proof condition, as appropriate. There shall be no loose or missing screws, gaskets, threaded connections, seals, or other impairments to a tight condition.

§ 1926.432 Environmental deterioration of equipment.

(a) **Deteriorating agents.**

   (1) Unless identified for use in the operating environment, no conductors or equipment shall be

   (i) located: In damp or wet locations;

   (ii) Where exposed to gases, fumes, vapors, liquids, or other agents having a deteriorating effect onthe conductors or equipment; or

   (iii) Where exposed to excessive temperatures.

   (2) Control equipment, utilization equipment, and busways approved for use in dry locations only shallbe protected against damage from the weather during building construction.

(b) **Protection against corrosion.** Metal raceways, cable armor, boxes, cable sheathing, cabinets, elbows, couplings, fittings, supports, and support hardware shall be of materials appropriate for the environment in which they are to be installed.

§§ 1926.433-1926.440 [Reserved]

SAFETY REQUIREMENTS FOR SPECIAL EQUIPMENT

§ 1926.441 Batteries and battery charging.

(a) *General requirements.*

(1) Batteries of the unsealed type shall be located in enclosures with outside vents or in well ventilatedrooms and shall be arranged so as to prevent the escape of fumes, gases, or electrolyte spray into other areas.

(2) Ventilation shall be provided to ensure diffusion of the gases from the battery and to prevent the accumulation of an explosive mixture.

(3) Racks and trays shall be substantial and shall be treated to make them resistant to the

(4) electrolyte.Floors shall be of acid resistant construction unless protected from acid

(5) accumulations.

Face shields, aprons, and rubber gloves shall be provided for workers handling acids or batteries.

(6) Facilities for quick drenching of the eyes and body shall be provided within 25 feet (7.62 m) of battery handling areas.

(7) Facilities shall be provided for flushing and neutralizing spilled electrolyte and for fire protection.

(b) *Charging.*

(2)

(1) Battery charging installations shall be located in areas designated for that

purpose.Charging apparatus shall be protected from damage by trucks.

(3) When batteries are being charged, the vent caps shall be kept in place to avoid electrolyte spray. Vent caps shall be maintained in functioning condition.

§§ 1926.442-1926.448 [Reserved]

DEFINITIONS

§ 1926.449 Definitions applicable to this subpart.

The definitions given in this section apply to the terms used in subpart K. The definitions given here for "approved"and "qualified person" apply, instead of the definitions given in § 1926.32, to the use of these terms in subpart K.

Acceptable. An installation or equipment is acceptable to the Assistant Secretary of Labor, and approved withinthe meaning of this subpart K:

(a) If it is accepted, or certified, or listed, or labeled, or otherwise determined to be safe by a qualified testing laboratory capable of determining the suitability of materials and equipment for installation and use in accordance with this standard; or

(b) With respect to an installation or equipment of a kind which no qualified testing laboratory accepts, certifies, lists, labels, or determines to be safe, if it is inspected or tested by another Federal agency, or by a State, municipal, or other local authority responsible for enforcing occupational safety provisions of the National Electrical Code, and found in compliance with those provisions; or

(c) With respect to custom-made equipment or related installations which are designed, fabricated for, and intended for use by a particular customer, if it is determined to be safe for its intended use by its manufacturer on the basis of test data which the employer keeps and makes available for inspection to the Assistant Secretary and his authorized representatives.

Accepted. An installation is "accepted" if it has been inspected and found to be safe by a qualified testing laboratory.

Accessible. (As applied to wiring methods.) Capable of being removed or exposed without damaging the building structure or finish, or not permanently closed in by the structure or finish of the building. (See *"concealed"* and *"exposed."*)

Accessible. (As applied to equipment.) Admitting close approach; not guarded by locked doors, elevation, or other effective means. (See *"Readily accessible."*)

Ampacity. The current in amperes a conductor can carry continuously under the conditions of use without exceeding its temperature rating.

Appliances. Utilization equipment, generally other than industrial, normally built in standardized sizes or types, which is installed or connecetcd as a unit to perform one or more functions.

Approved. Acceptable to the authority enforcing this subpart. The authority enforcing this subpart is the Assistant Secretary of Labor for Occupational Safety and Health. The definition of "acceptable" indicates what is acceptable to the Assistant Secretary of Labor, and therefore approved within the meaning of this subpart.

Askarel. A generic term for a group of nonflammable synthetic chlorinated hydrocarbons used as electrical insulating media. Askarels of various compositional types are used. Under arcing conditions the gases produced, while consisting predominantly of noncombustible hydrogen chloride, can include varying amounts of combustible gases depending upon the askarel type.

Attachment plug (Plug cap)(Cap). A device which, by insertion in a receptacle, establishes connection between the conductors of the attached flexible cord and the conductors connected permanently to the receptacle.

Automatic. Self-acting, operating by its own mechanism when actuated by some impersonal influence, as for example, a change in current strength, pressure, temperature, or mechanical configuration.

Bare conductor. See *"Conductor."*

Bonding. The permanent joining of metallic parts to form an electrically conductive path which will assure electrical continuity and the capacity to conduct safely any current likely to be imposed.

Bonding jumper. A reliable conductor to assure the required electrical conductivity between metal parts required to be electrically connected.

Branch circuit. The circuit conductors between the final overcurrent device protecting the circuit and the outlet(s).

Building. A structure which stands alone or which is cut off from adjoining structures by fire walls with all openings therein protected by approved fire doors.

Cabinet. An enclosure designed either for surface or flush mounting, and provided with a frame, mat, or trim in which a swinging door or doors are or may be hung.

Certified. Equipment is "certified" if it:

(a) Has been tested and found by a qualified testing laboratory to meet applicable test standards or to be safe for use in a specified manner, and

(b) Is of a kind whose production is periodically inspected by a qualified testing laboratory. Certified equipment must bear a label, tag, or other record of certification.

Circuit breaker

(a) (600 volts nominal, or less.) A device designed to open and close a circuit by nonautomatic means and to open the circuit automatically on a predetermined overcurrent without injury to itself when properly applied within its rating.

(b) (Over 600 volts, nominal.) A switching device capable of making, carrying, and breaking currents under normal circuit conditions, and also making, carrying for a specified time, and breaking currents under specified abnormal circuit conditions, such as those of short circuit.

Class I locations. Class I locations are those in which flammable gases or vapors are or may be present in the air in quantities sufficient to produce explosive or ignitible mixtures. Class I locations include the following:

(a) **Class I, Division 1.** A Class I, Division 1 location is a location:

(1) In which ignitible concentrations of flammable gases or vapors may exist under normal operating conditions; or

(2) In which ignitible concentrations of such gases or vapors may exist frequently because of repair or maintenance operations or because of leakage; or

(3) In which breakdown or faulty operation of equipment or processes might release ignitible concentrations of flammable gases or vapors, and might also cause simultaneous failure of electric equipment.

> Note: This classification usually includes locations where volatile flammable liquids or liquefied flammable gases are transferred from one container to another; interiors of spray booths and areas in the vicinity of spraying and painting operations where volatile flammable solvents are used; locations containing open tanks or vats of volatile flammable liquids; drying rooms or compartments for the evaporation of flammable solvents; inadequately ventilated pump rooms for flammable gas or for volatile flammable liquids; and all other locations where ignitible concentrations of flammable vapors or gases are likely to occur in the course of normal operations.

(b) **Class I, Division 2.** A Class I, Division 2 location is a location:

(1) In which volatile flammable liquids or flammable gases are handled, processed, or used, but in which the hazardous liquids, vapors, or gases will normally be confined within closed containers or closed systems from which they can escape only in case of accidental rupture or breakdown of such containers or systems, or in case of abnormal

operation of equipment; or

(2) In which ignitible concentrations of gases or vapors are normally prevented by positive mechanical ventilation, and which might become hazardous through failure or abnormal operations of the ventilating equipment; or

(3) That is adjacent to a Class I, Division 1 location, and to which ignitible concentrations of gasesor vapors might occasionally be communicated unless such communication is prevented by adequate positive-pressure ventilation from a source of clean air, and effective safeguards against ventilation failure are provided.

Note: This classification usually includes locations where volatile flammable liquids or flammablegases or vapors are used, but which would become hazardous only in case of an accident or of some unusual operating condition. The quantity of flammable material that might escape in case of accident, the adequacy of ventilating equipment, the total area involved, and the record of the industry or business with respect to explosions or fires are all factors that merit consideration indetermining the classification and extent of each location.

Piping without valves, checks, meters, and similar devices would not ordinarily introduce a hazardous condition even though used for flammable liquids or gases. Locations used for the storage of flammable liquids or of liquefied or compressed gases in sealed containers would notnormally be considered hazardous unless also subject to other hazardous conditions.

Electrical conduits and their associated enclosures separated from process fluids by a single sealor barrier are classed as a Division 2 location if the outside of the conduit and enclosures is a nonhazardous location.

Class II locations. Class II locations are those that are hazardous because of the presence of combustible dust.

Class II locations include the following:

(a) **Class II, Division 1.** A Class II, Division 1 location is a location:

(1) In which combustible dust is or may be in suspension in the air under normal operating conditions, in quantities sufficient to produce explosive or ignitible mixtures; or

(2) Where mechanical failure or abnormal operation of machinery or equipment might cause suchexplosive or ignitible mixtures to be produced, and might also provide a source of ignition through simultaneous failure of electric equipment, operation of protection devices, or from other causes, or

(3) In which combustible dusts of an electrically conductive nature may be present.

Note: Combustible dusts which are electrically nonconductive include dusts produced in the handling and processing of grain and grain products, pulverized sugar and cocoa, dried egg and milk powders, pulverized spices, starch and pastes, potato and woodflour, oil meal from beans and seed, dried hay, and other organic materials which may produce combustible dusts when processed or handled. Dusts containing magnesium or aluminum are particularly hazardous and the use of extreme caution is necessary to avoid ignition and explosion.

(b) *Class II, Division 2.* A Class II, Division 2 location is a location in which:

(1) Combustible dust will not normally be in suspension in the air in quantities sufficient to produceexplosive or ignitible mixtures, and dust accumulations are normally insufficient to interfere with the normal operation of electrical equipment or other apparatus; or

(2) Dust may be in suspension in the air as a result of infrequent malfunctioning of handling or processing equipment, and dust accumulations resulting therefrom may be ignitible by abnormal operation or failure of electrical equipment or other apparatus.

Note: This classification includes locations where dangerous concentrations of suspended dust would not be likely but where dust accumulations might form on or in the vicinity of electric equipment. These areas may contain equipment from which appreciable quantities of dust would escape under abnormal operating conditions or be adjacent to a Class II Division 1 location, as described above, into which an explosive or ignitible concentration of dust may be put into suspension under abnormal operating conditions.

Class III locations. Class III locations are those that are hazardous because of the presence of easily ignitible fibers or flyings but in which such fibers or flyings are not likely to be in suspension in the air in quantities sufficient to produce ignitible mixtures. Class 111 locations include the following:

(a) *Class III, Division 1.* A Class III, Division 1 location is a location in which easily ignitible fibers or materials producing combustible flyings are handled, manufactured, or used.

Note: Easily ignitible fibers and flyings include rayon, cotton (including cotton linters and cotton waste), sisal or henequen, istle, jute, hemp, tow, cocoa fiber, oakum, baled waste kapok, Spanish moss, excelsior, sawdust, woodchips, and other material of similar nature.

(b) *Class III, Division 2.* A Class III, Division 2 location is a location in which easily ignitible fibers arestored or handled, except in process of manufacture.

Collector ring. A collector ring is an assembly of slip rings for transferring electrical energy from a stationary to arotating member.

Concealed. Rendered inaccessible by the structure or finish of the building. Wires in concealed raceways are considered concealed, even though they may become accessible by withdrawing them. [See *"Accessible.* (As applied to wiring methods.)"]

Conductor

- (a) **Bare.** A conductor having no covering or electrical insulation whatsoever.
- (b) **Covered.** A conductor encased within material of composition or thickness that is not recognized as electrical insulation.
- (c) **Insulated.** A conductor encased within material of composition and thickness that is recognized as electrical insulation.

Controller. A device or group of devices that serves to govern, in some predetermined manner, the electric power delivered to the apparatus to which it is connected.

Covered conductor. See *"Conductor."*

Cutout. (Over 600 volts, nominal.) An assembly of a fuse support with either a fuseholder, fuse carrier, or disconnecting blade. The fuseholder or fuse carrier may include a conducting element (fuse link), or may act as the disconnecting blade by the inclusion of a nonfusible member.

Cutout box. An enclosure designed for surface mounting and having swinging doors or covers secured directly to and telescoping with the walls of the box proper. (See *"Cabinet."*)

Damp location. See *"Location."*

Dead front. Without live parts exposed to a person on the operating side of the equipment.

Device. A unit of an electrical system which is intended to carry but not utilize electric energy.

Disconnecting means. A device, or group of devices, or other means by which the conductors of a circuit can be disconnected from their source of supply.

Disconnecting (or Isolating) switch. (Over 600 volts, nominal.) A mechanical switching device used for isolating a circuit or equipment from a source of power.

Dry location. See *"Location."*

Enclosed. Surrounded by a case, housing, fence or walls which will prevent persons from accidentally contacting energized parts.

Enclosure. The case or housing of apparatus, or the fence or walls surrounding an installation to prevent personnel from accidentally contacting energized parts, or to protect the equipment from physical damage.

Equipment. A general term including material, fittings, devices, appliances, fixtures, apparatus, and the like, used as a part of, or in connection with, an electrical installation.

Equipment grounding conductor. See *"Grounding conductor, equipment."*

Explosion-proof apparatus. Apparatus enclosed in a case that is capable of withstanding an explosion of a specified gas or vapor which may occur within it and of preventing the ignition of a specified gas or vapor surrounding the enclosure by sparks, flashes, or explosion of the gas or vapor within, and which operates at such an external temperature that it will not ignite a surrounding flammable atmosphere.

Exposed. (As applied to live parts.) Capable of being inadvertently touched or approached nearer than a safe distance by a person. It is applied to parts not suitably guarded, isolated, or insulated. (See *"Accessible* and *"Concealed."*)

Exposed. (As applied to wiring methods.) On or attached to the surface or behind panels designed to allow access. [See *"Accessible.* (As applied to wiring methods.)"]

Exposed. (For the purposes of § 1926.408(d), Communications systems.) Where the circuit is in such a position that in case of failure of supports or insulation, contact with another circuit may result.

Externally operable. Capable of being operated without exposing the operator to contact with live parts.

Feeder. All circuit conductors between the service equipment, or the generator switchboard of an isolated plant, and the final branch-circuit overcurrent device.

Festoon lighting. A string of outdoor lights suspended between two points more than 15 feet (4.57 m) apart.

Fitting. An accessory such as a locknut, bushing, or other part of a wiring system that is intended primarily to perform a mechanical rather than an electrical function.

Fuse. (Over 600 volts, nominal.) An overcurrent protective device with a circuit opening fusible part that is heated and severed by the passage of overcurrent through it. A fuse comprises all the parts that form a unit capable of performing the prescribed functions. It may or may not be the complete device necessary to connect it into an electrical circuit.

Ground. A conducting connection, whether intentional or accidental, between an electrical circuit or equipment and the earth, or to some conducting body that serves in place of the earth.

Grounded. Connected to earth or to some conducting body that serves in place of the earth.

Grounded, effectively (Over 600 volts, nominal.) Permanently connected to earth through a ground connection of sufficiently low impedance and having sufficient ampacity that ground fault current which may occur cannot build up to voltages dangerous to personnel.

Grounded conductor. A system or circuit conductor that is intentionally grounded.

Grounding conductor. A conductor used to connect equipment or the grounded circuit of a wiring system to a grounding electrode or electrodes.

Grounding conductor, equipment. The conductor used to connect the noncurrent-carrying metal parts of equipment, raceways, and other enclosures to the system grounded conductor and/or the grounding electrode conductor at the service equipment or at the source of a separately derived system.

Grounding electrode conductor. The conductor used to connect the grounding electrode to the equipment grounding conductor and/or to the grounded conductor of the circuit at the service equipment or at the source of a separately derived system.

Ground-fault circuit interrupter. A device for the protection of personnel that functions to deenergize a circuit or portion thereof within an established period of time when a current to ground exceeds some predetermined value that is less than that required to operate the overcurrent protective device of the supply circuit.

Guarded. Covered, shielded, fenced, enclosed, or otherwise protected by means of suitable covers, casings, barriers, rails, screens, mats, or platforms to remove the likelihood of approach to a point of danger or contact by persons or objects.

Hoistway. Any shaftway, hatchway, well hole, or other vertical opening or space in which an elevator or dumbwaiter is designed to operate.

Identified (conductors or terminals). Identified, as used in reference to a conductor or its terminal, means that such conductor or terminal can be recognized as grounded.

Identified (for the use). Recognized as suitable for the specific purpose, function, use, environment, application, etc. where described as a requirement in this standard. Suitability of equipment for a specific purpose, environment, or application is determined by a qualified testing laboratory where such identification includes labeling or listing.

Insulated conductor. See "*Conductor.*"

Interrupter switch. (Over 600 volts, nominal.) A switch capable of making, carrying, and interrupting specified currents.

Intrinsically safe equipment and associated wiring. Equipment and associated wiring in which any spark or thermal effect, produced either normally or in specified fault conditions, is incapable, under certain prescribed test conditions, of causing ignition of a mixture of flammable or combustible material in air in its most easily ignitible concentration.

Isolated. Not readily accessible to persons unless special means for access are used.

Isolated power system. A system comprising an isolating transformer or its equivalent, a line isolation monitor, and its ungrounded circuit conductors.

Labeled. Equipment or materials to which has been attached a label, symbol or other identifying mark of a qualified testing laboratory which indicates compliance with appropriate standards or performance in a specified manner.

Lighting outlet. An outlet intended for the direct connection of a lampholder, a lighting fixture, or a pendant cord terminating in a lampholder.

Listed. Equipment or materials included in a list published by a qualified testing laboratory whose listing states either that the equipment or material meets appropriate standards or has been tested and found suitable for use in a specified manner.

Location

(a) **Damp location.** Partially protected locations under canopies, marquees, roofed open porches, and like locations, and interior locations subject to moderate degrees of moisture, such as some basements.

(b) **Dry location.** A location not normally subject to dampness or wetness. A location classified as dry may be temporarily subject to dampness or wetness, as in the case of a building under construction.

(c) **Wet location.** Installations underground or in concrete slabs or masonry in direct contact with the earth, and locations subject to saturation with water or other liquids, such as locations exposed to weather and unprotected.

Mobile X-ray. X-ray equipment mounted on a permanent base with wheels and/or casters for moving while completely assembled.

Motor control center. An assembly of one or more enclosed sections having a common power bus and principally containing motor control units.

Outlet. A point on the wiring system at which current is taken to supply utilization equipment.

Overcurrent. Any current in excess of the rated current of equipment or the ampacity of a conductor. It may result from overload (see definition), short circuit, or ground fault. A current in excess of rating may be accommodated by certain equipment and conductors for a given set of conditions. Hence the rules for overcurrent protection are specific for particular situations.

Overload. Operation of equipment in excess of normal, full load rating, or of a conductor in excess of *ratedampacity* which, when it persists for a sufficient length of time, would cause damage or dangerousoverheating. A fault, such as a short circuit or ground fault, is not an overload. (See *"Overcurrent."*)

Panelboard. A single panel or group of panel units designed for assembly in the form of a single panel; including buses, automatic overcurrent devices, and with or without switches for the control of light, heat, or power circuits; designed to be placed in a cabinet or cutout box placed in or against a wall or partition and accessible only from the front. (See *"Switchboard."*)

Portable X-ray. X-ray equipment designed to be hand-carried. (Over 600 volts, nominal.) See

Power fuse. "Fuse."

Power outlet. An enclosed assembly which may include receptacles, circuit breakers, fuseholders, fused switches, buses and watt-hour meter mounting means; intended to serve as a means for distributingpower required to operate mobile or temporarily installed equipment.

Premises wiring system. That interior and exterior wiring, including power, lighting, control, and signal circuitwiring together with all of its associated hardware, fittings, and wiring devices, both permanently and temporarily installed, which extends from the load end of the service drop, or load end of the servicelateral conductors to the outlet(s). Such wiring does not include wiring internal to appliances, fixtures, motors, controllers, motor control centers, and similar equipment.

Qualified person. One familiar with the construction and operation of the equipment and the hazards involved.

Qualified testing laboratory. A properly equipped and staffed testing laboratory which has capabilities for and which provides the following services:

(a) Experimental testing for safety of specified items of equipment and materials referred to in this standard to determine compliance with appropriate test standards or performance in a specified manner;

(b) Inspecting the run of such items of equipment and materials at factories for product evaluation toassure compliance with the test standards;

(c) Service-value determinations through field inspections to monitor the proper use of labels onproducts and with authority for recall of the label in the event a hazardous product is installed;

(d) Employing a controlled procedure for identifying the listed and/or labeled equipment or materials tested; and

(e) Rendering creditable reports or findings that are objective and without bias of the tests and test methods employed.

Raceway. A channel designed expressly for holding wires, cables, or busbars, with additional functions as permitted in this subpart. Raceways may be of metal or insulating material, and the term includes rigid metal conduit, rigid nonmetallic conduit, intermediate metal conduit, liquidtight flexible metal conduit,flexible metallic tubing, flexible metal conduit, electrical metallic tubing, underfloor raceways, cellular concrete floor raceways, cellular metal floor raceways, surface raceways, wireways, and busways.

Readily accessible. Capable of being reached quickly for operation, renewal, or inspections, without *requiring* those to whom ready access is requisite to climb over or remove obstacles or to resort to portable ladders, chairs, etc. (See *"Accessible."*)

Receptacle. A receptacle is a contact device installed at the outlet for the connection of a single attachment plug. A single receptacle is a single contact device with no other contact device on the same yoke. A multiple receptacle is a single device containing two or more receptacles.

Receptacle outlet. An outlet where one or more receptacles are installed.

Remote-control circuit. Any electric circuit that controls any other circuit through a relay or an equivalent device.

Sealable equipment. Equipment enclosed in a case or cabinet that is provided with a means of sealing or locking so that live parts cannot be made accessible without opening the enclosure. The equipment may or may not be operable without opening the enclosure.

Separately derived system. A premises wiring system whose power is derived from generator, transformer, or converter windings and has no direct electrical connection, including a solidly connected grounded circuit conductor, to supply conductors originating in another system.

Service. The conductors and equipment for delivering energy from the electricity supply system to the wiring system of the premises served.

Service conductors. The supply conductors that extend from the street main or from transformers to the service equipment of the premises supplied.

Service drop. The overhead service conductors from the last pole or other aerial support to and including the splices, if any, connecting to the service-entrance conductors at the building or other structure.

Service-entrance conductors, overhead system. The service conductors between the terminals of the service equipment and a point usually outside the building, clear of building walls, where joined by tap or splice to the service drop.

Service-entrance conductors, underground system. The service conductors between the terminals of the service equipment and the point of connection to the service lateral. Where service equipment is located outside the building walls, there may be no service-entrance conductors, or they may be entirely outside the building.

Service equipment. The necessary equipment, usually consisting of a circuit breaker or switch and fuses, and their accessories, located near the point of entrance of supply conductors to a building or other structure, or an otherwise defined area, and intended to constitute the main control and means of cutoff of the supply.

Service raceway. The raceway that encloses the service-entrance conductors.

Signaling circuit. Any electric circuit that energizes signaling equipment.

Switchboard. A large single panel, frame, or assembly of panels which have switches, buses, instruments, overcurrent and other protective devices mounted on the face or back or both. Switchboards are generally accessible from the rear as well as from the front and are not intended to be installed in cabinets. (See *"Panelboard."*)

Switches

    (a) **General-use switch.** A switch intended for use in general distribution and branch circuits. It is

(b) **General-use snap switch.** A form of general-use switch so constructed that it can be installed in flushdevice boxes or on outlet box covers, or otherwise used in conjunction with wiring systems recognized by this subpart.

(c) **Isolating switch.** A switch intended for isolating an electric circuit from the source of power. It has nointerrupting rating, and it is intended to be operated only after the circuit has been opened by someother means.

(d) **Motor-circuit switch.** A switch, rated in horsepower, capable of interrupting the maximum operatingoverload current of a motor of the same horsepower rating as the switch at the rated voltage.

Switching devices. (Over 600 volts, nominal.) Devices designed to close and/or open one or more electriccircuits. Included in this category are circuit breakers, cutouts, disconnecting (or isolating) switches, disconnecting means, and interrupter switches.

Transportable X-ray. X-ray equipment installed in a vehicle or that may readily be disassembled for transport in a vehicle.

Utilization equipment. Utilization equipment means equipment which utilizes electric energy for mechanical, chemical, heating, lighting, or similar useful purpose.

Utilization system. A utilization system is a system which provides electric power and light for employee workplaces, and includes the premises wiring system and utilization equipment.

Ventilated. Provided with a means to permit circulation of air sufficient to remove an excess of heat, fumes, or vapors.

Volatile flammable liquid. A flammable liquid having a flash point below 38 degrees C (100 degrees F) or whosetemperature is above its flash point, or a Class II combustible liquid having a vapor pressure not exceeding 40 psia (276 kPa) at 38 °C (100 °F) whose temperature is above its flash point.

Voltage. (Of a circuit.) The greatest root-mean-square (effective) difference of potential between any two conductors of the circuit concerned.

Voltage, nominal. A nominal value assigned to a circuit or system for the purpose of conveniently designating itsvoltage class (as 120/240, 480Y/277, 600, etc.). The actual voltage at which a circuit operates can vary from the nominal within a range that permits satisfactory operation of equipment.

Voltage to ground. For grounded circuits, the voltage between the given conductor and that point or conductor ofthe circuit that is grounded; for ungrounded circuits, the greatest voltage between the given conductor and any other conductor of the circuit.

Watertight. So constructed that moisture will not enter the enclosure.

Weatherproof. So constructed or protected that exposure to the weather will not interfere with successful operation. Rainproof, raintight, or watertight equipment can fulfill the requirements for weatherproof where varying weather conditions other than wetness, such as snow, ice, dust, or temperature extremes,are not a factor.

Wet location. See *"Location."*

Subpart L - Scaffolds

Authority: 40 U.S.C. 333; 29 U.S.C. 653, 655, 657; Secretary of Labor's Order Nos. 1-90 (55 FR 9033), 5-2007 (72 FR 31159), or 1-2012 (77 FR 3912); and 29 CFR part 1911.

Source: 61 FR 46104, Aug. 30, 1996, unless otherwise noted.

§ 1926.450 Scope, application and definitions applicable to this subpart.

(a) *Scope and application.* This subpart applies to all scaffolds used in workplaces covered by this part. It does not apply to crane or derrick suspended personnel platforms. The criteria for aerial lifts are set out exclusively in § 1926.453.

(b) *Definitions. Adjustable suspension scaffold* means a suspension scaffold equipped with a hoist(s) that can be operated by an employee(s) on the scaffold.

Bearer (putlog) means a horizontal transverse scaffold member (which may be supported by ledgers or runners) upon which the scaffold platform rests and which joins scaffold uprights, posts, poles, and similar members.

Boatswains' chair means a single-point adjustable suspension scaffold consisting of a seat or sling designed to support one employee in a sitting position.

Body belt (safety belt) means a strap with means both for securing it about the waist and for attaching it to a lanyard, lifeline, or deceleration device.

Body harness means a design of straps which may be secured about the employee in a manner to distribute the fall arrest forces over at least the thighs, pelvis, waist, chest and shoulders, with means for attaching it to other components of a personal fall arrest system.

Brace means a rigid connection that holds one scaffold member in a fixed position with respect to another member, or to a building or structure.

Bricklayers' square scaffold means a supported scaffold composed of framed squares which support a platform.

Carpenters' bracket scaffold means a supported scaffold consisting of a platform supported by brackets attached to building or structural walls.

Catenary scaffold means a suspension scaffold consisting of a platform supported by two essentially horizontal and parallel ropes attached to structural members of a building or other structure. Additional support may be provided by vertical pickups.

Chimney hoist means a multi-point adjustable suspension scaffold used to provide access to work inside chimneys. (See "Multi-point adjustable suspension scaffold".)

Cleat means a structural block used at the end of a platform to prevent the platform from slipping off its supports. Cleats are also used to provide footing on sloped surfaces such as crawling boards.

Competent person means one who is capable of identifying existing and predictable hazards in the surroundings or working conditions which are unsanitary, hazardous, or dangerous to employees, and who has authorization to take prompt corrective measures to eliminate them.

Continuous run scaffold (Run scaffold) means a two-point or multi-point adjustable suspension scaffold constructed using a series of interconnected braced scaffold members or supporting structures erected to form a continuous scaffold.

Coupler means a device for locking together the tubes of a tube and coupler scaffold.

Crawling board (chicken ladder) means a supported scaffold consisting of a plank with cleats spaced and secured to provide footing, for use on sloped surfaces such as roofs.

Deceleration device means any mechanism, such as a rope grab, rip-stitch lanyard, specially-woven lanyard, tearing or deforming lanyard, or automatic self-retracting lifeline lanyard, which dissipates a substantial amount of energy during a fall arrest or limits the energy imposed on an employee during fall arrest.

Double pole (independent pole) scaffold means a supported scaffold consisting of a platform(s) resting on cross beams (bearers) supported by ledgers and a double row of uprights independent of support (except ties, guys, braces) from any structure.

Equivalent means alternative designs, materials or methods to protect against a hazard which the employer can demonstrate will provide an equal or greater degree of safety for employees than the methods, materials or designs specified in the standard.

Exposed power lines means electrical power lines which are accessible to employees and which are not shielded from contact. Such lines do not include extension cords or power tool cords.

Eye or *Eye splice* means a loop with or without a thimble at the end of a wire rope.

Fabricated decking and planking means manufactured platforms made of wood (including laminated wood, and solid sawn wood planks), metal or other materials.

Fabricated frame scaffold (tubular welded frame scaffold) means a scaffold consisting of a platform(s) supported on fabricated end frames with integral posts, horizontal bearers, and intermediate members.

Failure means load refusal, breakage, or separation of component parts. Load refusal is the point where the ultimate strength is exceeded.

Float (ship) scaffold means a suspension scaffold consisting of a braced platform resting on two parallel bearers and hung from overhead supports by ropes of fixed length.

Form scaffold means a supported scaffold consisting of a platform supported by brackets attached to formwork.

Guardrail system means a vertical barrier, consisting of, but not limited to, toprails, midrails, and posts, erected to prevent employees from falling off a scaffold platform or walkway to lower levels.

Hoist means a manual or power-operated mechanical device to raise or lower a suspended scaffold.

Horse scaffold means a supported scaffold consisting of a platform supported by construction horses (saw horses). Horse scaffolds constructed of metal are sometimes known as trestle scaffolds.

Independent pole scaffold (see "Double pole scaffold").

Interior hung scaffold means a suspension scaffold consisting of a platform suspended from the ceiling or roof structure by fixed length supports.

Ladder jack scaffold means a supported scaffold consisting of a platform resting on brackets attached to ladders.

Ladder stand means a mobile, fixed-size, self-supporting ladder consisting of a wide flat tread ladder in the form of stairs.

Landing means a platform at the end of a flight of stairs.

Large area scaffold means a pole scaffold, tube and coupler scaffold, systems scaffold, or fabricated frame scaffold erected over substantially the entire work area. For example: a scaffold erected over the entire floor area of a room.

Lean-to scaffold means a supported scaffold which is kept erect by tilting it toward and resting it against a building or structure.

Lifeline means a component consisting of a flexible line that connects to an anchorage at one end to hang vertically (vertical lifeline), or that connects to anchorages at both ends to stretch horizontally (horizontal lifeline), and which serves as a means for connecting other components of a personal fall arrest system to the anchorage.

Lower levels means areas below the level where the employee is located and to which an employee can fall. Such areas include, but are not limited to, ground levels, floors, roofs, ramps, runways, excavations, pits, tanks, materials, water, and equipment.

Masons' adjustable supported scaffold (see "Self-contained adjustable scaffold").

Masons' multi-point adjustable suspension scaffold means a continuous run suspension scaffold designed and used for masonry operations.

Maximum intended load means the total load of all persons, equipment, tools, materials, transmitted loads, and other loads reasonably anticipated to be applied to a scaffold or scaffold component at any one time.

Mobile scaffold means a powered or unpowered, portable, caster or wheel-mounted supported scaffold.

Multi-level suspended scaffold means a two-point or multi-point adjustable suspension scaffold with a series of platforms at various levels resting on common stirrups.

Multi-point adjustable suspension scaffold means a suspension scaffold consisting of a platform(s) which is suspended by more than two ropes from overhead supports and equipped with means to raise and lower the platform to desired work levels. Such scaffolds include chimney hoists.

Needle beam scaffold means a platform suspended from needle beams.

Open sides and ends means the edges of a platform that are more than 14 inches (36 cm) away horizontally from a sturdy, continuous, vertical surface (such as a building wall) or a sturdy, continuous horizontal surface (such as a floor), or a point of access. Exception: For plastering and lathing operations the horizontal threshold distance is 18 inches (46 cm).

Outrigger means the structural member of a supported scaffold used to increase the base width of a scaffold in order to provide support for and increased stability of the scaffold.

Outrigger beam (Thrustout) means the structural member of a suspension scaffold or outrigger scaffold which provides support for the scaffold by extending the scaffold point of attachment to a point out and away from the structure or building.

Outrigger scaffold means a supported scaffold consisting of a platform resting on outrigger beams (thrustouts) projecting beyond the wall or face of the building or structure, the inboard ends of which are secured inside the building or structure.

Overhand bricklaying means the process of laying bricks and masonry units such that the surface of the wall to be jointed is on the opposite side of the wall from the mason, requiring the mason to lean over the wall to complete the work. It includes mason tending and electrical installation incorporated into the brick wall during the overhand bricklaying process.

Personal fall arrest system means a system used to arrest an employee's fall. It consists of an anchorage, connectors, a body belt or body harness and may include a lanyard, deceleration device, lifeline, or combinations of these.

Platform means a work surface elevated above lower levels. Platforms can be constructed using individual wood planks, fabricated planks, fabricated decks, and fabricated platforms.

Pole scaffold (see definitions for "Single-pole scaffold" and "Double (independent) pole scaffold").

Power operated hoist means a hoist which is powered by other than human energy.

Pump jack scaffold means a supported scaffold consisting of a platform supported by vertical poles and movable support brackets.

Qualified means one who, by possession of a recognized degree, certificate, or professional standing, or who by extensive knowledge, training, and experience, has successfully demonstrated his/her ability to solve or resolve problems related to the subject matter, the work, or the project.

Rated load means the manufacturer's specified maximum load to be lifted by a hoist or to be applied to a scaffold or scaffold component.

Repair bracket scaffold means a supported scaffold consisting of a platform supported by brackets which are secured in place around the circumference or perimeter of a chimney, stack, tank or other supporting structure by one or more wire ropes placed around the supporting structure.

Roof bracket scaffold means a rooftop supported scaffold consisting of a platform resting on angular-shaped supports.

Runner (ledger or ribbon) means the lengthwise horizontal spacing or bracing member which may support the bearers.

Scaffold means any temporary elevated platform (supported or suspended) and its supporting structure (including points of anchorage), used for supporting employees or materials or both.

Self-contained adjustable scaffold means a combination supported and suspension scaffold consisting of an adjustable platform(s) mounted on an independent supporting frame(s) not a part of the object being worked on, and which is equipped with a means to permit the raising and lowering of the platform(s). Such systems include rolling roof rigs, rolling outrigger systems, and some masons' adjustable supported scaffolds.

Shore scaffold means a supported scaffold which is placed against a building or structure and held inplace with props.

Single-point adjustable suspension scaffold   means a suspension scaffold consisting of a platform suspended by one rope from an overhead support and equipped with means to permit the movement of the platform to desired work levels.

Single-pole scaffold   means a supported scaffold consisting of a platform(s) resting on bearers, the outsideends of which are supported on runners secured to a single row of posts or uprights, and the inner ends of which are supported on or in a structure or building wall.

Stair tower (Scaffold stairway/tower)   means a tower comprised of scaffold components and which contains internal stairway units and rest platforms. These towers are used to provide access toscaffold platforms and other elevated points such as floors and roofs.

Stall load   means the load at which the prime-mover of a power-operated hoist stalls or the power to theprime-mover is automatically disconnected.

Step, platform, and trestle ladder scaffold   means a platform resting directly on the rungs of step ladders or trestle ladders.

Stilts   means a pair of poles or similar supports with raised footrests, used to permit walking above the ground or working surface.

Stonesetters' multi-point adjustable suspension scaffold   means a continuous run suspension scaffold designed and used for stonesetters' operations.

Supported scaffold   means one or more platforms supported by outrigger beams, brackets, poles, legs, uprights, posts, frames, or similar rigid support.

Suspension scaffold   means one or more platforms suspended by ropes or other non-rigid means from anoverhead structure(s).

System scaffold   means a scaffold consisting of posts with fixed connection points that accept runners, bearers, and diagonals that can be interconnected at predetermined levels.

Tank builders' scaffold   means a supported scaffold consisting of a platform resting on brackets that areeither directly attached to a cylindrical tank or attached to devices that are attached to such a tank.

Top plate bracket scaffold   means a scaffold supported by brackets that hook over or are attached to the top of a wall. This type of scaffold is similar to carpenters' bracket scaffolds and form scaffolds and is used in residential construction for setting trusses.

Tube and coupler scaffold   means a supported or suspended scaffold consisting of a platform(s) supportedby tubing, erected with coupling devices connecting uprights, braces, bearers, and runners.

Tubular welded frame scaffold   (see "Fabricated frame scaffold").

Two-point suspension scaffold (swing stage)   means a suspension scaffold consisting of a platform supported by hangers (stirrups) suspended by two ropes from overhead supports and equipped withmeans to permit the raising and lowering of the platform to desired work levels.

Unstable objects   means items whose strength, configuration, or lack of stability may allow them to becomedislocated and shift and therefore may not properly support the loads imposed on them. Unstable objects do not constitute a safe base support for scaffolds, platforms, or employees. Examples include, but are not limited to, barrels, boxes, loose brick, and concrete blocks.

Vertical pickup means a rope used to support the horizontal rope in catenary scaffolds.

Walkway means a portion of a scaffold platform used only for access and not as a work

Window jack scaffold

means a platform resting on a bracket or jack which projects through a window opening.

[61 FR 46104, Aug. 30, 1996, as amended at 75 FR 48133, Aug. 9, 2010]

§ 1926.451 General requirements.

This section does not apply to aerial lifts, the criteria for which are set out exclusively in § 1926.453.

(a) *Capacity.*

    (1) Except as provided in paragraphs (a)(2), (a)(3), (a)(4), (a)(5) and (g) of this section, each scaffold and scaffold component shall be capable of supporting, without failure, its own weight and at least 4 times the maximum intended load applied or transmitted to it.

    (2) Direct connections to roofs and floors, and counterweights used to balance adjustable suspensionscaffolds, shall be capable of resisting at least 4 times the tipping moment imposed by the scaffold operating at the rated load of the hoist, or 1.5 (minimum) times the tipping moment imposed by the scaffold operating at the stall load of the hoist, whichever is greater.

    (3) Each suspension rope, including connecting hardware, used on non-adjustable suspension scaffolds shall be capable of supporting, without failure, at least 6 times the maximum intended load appliedor transmitted to that rope.

    (4) Each suspension rope, including connecting hardware, used on adjustable suspension scaffolds shall be capable of supporting, without failure, at least 6 times the maximum intended load appliedor transmitted to that rope with the scaffold operating at either the rated load of the hoist, or 2 (minimum) times the stall load of the hoist, whichever is greater.

    (5) The stall load of any scaffold hoist shall not exceed 3 times its rated load.

    (6) Scaffolds shall be designed by a qualified person and shall be constructed and loaded in accordance with that design. Non-mandatory appendix A to this subpart contains examples of criteria that will enable an employer to comply with paragraph (a) of this section.

(b) *Scaffold platform construction.*

    (1) (ii) Each platform on all working levels of scaffolds shall be fully planked or decked between the front uprights and the guardrail supports as follows:

        (i) Each platform unit (e.g., scaffold plank, fabricated plank, fabricated deck, or fabricated platform) shall be installed so that the space between adjacent units and the space between the platform and the uprights is no more than 1 inch (2.5 cm) wide, except where the employer can demonstrate that a wider space is necessary (for example, to fit around uprights when sidebrackets are used to extend the width of the platform).

(ii) Where the employer makes the demonstration provided for in paragraph (b)(1)(i) of this section, the platform shall be planked or decked as fully as possible and the remaining open space between the platform and the uprights shall not exceed 9$^1/_2$ inches (24.1 cm).

Exception to paragraph (b)(1): The requirement in paragraph (b)(1) to provide full planking or decking does not apply to platforms used solely as walkways or solely by employees performing scaffold erection or dismantling. In these situations, only the planking that the employer establishes is necessary to provide safe working conditions is required.

(2) Except as provided in paragraphs (b)(2)(i) and (b)(2)(ii) of this section, each scaffold platform and walkway shall be at least 18 inches (46 cm) wide.

(i) Each ladder jack scaffold, top plate bracket scaffold, roof bracket scaffold, and pump jack scaffold shall be at least 12 inches (30 cm) wide. There is no minimum width requirement for boatswains' chairs.

Note to paragraph (b)(2)(i): Pursuant to an administrative stay effective November 29, 1996 and published in the FEDERAL REGISTER on November 25, 1996, the requirement in paragraph (b)(2)(i) that roof bracket scaffolds be at least 12 inches wide is stayed until November 25, 1997 or until rulemaking regarding the minimum width of roof bracket scaffolds has been completed, whichever is later.

Where scaffolds must be used in areas that the employer can demonstrate are so narrow that platforms and walkways cannot be at least 18 inches (46 cm) wide, such platforms and walkways shall be as wide as feasible, and employees on those platforms and walkways shall be protected from fall hazards by the use of guardrails and/or personal fall arrest systems.

(3) Except as provided in paragraphs (b)(3) (i) and (ii) of this section, the front edge of all platforms shall not be more than 14 inches (36 cm) from the face of the work, unless guardrail systems are erected along the front edge and/or personal fall arrest systems are used in accordance with paragraph (g) of this section to protect employees from falling.

(i) The maximum distance from the face for outrigger scaffolds shall be 3 inches (8 cm);

(ii) The maximum distance from the face for plastering and lathing operations shall be 18 inches (46 cm).

(4) Each end of a platform, unless cleated or otherwise restrained by hooks or equivalent means, shall extend over the centerline of its support at least 6 inches (15 cm).

(5)

(i) Each end of a platform 10 feet or less in length shall not extend over its support more than 12 inches (30 cm) unless the platform is designed and installed so that the cantilevered portion of the platform is able to support employees and/or materials without tipping, or has guardrails which block employee access to the cantilevered end.

(ii) Each platform greater than 10 feet in length shall not extend over its support more than 18 inches (46 cm), unless it is designed and installed so that the cantilevered portion of the platform is able to support employees without tipping, or has guardrails which block employee access to the cantilevered end.

(6) On scaffolds where scaffold planks are abutted to create a long platform, each abutted end shall rest on a separate support surface. This provision does not preclude the use of common support members, such as "T" sections, to support abutting planks, or hook on platforms designed to rest on common supports.

(7) On scaffolds where platforms are overlapped to create a long platform, the overlap shall occur only over supports, and shall not be less than 12 inches (30 cm) unless the platforms are nailed together or otherwise restrained to prevent movement.

(8) At all points of a scaffold where the platform changes direction, such as turning a corner, any platform that rests on a bearer at an angle other than a right angle shall be laid first, and platformswhich rest at right angles over the same bearer shall be laid second, on top of the first platform.

(9) Wood platforms shall not be covered with opaque finishes, except that platform edges may be covered or marked for identification. Platforms may be coated periodically with wood preservatives, fire-retardant finishes, and slip-resistant finishes; however, the coating may not obscure the top or bottom wood surfaces.

(10) Scaffold components manufactured by different manufacturers shall not be intermixed unless the components fit together without force and the scaffold's structural integrity is maintained by the user. Scaffold components manufactured by different manufacturers shall not be modified in order to intermix them unless a competent person determines the resulting scaffold is structurally sound.

(11) Scaffold components made of dissimilar metals shall not be used together unless a competent person has determined that galvanic action will not reduce the strength of any component to a levelbelow that required by paragraph (a)(1) of this section.

(c) **_Criteria for supported scaffolds._**

(1) Supported scaffolds with a height to base width (including outrigger supports, if used) ratio of more than four to one (4:1) shall be restrained from tipping by guying, tying, bracing, or equivalent means, as follows:

(i) Guys, ties, and braces shall be installed at locations where horizontal members support bothinner and outer legs.

(ii) Guys, ties, and braces shall be installed according to the scaffold manufacturer's recommendations or at the closest horizontal member to the 4:1 height and be repeated vertically at locations of horizontal members every 20 feet (6.1 m) or less thereafter for scaffolds 3 feet (0.91 m) wide or less, and every 26 feet (7.9 m) or less thereafter for scaffolds greater than 3 feet (0.91 m) wide. The top guy, tie or brace of completed scaffolds shall be placed no further than the 4:1 height from the top. Such guys, ties and braces shall be installedat each end of the scaffold and at horizontal intervals not to exceed 30 feet (9.1 m) (measured from one end [not both] towards the other).

(iii) Ties, guys, braces, or outriggers shall be used to prevent the tipping of supported scaffolds in all circumstances where an eccentric load, such as a cantilevered work platform, is applied or istransmitted to the scaffold.

(2) Supported scaffold poles, legs, posts, frames, and uprights shall bear on base plates and mud sills or other adequate firm foundation.

(i) Footings shall be level, sound, rigid, and capable of supporting the loaded scaffold withoutsettling or displacement.

- (ii) Unstable objects shall not be used to support scaffolds or platform
- (iii) units.Unstable objects shall not be used as working platforms.
- (iv) Front-end loaders and similar pieces of equipment shall not be used to support scaffold platforms unless they have been specifically designed by the manufacturer for such use.
- (v) Fork-lifts shall not be used to support scaffold platforms unless the entire platform is attached to the fork and the fork-lift is not moved horizontally while the platform is occupied.

(3) Supported scaffold poles, legs, posts, frames, and uprights shall be plumb and braced to preventswaying and displacement.

(d) *Criteria for suspension scaffolds.*

(1) All suspension scaffold support devices, such as outrigger beams, cornice hooks, parapet clamps, and similar devices, shall rest on surfaces capable of supporting at least 4 times the load imposedon them by the scaffold operating at the rated load of the hoist (or at least 1.5 times the load imposed on them by the scaffold at the stall capacity of the hoist, whichever is greater).

(2) Suspension scaffold outrigger beams, when used, shall be made of structural metal or equivalent strength material, and shall be restrained to prevent movement.

(3) The inboard ends of suspension scaffold outrigger beams shall be stabilized by bolts or other direct connections to the floor or roof deck, or they shall have their inboard ends stabilized by counterweights, except masons' multi-point adjustable suspension scaffold outrigger beams shall not be stabilized by counterweights.

- (i) Before the scaffold is used, direct connections shall be evaluated by a competent person who shall confirm, based on the evaluation, that the supporting surfaces are capable of supportingthe loads to be imposed. In addition, masons' multi-point adjustable suspension scaffold connections shall be designed by an engineer experienced in such scaffold design.
- (ii) Counterweights shall be made of non-flowable material. Sand, gravel and similar materials thatcan be easily dislocated shall not be used as counterweights.
- (iii) Only those items specifically designed as counterweights shall be used to counterweight scaffold systems. Construction materials such as, but not limited to, masonry units and rolls ofroofing felt, shall not be used as counterweights.
- (iv) Counterweights shall be secured by mechanical means to the outrigger beams to prevent accidental displacement.
- (v) Counterweights shall not be removed from an outrigger beam until the scaffold is disassembled.
- (vi) Outrigger beams which are not stabilized by bolts or other direct connections to the floor or roof deck shall be secured by tiebacks.
- (vii) Tiebacks shall be equivalent in strength to the suspension ropes.
- (viii) Outrigger beams shall be placed perpendicular to its bearing support (usually the face

of thebuilding or structure). However, where the employer can demonstrate that it is not possible toplace an outrigger beam perpendicular to the face of the building or structure because of obstructions that cannot be moved, the outrigger beam may be placed at some other angle, provided opposing angle tiebacks are used.

(ix) Tiebacks shall be secured to a structurally sound anchorage on the building or structure. Soundanchorages include structural members, but do not include standpipes, vents, other piping systems, or electrical conduit.

(x) Tiebacks shall be installed perpendicular to the face of the building or structure, or opposing angle tiebacks shall be installed. Single tiebacks installed at an angle are prohibited.

(4) Suspension scaffold outrigger beams shall be:

(i) Provided with stop bolts or shackles at both

(ii) ends;

Securely fastened together with the flanges turned out when channel iron beams are used inplace of I-beams;

(iii) Installed with all bearing supports perpendicular to the beam center

(iv) line;Set and maintained with the web in a vertical position; and

(v) When an outrigger beam is used, the shackle or clevis with which the rope is attached to the outrigger beam shall be placed directly over the center line of the stirrup.

(5) Suspension scaffold support devices such as cornice hooks, roof hooks, roof irons, parapet clamps,or similar devices shall be:

(i) Made of steel, wrought iron, or materials of equivalent strength;

(ii) Supported by bearing blocks; and

(iii) Secured against movement by tiebacks installed at right angles to the face of the building or structure, or opposing angle tiebacks shall be installed and secured to a structurally sound point of anchorage on the building or structure. Sound points of anchorage include structuralmembers, but do not include standpipes, vents, other piping systems, or electrical conduit.

(iv) Tiebacks shall be equivalent in strength to the hoisting rope.

(6) When winding drum hoists are used on a suspension scaffold, they shall contain not less than fourwraps of the suspension rope at the lowest point of scaffold travel. When other types of hoists are used, the suspension ropes shall be long enough to allow the scaffold to be lowered to the level below without the rope end passing through the hoist, or the rope end shall be configured or provided with means to prevent the end from passing through the hoist.

(7) The use of repaired wire rope as suspension rope is prohibited.

(8) Wire suspension ropes shall not be joined together except through the use of eye splice thimbles connected with shackles or coverplates and bolts.

(9) The load end of wire suspension ropes shall be equipped with proper size thimbles and secured

byeyesplicing or equivalent means.

(10) Ropes shall be inspected for defects by a competent person prior to each workshift and after everyoccurrence which could affect a rope's integrity. Ropes shall be replaced if any of the following conditions exist:

   (i) Any physical damage which impairs the function and strength of the rope.

   (ii) Kinks that might impair the tracking or wrapping of rope around the drum(s) or sheave(s).

   (iii) Six randomly distributed broken wires in one rope lay or three broken wires in one strand in onerope lay.

   (iv) Abrasion, corrosion, scrubbing, flattening or peening causing loss of more than one-third of theoriginal diameter of the outside wires.

   (v) Heat damage caused by a torch or any damage caused by contact with electrical wires.

   (vi) Evidence that the secondary brake has been activated during an overspeed condition and hasengaged the suspension rope.

(11) Swaged attachments or spliced eyes on wire suspension ropes shall not be used unless they aremade by the wire rope manufacturer or a qualified person.

(12) When wire rope clips are used on suspension scaffolds:

   (i) There shall be a minimum of 3 wire rope clips installed, with the clips a minimum of 6 rope diameters apart;

   (ii) Clips shall be installed according to the manufacturer's recommendations;

   (iii) Clips shall be retightened to the manufacturer's recommendations after the initial loading;

   (iv) Clips shall be inspected and retightened to the manufacturer's recommendations at the start ofeach workshift thereafter;

   (v) U-bolt clips shall not be used at the point of suspension for any scaffold hoist;

   (vi) When U-bolt clips are used, the U-bolt shall be placed over the dead end of the rope, and the saddle shall be placed over the live end of the rope.

(13) Suspension scaffold power-operated hoists and manual hoists shall be tested by a qualified testinglaboratory.

(14) Gasoline-powered equipment and hoists shall not be used on suspension scaffolds.

(15) Gears and brakes of power-operated hoists used on suspension scaffolds shall be enclosed.

(16) In addition to the normal operating brake, suspension scaffold power-operated hoists and manuallyoperated hoists shall have a braking device or locking pawl which engages automatically when a hoist makes either of the following uncontrolled movements: an instantaneous change in momentum or an accelerated overspeed.

(17) Manually operated hoists shall require a positive crank force to descend.

(18) Two-point and multi-point suspension scaffolds shall be tied or otherwise secured to prevent them from swaying, as determined to be necessary based on an evaluation by a competent person.
Window cleaners' anchors shall not be used for this purpose.

(19) Devices whose sole function is to provide emergency escape and rescue shall not be used as working platforms. This provision does not preclude the use of systems which are designed to function both as suspension scaffolds and emergency systems.

(e) **Access.** This paragraph applies to scaffold access for all employees. Access requirements for employees erecting or dismantling supported scaffolds are

(1) When scaffold platforms are more than 2 feet (0.6 m) above or below a point of access, portable ladders, hook-on ladders, attachable ladders, stair towers (scaffold stairways/towers), stairway-type ladders (such as ladder stands), ramps, walkways, integral prefabricated scaffold access, or direct access from another scaffold, structure, personnel hoist, or similar surface shall be used.
Crossbraces shall not be used as a means of access.

(2) Portable, hook-on, and attachable ladders (Additional requirements for the proper construction anduse of portable ladders are contained in subpart X of this part - Stairways and Ladders):

   (i) Portable, hook-on, and attachable ladders shall be positioned so as not to tip the scaffold;

   (ii) Hook-on and attachable ladders shall be positioned so that their bottom rung is not more than24 inches (61 cm) above the scaffold supporting level;

   (iii) When hook-on and attachable ladders are used on a supported scaffold more than 35 feet (10.7m) high, they shall have rest platforms at 35-foot (10.7 m) maximum vertical intervals.

   (iv) Hook-on and attachable ladders shall be specifically designed for use with the type of scaffoldused;

   (v) Hook-on and attachable ladders shall have a minimum rung length of $11^1/_2$ inches (29 cm); and

   (vi) Hook-on and attachable ladders shall have uniformly spaced rungs with a maximum spacing between rungs of $16^3/_4$ inches.

(3) Stairway-type ladders shall:

   (i) Be positioned such that their bottom step is not more than 24 inches (61 cm) above the scaffold supporting level;

   (ii) Be provided with rest platforms at 12 foot (3.7 m) maximum vertical intervals;

   (iii) Have a minimum step width of 16 inches (41 cm), except that mobile scaffold stairway-type ladders shall have a minimum step width of $11^1/_2$ inches (30 cm); and

   (iv) Have slip-resistant treads on all steps and landings.

(4) Stairtowers (scaffold stairway/towers) shall be positioned such that their bottom step is not morethan 24 inches (61 cm.) above the scaffold supporting level.

   (i) A stairrail consisting of a toprail and a midrail shall be provided on each side of each scaffoldstairway.

   (ii) The toprail of each stairrail system shall also be capable of serving as a handrail, unless a separate handrail is provided.

   (iii) Handrails, and toprails that serve as handrails, shall provide an adequate handhold for employees grasping them to avoid falling.

(iv) Stairrail systems and handrails shall be surfaced to prevent injury to employees from punctures or lacerations, and to prevent snagging of clothing.

(v) The ends of stairrail systems and handrails shall be constructed so that they do not constitute a projection hazard.

(vi) Handrails, and toprails that are used as handrails, shall be at least 3 inches (7.6 cm) from other objects.

(vii) Stairrails shall be not less than 28 inches (71 cm) nor more than 37 inches (94 cm) from the upper surface of the stairrail to the surface of the tread, in line with the face of the riser at the forward edge of the tread.

(viii) A landing platform at least 18 inches (45.7 cm) wide by at least 18 inches (45.7 cm) long shall be provided at each level.

(ix) Each scaffold stairway shall be at least 18 inches (45.7 cm) wide between stairrails.

(x) Treads and landings shall have slip-resistant surfaces.

(xi) Stairways shall be installed between 40 degrees and 60 degrees from the horizontal.

(xii) Guardrails meeting the requirements of paragraph (g)(4) of this section shall be provided on the open sides and ends of each landing.

(xiii) Riser height shall be uniform, within $1/4$ inch, (0.6 cm) for each flight of stairs. Greater variations in riser height are allowed for the top and bottom steps of the entire system, not for each flight of stairs.

(xiv) Tread depth shall be uniform, within $1/4$ inch, for each flight of stairs.

(5) Ramps and walkways.

(i) Ramps and walkways 6 feet (1.8 m) or more above lower levels shall have guardrail systems which comply with subpart M of this part - Fall Protection;

(f) *Use.*

(1)

(ii) No ramp or walkway shall be inclined more than a slope of one (1) vertical to three (3) horizontal (20 degrees above the horizontal).

(iii) If the slope of a ramp or a walkway is steeper than one (1) vertical in eight (8) horizontal, the ramp or walkway shall have cleats not more than fourteen (14) inches (35 cm) apart which are securely fastened to the planks to provide footing.

Integral prefabricated scaffold access frames shall:

(i) Be specifically designed and constructed for use as ladder rungs; Have a rung

(ii) length at least 8 inches (20 cm);

(iii) Not be used as work platforms when rungs are less than $11^{1}/_{2}$ inches in length, unless each affected employee uses fall protection, or a positioning device, which complies with § 1926.502;

(iv) Be uniformly spaced within each frame section;

(v) Be provided with rest platforms at 35-foot (10.7 m) maximum vertical intervals on all supported scaffolds more than 35 feet (10.7 m) high; and

Have a (vi) mum spacing between rungs of 16³⁄₄ inches (43 cm). Non-uniform rung spacingcaused by joining end frames together is allowed, provided the resulting spacing does not exceed 16³⁄₄ inches (43 cm).

Steps and rungs of ladder and stairway type access shall line up vertically with each other betweenrest platforms.

Direct access to or from another surface shall be used only when the scaffold is not more than 14 inches (36 cm) horizontally and not more than 24 inches (61 cm) vertically from the other surface.

Effective September 2, 1997, access for employees erecting or dismantling supported scaffolds shallbe in accordance with the following:

The en (i) er shall provide safe means of access for each employee erecting or dismantling ascaffold where the provision of safe access is feasible and does not create a greater hazard. The employer shall have a competent person determine whether it is feasible or would pose a greater hazard to provide, and have employees use a safe means of access. This determinationshall be based on site conditions and the type of scaffold being erected or dismantled.

Hook-c (ii) ttachable ladders shall be installed as soon as scaffold erection has progressed toa point that permits safe installation and use.

When (iii) ng or dismantling tubular welded frame scaffolds, (end) frames, with horizontal members that are parallel, level and are not more than 22 inches apart vertically may be usedas climbing devices for access, provided they are erected in a manner that creates a usable ladder and provides good hand hold and foot space.

Cross (iv) s on tubular welded frame scaffolds shall not be used as a means of access oregress.

(2) The use of shore or lean-to scaffolds is prohibited.

(3) Scaffolds and scaffold components shall be inspected for visible defects by a competent person before each work shift, and after any occurrence which could affect a scaffold's structural integrity.

(4) Any part of a scaffold damaged or weakened such that its strength is less than that required by paragraph (a) of this section shall be immediately repaired or replaced, braced to meet those provisions, or removed from service until repaired.

(5) Scaffolds shall not be moved horizontally while employees are on them, unless they have been designed by a registered professional engineer specifically for such movement or, for mobile scaffolds, where the provisions of § 1926.452(w) are followed.

(6) The clearance between scaffolds and power lines shall be as follows: Scaffolds shall not be erected, used, dismantled, altered, or moved such that they or any conductive material handled on them might come closer to exposed and energized power lines than as follows:

Insulated lines voltage	Minimum distance	Alternatives

Less than 300 volts	3 feet (0.9 m)	
300 volts to 50 kv	10 feet (3.1m)	
More than 50 kv	10 feet (3.1 m) plus 0.4 inches (1.0 cm) for each 1 kv over 50 kv	2 times the length of the line insulator, but never less than 10 feet (3.1 m).

Uninsulated lines voltage	Minimum distance	Alternatives
Less than 50 kv	10 feet (3.1 m)	
More than 50 kv	10 feet (3.1 m) plus 0.4 inches (1.0 cm) for each 1 kv over 50 kv	2 times the length of the line insulator, but never less than 10 feet (3.1 m).

Exception to paragraph (f)(6): Scaffolds and materials may be closer to power lines than specified above where such clearance is necessary for performance of work, and only after the utility company, or electrical system operator, has been notified of the need to work closer and the utility company, or electrical system operator, has deenergized the lines, relocated the lines, or installed protective coverings to prevent accidental contact with the lines.

> Scaffolds shall be erected, moved, dismantled, or altered only under the supervision and direction of a competent person qualified in scaffold erection, moving, dismantling or alteration. Such activities shall be performed only by experienced and trained employees selected for such work by the competent person.

(8) Employees shall be prohibited from working on scaffolds covered with snow, ice, or other slippery material except as necessary for removal of such materials.

(9) Where swinging loads are being hoisted onto or near scaffolds such that the loads might contact the scaffold, tag lines or equivalent measures to control the loads shall be used.

(10) Suspension ropes supporting adjustable suspension scaffolds shall be of a diameter large enough to provide sufficient surface area for the functioning of brake and hoist mechanisms.

(11) Suspension ropes shall be shielded from heat-producing processes. When acids or other corrosive substances are used on a scaffold, the ropes shall be shielded, treated to protect against the corrosive substances, or shall be of a material that will not be damaged by the substance being used.

(12) Work on or from scaffolds is prohibited during storms or high winds unless a competent person has determined that it is safe for employees to be on the scaffold and those employees are protected by a personal fall arrest system or wind screens. Wind screens shall not be used unless the scaffold is secured against the anticipated wind forces imposed.

(13) Debris shall not be allowed to accumulate on platforms.

(14) Makeshift devices, such as but not limited to boxes and barrels, shall not be used on top of scaffold platforms to increase the working level height of employees.

(15) Ladders shall not be used on scaffolds to increase the working level height of employees, except on large area scaffolds where employers have satisfied the following criteria:

    (i) When the ladder is placed against a structure which is not a part of the scaffold, the scaffoldshall be secured against the sideways thrust exerted by the ladder;

    (ii) The platform units shall be secured to the scaffold to prevent their movement;

    (iii) The ladder legs shall be on the same platform or other means shall be provided to stabilize theladder against unequal platform deflection, and

    (iv) The ladder legs shall be secured to prevent them from slipping or being pushed off the platform.

(16) Platforms shall not deflect more than $1/60$ of the span when loaded.

(17) To reduce the possibility of welding current arcing through the suspension wire rope when performing welding from suspended scaffolds, the following precautions shall be taken, as applicable:

    (i) An insulated thimble shall be used to attach each suspension wire rope to its hanging support(such as cornice hook or outrigger). Excess suspension wire rope and any additional independent lines from grounding shall be insulated;

    (ii) The suspension wire rope shall be covered with insulating material extending at least 4 feet (1.2m) above the hoist. If there is a tail line below the hoist, it shall be insulated to prevent contactwith the platform. The portion of the tail line that hangs free below the scaffold shall be guided or retained, or both, so that it does not become grounded;

    (iii) Each hoist shall be covered with insulated protective covers;

    (iv) In addition to a work lead attachment required by the welding process, a grounding conductorshall be connected from the scaffold to the structure. The size of this conductor shall be at least the size of the welding process work lead, and this conductor shall not be in series with the welding process or the work piece;

    (v) If the scaffold grounding lead is disconnected at any time, the welding machine shall be shutoff; and

    (vi) An active welding rod or uninsulated welding lead shall not be allowed to contact the scaffoldor its suspension system.

(g) *Fall protection.*

    (1) Each employee on a scaffold more than 10 feet (3.1 m) above a lower level shall be protected from falling to that lower level. Paragraphs (g)(1) (i) through (vii) of this section establish the types of fall protection to be provided to the employees on each type of scaffold. Paragraph (g)(2) of this sectionaddresses fall protection for scaffold erectors and dismantlers.

    Note to paragraph (g)(1): The fall protection requirements for employees installing suspension scaffold support systems on floors, roofs, and other elevated surfaces are setforth in subpart M of this part.

    Each employee on a boatswains' chair, catenary scaffold, float scaffold, needle beam scaffold, or ladder jack scaffold shall be protected by a personal fall arrest system;

(i)

(ii) Each employee on a single-point or two-point adjustable suspension scaffold shall be protected by both a personal fall arrest system and guardrail system;

(iii) Each employee on a crawling board (chicken ladder) shall be protected by a personal fall arrest system, a guardrail system (with minimum 200 pound toprail capacity), or by a three-fourth inch(1.9 cm) diameter grabline or equivalent handhold securely fastened beside each crawling board;

(iv) Each employee on a self-contained adjustable scaffold shall be protected by a guardrail system(with minimum 200 pound toprail capacity) when the platform is supported by the frame structure, and by both a personal fall arrest system and a guardrail system (with minimum 200 pound toprail capacity) when the platform is supported by ropes;

(v) Each employee on a walkway located within a scaffold shall be protected by a guardrail system(with minimum 200 pound toprail capacity) installed within $9^1/_2$ inches (24.1 cm) of and alongat least one side of the walkway.

(vi) Each employee performing overhand bricklaying operations from a supported scaffold shall beprotected from falling from all open sides and ends of the scaffold (except at the side next to the wall being laid) by the use of a personal fall arrest system or guardrail system (with minimum 200 pound toprail capacity).

(vii) For all scaffolds not otherwise specified in paragraphs (g)(1)(i) through (g)(1)(vi) of this section, each employee shall be protected by the use of personal fall arrest systems or guardrail systems meeting the requirements of paragraph (g)(4) of this section.

(2) Effective September 2, 1997, the employer shall have a competent person determine the feasibility and safety of providing fall protection for employees erecting or dismantling supported scaffolds. Employers are required to provide fall protection for employees erecting or dismantling supported scaffolds where the installation and use of such protection is feasible and does not create a greaterhazard.

(3) In addition to meeting the requirements of § 1926.502(d), personal fall arrest systems used on scaffolds shall be attached by lanyard to a vertical lifeline, horizontal lifeline, or scaffold structural member. Vertical lifelines shall not be used when overhead components, such as overhead protection or additional platform levels, are part of a single-point or two-point adjustable suspension scaffold.

   (i) When vertical lifelines are used, they shall be fastened to a fixed safe point of anchorage, shall be independent of the scaffold, and shall be protected from sharp edges and abrasion. Safe points of anchorage include structural members of buildings, but do not include standpipes, vents, other piping systems, electrical conduit, outrigger beams, or counterweights.

   (ii) When horizontal lifelines are used, they shall be secured to two or more structural members of the scaffold, or they may be looped around both suspension and independent suspension lines(on scaffolds so equipped) above the hoist and brake attached to the end of the scaffold.
   Horizontal lifelines shall not be attached only to the suspension ropes.

   (iii) When lanyards are connected to horizontal lifelines or structural members on a single-point ortwo-point adjustable suspension scaffold, the scaffold shall be equipped with additional independent support lines and automatic locking devices capable of stopping the fall of the scaffold in the event one or both of the suspension ropes fail. The independent support linesshall be equal in number and strength to the suspension ropes.

(iv) Vertical lifelines, independent support lines, and suspension ropes shall not be attached to eachother, nor shall they be attached to or use the same point of anchorage, nor shall they be attached to the same point on the scaffold or personal fall arrest system.

(4) Guardrail systems installed to meet the requirements of this section shall comply with the followingprovisions (guardrail systems built in accordance with appendix A to this subpart will be deemed to meet the requirements of paragraphs (g)(4) (vii), (viii), and (ix) of this section):

(i) Guardrail systems shall be installed along all open sides and ends of platforms. Guardrail systems shall be installed before the scaffold is released for use by employees other thanerection/dismantling crews.

(ii) The top edge height of toprails or equivalent member on supported scaffolds manufactured or placed in service after January 1, 2000 shall be installed between 38 inches (0.97 m) and 45 inches (1.2 m) above the platform surface. The top edge height on supported scaffolds manufactured and placed in service before January 1, 2000, and on all suspended scaffolds where both a guardrail and a personal fall arrest system are required shall be between 36 inches (0.9 m) and 45 inches (1.2 m). When conditions warrant, the height of the top edge mayexceed the 45-inch height, provided the guardrail system meets all other criteria of paragraph (g)(4).

(iii) When midrails, screens, mesh, intermediate vertical members, solid panels, or equivalent structural members are used, they shall be installed between the top edge of the guardrail system and the scaffold platform.

(iv) When midrails are used, they shall be installed at a height approximately midway between thetop edge of the guardrail system and the platform surface.

(v) When screens and mesh are used, they shall extend from the top edge of the guardrail systemto the scaffold platform, and along the entire opening between the supports.

(vi) When intermediate members (such as balusters or additional rails) are used, they shall not bemore than 19 inches (48 cm) apart.

(vii) Each toprail or equivalent member of a guardrail system shall be capable of withstanding, without failure, a force applied in any downward or horizontal direction at any point along its topedge of at least 100 pounds (445 n) for guardrail systems installed on single-point adjustable suspension scaffolds or two-point adjustable suspension scaffolds, and at least 200 pounds (890 n) for guardrail systems installed on all other scaffolds.

(viii) When the loads specified in paragraph (g)(4)(vii) of this section are applied in a downward direction, the top edge shall not drop below the height above the platform surface that is prescribed in paragraph (g)(4)(ii) of this section.

(ix) Midrails, screens, mesh, intermediate vertical members, solid panels, and equivalent structuralmembers of a guardrail system shall be capable of withstanding, without failure, a force appliedin any downward or horizontal direction at any point along the midrail or other member of at least 75 pounds (333 n) for guardrail systems with a minimum 100 pound toprail capacity, and at least 150 pounds (666 n) for guardrail systems with a minimum 200 pound toprail capacity.

(x) Suspension scaffold hoists and non-walk-through stirrups may be used as end guardrails, if the space between the hoist or stirrup and the side guardrail or structure does not allow passage of an employee to the end of the scaffold.

(xi) Guardrails shall be surfaced to prevent injury to an employee from punctures or lacerations, and to prevent snagging of clothing.

(xii) The ends of all rails shall not overhang the terminal posts except when such overhang does not constitute a projection hazard to employees.

(xiii) Steel or plastic banding shall not be used as a toprail or midrail.

(xiv) Manila or plastic (or other synthetic) rope being used for toprails or midrails shall be inspected by a competent person as frequently as necessary to ensure that it continues to meet the strength requirements of paragraph (g) of this section.

(xv) Crossbracing is acceptable in place of a midrail when the crossing point of two braces is between 20 inches (0.5 m) and 30 inches (0.8 m) above the work platform or as a toprail when the crossing point of two braces is between 38 inches (0.97 m) and 48 inches (1.3 m) above the work platform. The end points at each upright shall be no more than 48 inches (1.3 m) apart.

(h) **Falling object protection.**

(1) In addition to wearing hardhats each employee on a scaffold shall be provided with additional protection from falling hand tools, debris, and other small objects through the installation of toeboards, screens, or guardrail systems, or through the erection of debris nets, catch platforms, or canopy structures that contain or deflect the falling objects. When the falling objects are too large,

heavy or massive to be contained or deflected by any of the above-listed measures, the employer shall place such potential falling objects away from the edge of the surface from which they could fall and shall secure those materials as necessary to prevent their falling.

(2) Where there is a danger of tools, materials, or equipment falling from a scaffold and striking employees below, the following provisions apply:

(i) The area below the scaffold to which objects can fall shall be barricaded, and employees shall not be permitted to enter the hazard area; or

(ii) A toeboard shall be erected along the edge of platforms more than 10 feet (3.1 m) above lower levels for a distance sufficient to protect employees below, except on float (ship) scaffolds where an edging of $3/4 \times 1 1/2$ inch (2 × 4 cm) wood or equivalent may be used in lieu of toeboards;

(iii) Where tools, materials, or equipment are piled to a height higher than the top edge of the toeboard, paneling or screening extending from the toeboard or platform to the top of the guardrail shall be erected for a distance sufficient to protect employees below; or

(iv) A guardrail system shall be installed with openings small enough to prevent passage of potential falling objects; or

(v) A canopy structure, debris net, or catch platform strong enough to withstand the impact forces of the potential falling objects shall be erected over the employees below.

(3) Canopies, when used for falling object protection, shall comply with the following

(i) criteria: Canopies shall be installed between the falling object hazard and the

(ii)

employees.

When canopies are used on suspension scaffolds for falling object protection, the scaffold shall be equipped with additional independent support lines equal in number to the number of points supported, and equivalent in strength to the strength of the suspension ropes.

(iii) Independent support lines and suspension ropes shall not be attached to the same points of anchorage.

(4) Where used, toeboards shall be:

(i) Capable of withstanding, without failure, a force of at least 50 pounds (222 n) applied in any downward or horizontal direction at any point along the toeboard (toeboards built in accordance with appendix A to this subpart will be deemed to meet this requirement); and

(ii) At least three and one-half inches (9 cm) high from the top edge of the toeboard to the level of the walking/working surface. Toeboards shall be securely fastened in place at the outermost edge of the platform and have not more than $1/4$ inch (0.7 cm) clearance above the walking/ working surface. Toeboards shall be solid or with openings not over one inch (2.5 cm) in the greatest dimension.

[61 FR 46107, Aug. 30, 1996, as corrected and amended at 61 FR 59831, 59832, Nov. 25, 1996]

Effective Date Note: At 61 FR 59832, Nov. 25, 1996, § 1926.451(b)(2)(i) was amended and certain requirements stayed until Nov. 25, 1997, or until further rulemaking has been completed, whichever is later.

§ 1926.452 Additional requirements applicable to specific types of scaffolds.

In addition to the applicable requirements of § 1926.451, the following requirements apply to the specific types of scaffolds indicated. Scaffolds not specifically addressed by § 1926.452, such as but not limited to systems scaffolds, must meet the requirements of § 1926.451.

(a) *Pole scaffolds.*

(1) When platforms are being moved to the next level, the existing platform shall be left undisturbed until the new bearers have been set in place and braced, prior to receiving the new platforms.

(2) Crossbracing shall be installed between the inner and outer sets of poles on double pole scaffolds.

(3) Diagonal bracing in both directions shall be installed across the entire inside face of double-pole scaffolds used to support loads equivalent to a uniformly distributed load of 50 pounds (22.7 kg) or more per square foot (929 square cm).

(4) Diagonal bracing in both directions shall be installed across the entire outside face of all double- and single-pole scaffolds.

(5) Runners and bearers shall be installed on edge.

(6) Bearers shall extend a minimum of 3 inches (7.6 cm) over the outside edges of runners.

(7) Runners shall extend over a minimum of two poles, and shall be supported by bearing blocks securely attached to the poles.

(8) Braces, bearers, and runners shall not be spliced between poles.

(9) Where wooden poles are spliced, the ends shall be squared and the upper section shall rest squarely on the lower section. Wood splice plates shall be provided on at least two adjacent sides, and shall extend at least 2 feet (0.6 m) on either side of the splice, overlap the abutted ends equally, and have at least the same cross-sectional areas as the pole. Splice plates of other materials of equivalent strength may be used.

(10) Pole scaffolds over 60 feet in height shall be designed by a registered professional engineer, and shall be constructed and loaded in accordance with that design. Non-mandatory appendix A to this subpart contains examples of criteria that will enable an employer to comply with design and loading requirements for pole scaffolds under 60 feet in height.

(b) *Tube and coupler scaffolds.*

(1) When platforms are being moved to the next level, the existing platform shall be left undisturbed until the new bearers have been set in place and braced prior to receiving the new platforms.

(2) Transverse bracing forming an "X" across the width of the scaffold shall be installed at the scaffold ends and at least at every third set of posts horizontally (measured from only one end) and every fourth runner vertically. Bracing shall extend diagonally from the inner or outer posts or runners upward to the next outer or inner posts or runners. Building ties shall be installed at the bearer levels between the transverse bracing and shall conform to the requirements of § 1926.451(c)(1).

(3) On straight run scaffolds, longitudinal bracing across the inner and outer rows of posts shall be installed diagonally in both directions, and shall extend from the base of the end posts upward to the top of the scaffold at approximately a 45 degree angle. On scaffolds whose length is greater than their height, such bracing shall be repeated beginning at least at every fifth post. On scaffolds whose

length is less than their height, such bracing shall be installed from the base of the end posts upward to the opposite end posts, and then in alternating directions until reaching the top of the scaffold.
Bracing shall be installed as close as possible to the intersection of the bearer and post or runner and post.

(4) Where conditions preclude the attachment of bracing to posts, bracing shall be attached to the runners as close to the post as possible.

(5) Bearers shall be installed transversely between posts, and when coupled to the posts, shall have the inboard coupler bear directly on the runner coupler. When the bearers are coupled to the runners, the couplers shall be as close to the posts as possible.

(6) Bearers shall extend beyond the posts and runners, and shall provide full contact with the coupler.

(7) Runners shall be installed along the length of the scaffold, located on both the inside and outside posts at level heights (when tube and coupler guardrails and midrails are used on outside posts, they may be used in lieu of outside runners).

(8) Runners shall be interlocked on straight runs to form continuous lengths, and shall be coupled to each post. The bottom runners and bearers shall be located as close to the base as possible.

(9) Couplers shall be of a structural metal, such as drop-forged steel, malleable iron, or structural gradealuminum. The use of gray cast iron is prohibited.

(10) Tube and coupler scaffolds over 125 feet in height shall be designed by a registered professional engineer, and shall be constructed and loaded in accordance with such design. Non-mandatory appendix A to this subpart contains examples of criteria that will enable an employer to comply withdesign and loading requirements for tube and coupler scaffolds under 125 feet in height.

(c) *Fabricated frame scaffolds* (tubular welded frame scaffolds).

(1) When moving platforms to the next level, the existing platform shall be left undisturbed until the new end frames have been set in place and braced prior to receiving the new platforms.

(2) Frames and panels shall be braced by cross, horizontal, or diagonal braces, or combination thereof, which secure vertical members together laterally. The cross braces shall be of such length as will automatically square and align vertical members so that the erected scaffold is always plumb, level,and square. All brace connections shall be secured.

(3) Frames and panels shall be joined together vertically by coupling or stacking pins or equivalent means.

(4) Where uplift can occur which would displace scaffold end frames or panels, the frames or panelsshall be locked together vertically by pins or equivalent means.

(5) Brackets used to support cantilevered loads shall:

(i) Be seated with side-brackets parallel to the frames and end-brackets at 90 degrees to the frames;

(ii) Not be bent or twisted from these positions; and

(iii) Be used only to support personnel, unless the scaffold has been designed for other loads by aqualified engineer and built to withstand the tipping forces caused by those other loads being placed on the bracket-supported section of the scaffold.

(6) Scaffolds over 125 feet (38.0 m) in height above their base plates shall be designed by a registeredprofessional engineer, and shall be constructed and loaded in accordance with such design.

(d) *Plasterers', decorators', and large area scaffolds.* Scaffolds shall be constructed in accordance withparagraphs (a), (b), or (c) of this section, as appropriate.

(e) *Bricklayers' square scaffolds (squares).*

(1) Scaffolds made of wood shall be reinforced with gussets on both sides of each

(2) corner.Diagonal braces shall be installed on all sides of each square.

(3) Diagonal braces shall be installed between squares on the rear and front sides of the scaffold, andshall extend from the bottom of each square to the top of the next square.

(4) Scaffolds shall not exceed three tiers in height, and shall be so constructed and arranged that onesquare rests directly above the other. The upper tiers shall stand on a continuous row of planks laidacross the next lower tier, and shall be nailed down or otherwise secured to prevent displacement.

(f) *Horse scaffolds.*

    (1) Scaffolds shall not be constructed or arranged more than two tiers or 10 feet (3.0 m) in height, whichever is less.

    (2) When horses are arranged in tiers, each horse shall be placed directly over the horse in the tier below.

    (3) When horses are arranged in tiers, the legs of each horse shall be nailed down or otherwise secured to prevent displacement.

    (4) When horses are arranged in tiers, each tier shall be crossbraced.

(g) *Form scaffolds and carpenters' bracket scaffolds.*

    (1) Each bracket, except those for wooden bracket-form scaffolds, shall be attached to the supporting formwork or structure by means of one or more of the following: nails; a metal stud attachment device; welding; hooking over a secured structural supporting member, with the form wales either bolted to the form or secured by snap ties or tie bolts extending through the form and securely anchored; or, for carpenters' bracket scaffolds only, by a bolt extending through to the opposite side of the structure's wall.

    (2) Wooden bracket-form scaffolds shall be an integral part of the form panel.

    (3) Folding type metal brackets, when extended for use, shall be either bolted or secured with a locking-type pin.

(h) *Roof bracket scaffolds.*

    (1) Scaffold brackets shall be constructed to fit the pitch of the roof and shall provide a level support for the platform.

    (2) Brackets (including those provided with pointed metal projections) shall be anchored in place by nails unless it is impractical to use nails. When nails are not used, brackets shall be secured in place with first-grade manila rope of at least three-fourth inch (1.9 cm) diameter, or equivalent.

(i) *Outrigger scaffolds.*

    (1) The inboard end of outrigger beams, measured from the fulcrum point to the extreme point of anchorage, shall be not less than one and one-half times the outboard end in length.

    (2) Outrigger beams fabricated in the shape of an I-beam or channel shall be placed so that the web section is vertical.

    (3) The fulcrum point of outrigger beams shall rest on secure bearings at least 6 inches (15.2 cm) in each horizontal dimension.

    (4) Outrigger beams shall be secured in place against movement, and shall be securely braced at the fulcrum point against tipping.

    (5) The inboard ends of outrigger beams shall be securely anchored either by means of braced struts bearing against sills in contact with the overhead beams or ceiling, or by means of tension members secured to the floor joists underfoot, or by both.

    (6) The entire supporting structure shall be securely braced to prevent any horizontal movement.

    (7) To prevent their displacement, platform units shall be nailed, bolted, or otherwise secured to outriggers.

(8) Scaffolds and scaffold components shall be designed by a registered professional engineer andshall be constructed and loaded in accordance with such design.

(j) **Pump jack scaffolds.**

(1) Pump jack brackets, braces, and accessories shall be fabricated from metal plates and angles. Eachpump jack bracket shall have two positive gripping mechanisms to prevent any failure or slippage.

(2) Poles shall be secured to the structure by rigid triangular bracing or equivalent at the bottom, top, and other points as necessary. When the pump jack has to pass bracing already installed, an additional brace shall be installed approximately 4 feet (1.2 m) above the brace to be passed, andshall be left in place until the pump jack has been moved and the original brace reinstalled.

(3) When guardrails are used for fall protection, a workbench may be used as the toprail only if it meetsall the requirements in paragraphs (g)(4) (ii), (vii), (viii), and (xiii) of § 1926.451.

(4) Work benches shall not be used as scaffold platforms.

(5) When poles are made of wood, the pole lumber shall be straight-grained, free of shakes, large loose or dead knots, and other defects which might impair strength.

(6) When wood poles are constructed of two continuous lengths, they shall be joined together with the seam parallel to the bracket.

(7) When two by fours are spliced to make a pole, mending plates shall be installed at all splices todevelop the full strength of the member.

(k) **Ladder jack scaffolds.**

(1) Platforms shall not exceed a height of 20 feet (6.1 m).

(2) All ladders used to support ladder jack scaffolds shall meet the requirements of subpart X of this part - Stairways and Ladders, except that job-made ladders shall not be used to support ladder jackscaffolds.

(3) The ladder jack shall be so designed and constructed that it will bear on the side rails and ladder rungs or on the ladder rungs alone. If bearing on rungs only, the bearing area shall include a length ofat least 10 inches (25.4 cm) on each rung.

(4) Ladders used to support ladder jacks shall be placed, fastened, or equipped with devices to prevent slipping.

(5) Scaffold platforms shall not be bridged one to another.

(l) **Window jack scaffolds.**

(1) Scaffolds shall be securely attached to the window opening.

(2) Scaffolds shall be used only for the purpose of working at the window opening through which the jack is placed.

(3) Window jacks shall not be used to support planks placed between one window jack and another, orfor other elements of scaffolding.

(m) **Crawling boards (chicken ladders).**

(1) Crawling boards shall extend from the roof peak to the eaves when used in connection with roof construction, repair, or maintenance.

(2) Crawling boards shall be secured to the roof by ridge hooks or by means that meet equivalent criteria(e.g., strength and durability).

(n) *Step, platform, and trestle ladder scaffolds.*

(1) Scaffold platforms shall not be placed any higher than the second highest rung or step of the ladder supporting the platform.

(2) All ladders used in conjunction with step, platform and trestle ladder scaffolds shall meet the pertinent requirements of subpart X of this part - Stairways and Ladders, except that job-made ladders shall not be used to support such scaffolds.

(3) Ladders used to support step, platform, and trestle ladder scaffolds shall be placed, fastened, or equipped with devices to prevent slipping.

(4) Scaffolds shall not be bridged one to another.

(o) *Single-point adjustable suspension scaffolds.*

(1) When two single-point adjustable suspension scaffolds are combined to form a two-point adjustable suspension scaffold, the resulting two-point scaffold shall comply with the requirements for two- point adjustable suspension scaffolds in paragraph (p) of this section.

(2) The supporting rope between the scaffold and the suspension device shall be kept vertical unless all of the following conditions are met:

   (i) The rigging has been designed by a qualified person,

   (ii) andThe scaffold is accessible to rescuers, and

   (iii) The supporting rope is protected to ensure that it will not chafe at any point where a change indirection occurs, and

   (iv) The scaffold is positioned so that swinging cannot bring the scaffold into contact with anothersurface.

(3) Boatswains' chair tackle shall consist of correct size ball bearings or bushed blocks containing safety hooks and properly "eye-spliced" minimum five-eighth ($^5/_8$) inch (1.6 cm) diameter first-grade manila rope, or other rope which will satisfy the criteria (e.g., strength and durability) of manila rope.

(4) Boatswains' chair seat slings shall be reeved through four corner holes in the seat; shall cross eachother on the underside of the seat; and shall be rigged so as to prevent slippage which could cause an out-of-level condition.

(5) Boatswains' chair seat slings shall be a minimum of five-eight ($^5/_8$) inch (1.6 cm) diameter fiber, synthetic, or other rope which will satisfy the criteria (e.g., strength, slip resistance, durability, etc.) of first grade manila rope.

(6) When a heat-producing process such as gas or arc welding is being conducted, boatswains' chair seat slings shall be a minimum of three-eight ($^3/_8$) inch (1.0 cm) wire rope.

(7) Non-cross-laminated wood boatswains' chairs shall be reinforced on their underside by cleats securely fastened to prevent the board from splitting.

(p) **Two-point adjustable suspension scaffolds (swing stages).** The following requirements do not apply to two-point adjustable suspension scaffolds used as masons' or stonesetters' scaffolds. Such scaffolds are covered by paragraph (q) of this section.

   (1) Platforms shall not be more than 36 inches (0.9 m) wide unless designed by a qualified person to prevent unstable conditions.

   (2) The platform shall be securely fastened to hangers (stirrups) by U-bolts or by other means which satisfy the requirements of § 1926.451(a).

   (3) The blocks for fiber or synthetic ropes shall consist of at least one double and one single block. The sheaves of all blocks shall fit the size of the rope used.

   (4) Platforms shall be of the ladder-type, plank-type, beam-type, or light-metal type. Light metal-type platforms having a rated capacity of 750 pounds or less and platforms 40 feet (12.2 m) or less in length shall be tested and listed by a nationally recognized testing laboratory.

   (5) Two-point scaffolds shall not be bridged or otherwise connected one to another during raising and lowering operations unless the bridge connections are articulated (attached), and the hoists properly sized.

   (6) Passage may be made from one platform to another only when the platforms are at the same height, are abutting, and walk-through stirrups specifically designed for this purpose are used.

(q) **Multi-point adjustable suspension scaffolds, stonesetters' multi-point adjustable suspension scaffolds, and masons' multi-point adjustable suspension scaffolds.**

   (1) When two or more scaffolds are used they shall not be bridged one to another unless they are designed to be bridged, the bridge connections are articulated, and the hoists are properly sized.

   (2) If bridges are not used, passage may be made from one platform to another only when the platforms are at the same height and are abutting.

   (3) Scaffolds shall be suspended from metal outriggers, brackets, wire rope slings, hooks, or means that meet equivalent criteria (e.g., strength, durability).

(r) **Catenary scaffolds.**

   (1) No more than one platform shall be placed between consecutive vertical pickups, and no more than two platforms shall be used on a catenary scaffold.

   (2) Platforms supported by wire ropes shall have hook-shaped stops on each end of the platforms to prevent them from slipping off the wire ropes. These hooks shall be so placed that they will prevent the platform from falling if one of the horizontal wire ropes breaks.

   (3) Wire ropes shall not be tightened to the extent that the application of a scaffold load will overstress them.

   (4) Wire ropes shall be continuous and without splices between anchors.

(s) **Float (ship) scaffolds.**

   (1) The platform shall be supported by a minimum of two bearers, each of which shall project a minimum of 6 inches (15.2 cm) beyond the platform on both sides. Each bearer shall be securely fastened to the platform.

   (2) Rope connections shall be such that the platform cannot shift or

   (3)

(i) slip. When only two ropes are used with each float:

> They shall be arranged so as to provide four ends which are securely fastened to overhead supports.

(ii) Each supporting rope shall be hitched around one end of the bearer and pass under the platform to the other end of the bearer where it is hitched again, leaving sufficient rope at eachend for the supporting ties.

(t) *Interior hung scaffolds.*

(1) Scaffolds shall be suspended only from the roof structure or other structural member such as ceiling beams.

(2) Overhead supporting members (roof structure, ceiling beams, or other structural members) shall be inspected and checked for strength before the scaffold is erected.

(3) Suspension ropes and cables shall be connected to the overhead supporting members by shackles, clips, thimbles, or other means that meet equivalent criteria (e.g., strength, durability).

(u) *Needle beam scaffolds.*

(1) Scaffold support beams shall be installed on edge.

(2) Ropes or hangers shall be used for supports, except that one end of a needle beam scaffold may besupported by a permanent structural member.

(3) The ropes shall be securely attached to the needle beams.

(4) The support connection shall be arranged so as to prevent the needle beam from rolling or becomingdisplaced.

(5) Platform units shall be securely attached to the needle beams by bolts or equivalent means. Cleatsand overhang are not considered to be adequate means of attachment.

(v) *Multi-level suspended scaffolds.*

(1) Scaffolds shall be equipped with additional independent support lines, equal in number to thenumber of points supported, and of equivalent strength to the suspension ropes, and rigged tosupport the scaffold in the event the suspension rope(s) fail.

(2) Independent support lines and suspension ropes shall not be attached to the same points of anchorage.

(3) Supports for platforms shall be attached directly to the support stirrup and not to any other platform.

(w) *Mobile scaffolds.*

(1) Scaffolds shall be braced by cross, horizontal, or diagonal braces, or combination thereof, to preventracking or collapse of the scaffold and to secure vertical members together laterally so as to automatically square and align the vertical members. Scaffolds shall be plumb, level, and squared. All brace connections shall be secured.

(i) Scaffolds constructed of tube and coupler components shall also comply with the requirementsof paragraph (b) of this section;

(ii) Scaffolds constructed of fabricated frame components shall also comply with the requirementsof paragraph (c) of this section.

(2) Scaffold casters and wheels shall be locked with positive wheel and/or wheel and swivel locks, orequivalent means, to prevent movement of the scaffold while the scaffold is used in a stationarymanner.

(3) Manual force used to move the scaffold shall be applied as close to the base as practicable, but notmore than 5 feet (1.5 m) above the supporting surface.

(4) Power systems used to propel mobile scaffolds shall be designed for such use. Forklifts, trucks, similar motor vehicles or add-on motors shall not be used to propel scaffolds unless the scaffold isdesigned for such propulsion systems.

(5) Scaffolds shall be stabilized to prevent tipping during movement.

(6) Employees shall not be allowed to ride on scaffolds unless the following conditions exist:

   (i) The surface on which the scaffold is being moved is within 3 degrees of level, and free of pits,holes, and obstructions;

   (ii) The height to base width ratio of the scaffold during movement is two to one or less, unless thescaffold is designed and constructed to meet or exceed nationally recognized stability test requirements such as those listed in paragraph 2.(w) of appendix A to this subpart;

   (iii) Outrigger frames, when used, are installed on both sides of the scaffold;

   (iv) When power systems are used, the propelling force is applied directly to the wheels, and doesnot produce a speed in excess of 1 foot per second (.3 mps); and

   (v) No employee is on any part of the scaffold which extends outward beyond the wheels, casters,or other supports.

(7) Platforms shall not extend outward beyond the base supports of the scaffold unless outrigger frames or equivalent devices are used to ensure stability.

(8) Where leveling of the scaffold is necessary, screw jacks or equivalent means shall be used.

(9) Caster stems and wheel stems shall be pinned or otherwise secured in scaffold legs or adjustment screws.

(10) Before a scaffold is moved, each employee on the scaffold shall be made aware of the move.

(x) **Repair bracket scaffolds.**

(1) Brackets shall be secured in place by at least one wire rope at least $1/2$ inch (1.27 cm) in diameter.

(2) Each bracket shall be attached to the securing wire rope (or ropes) by a positive locking device capable of preventing the unintentional detachment of the bracket from the rope, or by equivalent means.

(3) Each bracket, at the contact point between the supporting structure and the bottom of the bracket, shall be provided with a shoe (heel block or foot) capable of preventing the lateral movement of thebracket.

(4) Platforms shall be secured to the brackets in a manner that will prevent the separation of the platforms from the brackets and the movement of the platforms or the brackets on a completedscaffold.

(5) When a wire rope is placed around the structure in order to provide a safe anchorage for personal fall arrest systems used by employees erecting or dismantling scaffolds, the wire rope shall

meet the requirements of subpart M of this part, but shall be at least $^5/_{16}$ inch (0.8 cm) in diameter.

(6) Each wire rope used for securing brackets in place or as an anchorage for personal fall arrest systems shall be protected from damage due to contact with edges, corners, protrusions, or otherdiscontinuities of the supporting structure or scaffold components.

(7) Tensioning of each wire rope used for securing brackets in place or as an anchorage for personal fall arrest systems shall be by means of a turnbuckle at least 1 inch (2.54 cm) in diameter, or by equivalent means.

(8) Each turnbuckle shall be connected to the other end of its rope by use of an eyesplice thimble of asize appropriate to the turnbuckle to which it is attached.

(9) U-bolt wire rope clips shall not be used on any wire rope used to secure brackets or to serve as an anchor for personal fall arrest systems.

(10) The employer shall ensure that materials shall not be dropped to the outside of the supporting structure.

(11) Scaffold erection shall progress in only one direction around any structure.

(y) **Stilts.** Stilts, when used, shall be used in accordance with the following

(1) requirements:An employee may wear stilts on a scaffold only if it is a large

(2) area scaffold.

When an employee is using stilts on a large area scaffold where a guardrail system is used to providefall protection, the guardrail system shall be increased in height by an amount equal to the height of the stilts being used by the employee.

(3) Surfaces on which stilts are used shall be flat and free of pits, holes and obstructions, such as debris, as well as other tripping and falling hazards.

(4) Stilts shall be properly maintained. Any alteration of the original equipment shall be approved by themanufacturer.

[61 FR 46104, Aug. 30, 1996, as amended at 85 FR 8736, Feb. 18, 2020]

§ 1926.453 Aerial lifts.

(a) *General requirements.*

(1) Unless otherwise provided in this section, aerial lifts acquired for use on or after January 22, 1973 shall be designed and constructed in conformance with the applicable requirements of the AmericanNational Standards for "Vehicle Mounted Elevating and Rotating Work Platforms," ANSI A92.2-1969,including appendix. Aerial lifts acquired before January 22, 1973 which do not meet the requirements of ANSI A92.2-1969, may not be used after January 1, 1976, unless they shall have been modified so as to conform with the applicable design and construction requirements of ANSI A92.2-1969. Aerial lifts include the following types of vehicle-mounted aerial devices used to elevate personnel to job-sites above ground:

(i) Extensible boom

(ii) platforms;Aerial

(iii)

ladders;

Articulating boom platforms;

(iv) Vertical towers; and

(v) A combination of any such devices. Aerial equipment may be made of metal, wood, fiberglassreinforced plastic (FRP), or other material; may be powered or manually operated; and are deemed to be aerial lifts whether or not they are capable of rotating about a substantially vertical axis.

(2) Aerial lifts may be "field modified" for uses other than those intended by the manufacturer provided the modification has been certified in writing by the manufacturer or by any other equivalent entity, such as a nationally recognized testing laboratory, to be in conformity with all applicable provisionsof ANSI A92.2-1969 and this section and to be at least as safe as the equipment was before modification.

(b) *Specific requirements* -

(1) **Ladder trucks and tower trucks.** Aerial ladders shall be secured in the lower traveling position by thelocking device on top of the truck cab, and the manually operated device at the base of the ladder before the truck is moved for highway travel.

(2) **Extensible and articulating boom platforms.**

(i) Lift controls shall be tested each day prior to use to determine that such controls are in safe working condition.

(ii) Only authorized persons shall operate an aerial lift.

(iii) Belting off to an adjacent pole, structure, or equipment while working from an aerial lift shall notbe permitted.

(iv) Employees shall always stand firmly on the floor of the basket, and shall not sit or climb on theedge of the basket or use planks, ladders, or other devices for a work position.

(v) A body belt shall be worn and a lanyard attached to the boom or basket when working from anaerial lift.

Note to paragraph (b)(2)(v): As of January 1, 1998, subpart M of this part (§ 1926.502(d)) provides that body belts are not acceptable as part of a personal fall arrest system. The use of a body belt in a tethering system or in a restraint system is acceptable and is regulated under § 1926.502(e).

(vi) Boom and basket load limits specified by the manufacturer shall not be exceeded.

(vii) The brakes shall be set and when outriggers are used, they shall be positioned on pads or a solid surface. Wheel chocks shall be installed before using an aerial lift on an incline, providedthey can be safely installed.

(viii) An aerial lift truck shall not be moved when the boom is elevated in a working position with menin the basket, except for equipment which is specifically designed for this type of operation in accordance with the provisions of paragraphs (a) (1) and (2) of this section.

(ix) Articulating boom and extensible boom platforms, primarily designed as personnel carriers, shall have both platform (upper) and lower controls. Upper controls shall be in or beside the platform within easy reach of the operator. Lower controls shall provide for overriding the uppercontrols. Controls shall be plainly marked as to their function. Lower level controls shall not be operated unless permission has been obtained from the employee in the lift, except in case of emergency.

(x) Climbers shall not be worn while performing work from an aerial lift.

(xi) The insulated portion of an aerial lift shall not be altered in any manner that might reduce its insulating value.

(xii) Before moving an aerial lift for travel, the boom(s) shall be inspected to see that it is properly cradled and outriggers are in stowed position except as provided in paragraph (b)(2)(viii) of thissection.

(3) **Electrical tests.** All electrical tests shall conform to the requirements of ANSI A92.2-1969 section 5. However equivalent d.c.; voltage tests may be used in lieu of the a.c. voltage specified in A92.2-1969;
d.c. voltage tests which are approved by the equipment manufacturer or equivalent entity shall be considered an equivalent test for the purpose of this paragraph (b)(3).

(4) **Bursting safety factor.** The provisions of the American National Standards Institute standard ANSI A92.2-1969, section 4.9 Bursting Safety Factor shall apply to all critical hydraulic and pneumatic components. Critical components are those in which a failure would result in a free fall or free rotation of the boom. All noncritical components shall have a bursting safety factor of at least 2 to 1.

(5) **Welding standards.** All welding shall conform to the following standards as applicable:

  (i) Standard Qualification Procedure, AWS B3.0-41.

  (ii) Recommended Practices for Automotive Welding Design, AWS D8.4-61.

  (iii) Standard Qualification of Welding Procedures and Welders for Piping and Tubing, AWSD10.9-69.

  (iv) Specifications for Welding Highway and Railway Bridges, AWS D2.0-69.

Note to § 1926.453: Non-mandatory appendix C to this subpart lists examples of national consensus standards that are considered to provide employee protection equivalent to that provided through the application of ANSI A92.2-1969, where appropriate. This incorporation by reference was approved by the Director of the Federal Register in accordance with 5 U.S.C. 552(a) and 1 CFR part 51. Copies may be obtained from the American National Standards Institute. Copies may be inspected at the Docket Office, Occupational Safety and Health Administration, U.S. Department of Labor, 200 Constitution Avenue, NW., room N2634, Washington, DC or at the National Archives and Records Administration (NARA). For information on the availability of this material at NARA, call 202-741-6030, or go to: *http://www.archives.gov/federal_register/ code_of_federal_regulations/ibr_locations.html.*

[61 FR 46116, Aug. 30, 1996; 61 FR 59832, Nov. 25, 1996, as amended at 69 FR 18803, Apr. 9, 2004]

§ 1926.454 Training requirements.

This section supplements and clarifies the requirements of § 1926.21(b)(2) as these relate to the hazards of work on scaffolds.

(a) The employer shall have each employee who performs work while on a scaffold trained by a person qualified in the subject matter to recognize the hazards associated with the type of scaffold being used and to understand the procedures to control or minimize those hazards. The training shall include the following areas, as applicable:

(1) The nature of any electrical hazards, fall hazards and falling object hazards in the work area;

(2) The correct procedures for dealing with electrical hazards and for erecting, maintaining, and disassembling the fall protection systems and falling object protection systems being used;

(3) The proper use of the scaffold, and the proper handling of materials on the

(4) scaffold; The maximum intended load and the load-carrying capacities of the

(5) scaffolds used; and Any other pertinent requirements of this subpart.

(b) The employer shall have each employee who is involved in erecting, disassembling, moving, operating, repairing, maintaining, or inspecting a scaffold trained by a competent person to recognize any hazards associated with the work in question. The training shall include the following topics, as applicable:

(1) The nature of scaffold hazards;

(2) The correct procedures for erecting, disassembling, moving, operating, repairing, inspecting, and maintaining the type of scaffold in question;

(3) The design criteria, maximum intended load-carrying capacity and intended use of the scaffold;

(4) Any other pertinent requirements of this subpart.

(c) When the employer has reason to believe that an employee lacks the skill or understanding needed for safe work involving the erection, use or dismantling of scaffolds, the employer shall retrain each such employee so that the requisite proficiency is regained. Retraining is required in at least the following situations:

(1) Where changes at the worksite present a hazard about which an employee has not been previously trained; or

(2) Where changes in the types of scaffolds, fall protection, falling object protection, or other equipment present a hazard about which an employee has not been previously trained; or

(3) Where inadequacies in an affected employee's work involving scaffolds indicate that the employee has not retained the requisite proficiency.

Made in the USA
Coppell, TX
27 September 2023